Secure Communication in Internet of Things

The book *Secure Communication in Internet of Things: Emerging Technologies, Challenges, and Mitigation* will be of value to the readers in understanding the key theories, standards, various protocols, and techniques for the security of Internet of Things hardware, software, and data, and explains how to design a secure Internet of Things system. It presents the regulations, global standards, and standardization activities with an emphasis on ethics, legal, and social considerations about Internet of Things security.

Features:

- Explores the new Internet of Things security challenges, threats, and future regulations to end-users.
- Presents authentication, authorization, and anonymization techniques in the Internet of Things.
- Illustrates security management through emerging technologies such as blockchain and artificial intelligence.
- Highlights the theoretical and architectural aspects, foundations of security, and privacy of the Internet of Things framework.
- Discusses artificial-intelligence-based security techniques, and cloud security for the Internet of Things.

It will be a valuable resource for senior undergraduates, graduate students, and academic researchers in fields such as electrical engineering, electronics and communications engineering, computer engineering, and information technology.

Secure Communication in Internet of Things

Emerging Technologies, Challenges, and Mitigation

Edited by
T. Kavitha
M. K. Sandhya
V. J. Subashini
Prasidh Srikanth

CRC Press
Taylor & Francis Group
Boca Raton London New York

CRC Press is an imprint of the
Taylor & Francis Group, an **informa** business

Designed cover image: © Sam Daniel A

First edition published 2024
by CRC Press
2385 NW Executive Center Drive, Suite 320, Boca Raton FL 33431

and by CRC Press
4 Park Square, Milton Park, Abingdon, Oxon, OX14 4RN

CRC Press is an imprint of Taylor & Francis Group, LLC

© 2024 selection and editorial matter, T. Kavitha, M. K. Sandhya, V. J. Subashini, and Prasidh Srikanth; individual chapters, the contributors

ISBN: 978-1-032-43573-2 (hbk)
ISBN: 978-1-032-76157-2 (pbk)
ISBN: 978-1-003-47732-7 (ebk)

DOI: 10.1201/9781003477327

Typeset in Times
by SPi Technologies India Pvt Ltd (Straive)

Editorial Advisory Board and Reviewers

Contents

Preface

The book *Secure Communication in the Internet of Things: Standards, Emerging Technologies, Challenges, and Mitigation* explores the various security aspects of the Internet of Things (IoT). This book analyzes the effective practical applications of IoT which integrates secure communication. The advancements in IoT offer a wide range of opportunities for manufacturers and consumers. It poses major risks in terms of security as more devices get interconnected, securing all these devices will be the biggest challenge. This book contains 25 chapters exploring the various facets of secure communication in the IoT.

In **Chapter 1**, a comprehensive and in-depth reference to the security aspects of the IoT with a focus on regulation for IoT security, standards, security life cycle, and security-layered model is presented.

Chapter 2 provides insight into the building blocks of IoT and their characteristics, the platforms of IoT, and specialized aspects of IoT protocols for device security. It is mainly designed for educational programs at colleges and universities, and also for IoT vendors and service providers who may be interested in offering a broader perspective of the IoT to accompany their customers and for developer training programs.

Chapter 3 introduces the concepts of secure device management and device attestation, and describes how they can be used to increase the safety of IoT gadgets. The chapter presents an architecture that uses a combination of encryption, digital signatures, and secure bootstrapping to provide end-to-end security for IoT devices. Further, it explores the various device attestation techniques.

Chapter 4 emphasizes and focuses on secure operating systems for IoT devices and their current and future trends. The chapter discusses the anatomy and architectural design of operating systems. It also focuses on programming support and modern operating systems as it is crucial for developing apps required for the devices. The chapter also focuses on resource management principles and strategies. Operating systems designed for IoT require exclusive process, memory, and resource management while focusing on communication and file management. The chapter also deals with the security challenges while designing an operating system.

Chapter 5 presents a comprehensive study with an overview of wireless networks and delves into physical-layer techniques for improved security. It addresses potential attacks and vulnerabilities while acknowledging the inherent limitations of these security measures. The chapter concludes by discussing future research possibilities in this domain and the need for ongoing innovation to address evolving digital security concerns.

In **Chapter 6**, a case study about the design and implementation of a two-level cryptosystem for high-security-enabled applications is presented. The implementation presented in the chapter is coded in Verilog, verified for functionality, and synthesized and developed as an Intellectual Property, which can be easily embedded into the IoT design to provide high security.

Chapter 7 discusses secure data collection, encryption, and authentication techniques that can be used to protect the data from interception and unauthorized access. Access-control mechanisms implemented to regulate the data access and the data integrity mechanisms are summarized.

Chapter 8 enumerates the vulnerabilities across IoT devices, network systems, user interfaces, and customer back-end systems. It also addresses strategies for mitigating these vulnerabilities throughout the entire IoT life cycle, particularly as it expands across geographical boundaries. It delves into authentication techniques and protocols employed in IoT to bolster security among IoT devices. The book chapter extensively explores various forms of access control and provides a comprehensive methodology for effectively integrating authentication, authorization, and anonymization techniques within IoT subsystems. Additionally, it analyzes a case study, examining the management of authentication within AWS Core, with detailed discussions presented across different sections.

Chapter 9 focuses on secure identity, access, and mobility management solutions that can help mitigate risks by providing device identity, authentication, access control, mobility management, improved efficiency, and manageability of IoT networks. The solutions are essential for protecting the security, privacy, and reliability of IoT networks. By implementing such robust solutions, organizations can reduce the risk of cyberattacks and ensure that their IoT devices and data are safe.

Chapter 10 focuses on privacy protection systems for the IoT. Privacy protection policies protect the privacy of individuals by limiting the collection, storage, and use of personal data. The chapter will give readers a better understanding of the importance of privacy in the IoT, and the different approaches to protecting individual privacy in this context.

Chapter 11 illustrates the various ways in which data security can be assessed and ensured in an IoT environment. This is an emerging research area which finds prominence in the modern cyber world. The application of IoT provides enormous possibilities to make new concepts emerge in industries also. Industry 4.0, which is said to be the fourth industrial revolution, is paving the way for more efficient industrial operations through the extensive use of IoT. In all these activities mentioned so far, the reliability and quality of service is a factor of utmost concern. Since these operations are carried out over vast, massively distributed digital data networks, there is always a concern for data security.

Chapter 12 enumerates the various security mechanisms, access controls, and other security measures. It investigates the need for adopting secure communication protocols by which the IoT ecosystem can continue to grow and evolve, without compromising on security and privacy. Furthermore, deep learning algorithms can significantly boost security levels to provide safer and more secure communication between these devices.

Chapter 13 unfolds the concept of adaptive security within the IoT context, highlighting key challenges and exploring potential solutions for establishing resilient and self-defending IoT systems.

Chapter 14 delves into the fascinating realm of covert communication security methods. It explores the ingenious techniques and technologies developed to conceal information, ensuring that messages remain hidden from prying eyes while also guaranteeing their integrity and authenticity. This journey will take the readers through the history of covert communication, from ancient ciphers to cutting-edge steganography and encryption methods. The chapter navigates through the intricate landscape of encryption, authentication, secure protocols, and hardware-based security mechanisms. Additionally, emerging technologies such as blockchain and privacy-enhancing techniques that hold promise in fortifying IoT security are discussed.

In **Chapter 15**, a systematic approach to understanding the security requirements for the IoT and different scenarios of potential threats and attacks within the IoT are presented. The existing security requirements for IoT in the literature are presented and the proposed approach for security requirements for the IoT is discussed with regard to some of the security designs like cross-layer designs for IoT systems.

Chapter 16 focuses on the survey of various levels of security and the risk of edge computing, followed by a list of algorithms at the level of security available on IoT. A detailed analysis of security using edge computing for the IoT is highlighted. The chapter concludes with challenges in edge computing.

In **Chapter 17**, a journey through the realm of cloud security for the IoT is embarked upon. The chapter delves into the multifaceted domain of cloud security in the context of the IoT, offering a comprehensive exploration of the key principles, emerging threats, and best practices necessary to safeguard the integrity, confidentiality, and availability of IoT systems.

Chapter 18 explores the symbiotic relationship between the IoT and blockchain. The chapter highlights the need for security, the possibilities of creative thinking, and the opportunity to create a world where connections and privacy coexist peacefully. The chapter sets off on an intriguing trip into the nexus of two transformational forces: blockchain technology and the IoT. The chapter explores these subjects in-depth with a strong focus on security, revealing how blockchain technology has the ability to defend IoT devices from a variety of security threats.

In **Chapter 19**, an IoT-based weather-monitoring system for irrigation using case studies and relevant research under the IoT ecosystem-level security auditing, analysis, and recovery is discussed. The chapter also emphasizes in detail the issues related to challenges faced in IoT ecosystems analysis and recovery including security auditing under different sections.

Chapter 20 presents an insight into risk assessment and vulnerability analysis. The chapter aims to foster a more secure and resilient IoT environment, enabling the continued evolution of this transformative technology.

In **Chapter 21**, the security of IoT systems and threat models are utilized to recognize potential threats and vulnerabilities. Furthermore, attackers may resort to social engineering tactics to access sensitive data or manipulate user actions. Implementing security measures such as regular patching, vigilant monitoring, and encryption can be instrumental in mitigating these risks. The book chapter also emphasizes in detail the issues related to challenges faced in IoT threat models and different types of attacks in IoT systems and how to mitigate those risks are described under different sections.

Chapter 22 delves into the complex and multifaceted landscape of IoT security and privacy. It is a comprehensive exploration of the challenges, risks, and solutions that surround this rapidly evolving field.

Chapter 23 introduces a comprehensive exploration of mitigation techniques within the context of SDN. This delves into a diverse array of strategies and approaches aimed at safeguarding SDN infrastructures against cyber threats and attacks. From analyzing the fundamental principles of SDN and its inherent security implications to exploring cutting-edge solutions and best practices, this work endeavors to equip network administrators, security professionals, and researchers with the knowledge needed to fortify SDN environments.

In **Chapter 24**, a detailed review of current malware technologies used for detection in IoT is presented. Further, the preventive measures that need to be looked into to avoid such malware attacks are also presented. It is emphasized that adopting futuristic approaches for the development of malware detection applications shall provide significant advantages.

Chapter 25 encompasses the critical aspects of the security life cycle for 5G and beyond, providing guidance on constructing robust security architectures that accommodate the unique challenges posed by IoT devices in the 5G and 6G landscapes. It addresses mobile network security from multiple angles, including network-centric, device-centric, information-centric, and, most crucially, people-centric perspectives. This comprehensive resource caters to the security concerns of various stakeholders within the IoT ecosystem of 5G and 6G, including mobile network operators (MNOs), mobile virtual network operators (MVNOs), IoT device manufacturers and users, wireless service providers, as well as experts in cybersecurity, security researchers, and engineers.

Editors

Dr. T. Kavitha is working as a Professor in the Department of Computer Engineering, New Horizon College of Engineering. She completed her Ph.D. in the Faculty of Information and Communication Engineering, Anna University, Chennai, India, in the year 2014. She received her M.E. degree in Systems Engineering and Operations Research from Anna University in the year 2006 and her B.E. in Electronics and Communication Engineering from Bharathidasan University, India, in the year 2000. She has 20+ years of experience in teaching and research from reputed engineering colleges. She is an Anna University and VTU-recognized supervisor for guiding the Ph.D. and M.S. (by Research) programs. Under her guidance, a scholar had completed a Ph.D. at Anna University. She has received funds from various agencies such as ISTE-SRM, VTU-TEQIP, IE, VTU, AICTE-ISTE, and AICTE to organize FDPs, workshops, training, and conferences. She is also a mentor for projects that received funds from VTU and KSCKT. Presently, she is guiding three scholars under VTU. She has filed three Indian patents to her credit. She has published 30 papers in national/international conferences, two in book chapters, and 15 in international journals. She is a lifetime member of ISTE, IETE, and IE. Her fields of interest include wireless networks, wireless sensor networks, information security, the Internet of Things, deep learning, and machine learning.

Dr. M. K. Sandhya received her Bachelor of Engineering in Computer Science and Engineering from Easwari Engineering College for Women, affiliated to the University of Madras, in 2003 and Master of Engineering in Computer Science and Engineering from Sri Venkateswara College of Engineering, affiliated to Anna University, Chennai, India, in 2006. She received her Ph.D. under the faculty of Information and Communication Engineering from Anna University in 2014. She is presently working as a Professor in the Department of Computer Science and Engineering at Meenakshi Sundararajan Engineering College, Chennai. She has 18 years of teaching and 9 years of research experience. She received her supervisor recognition in the Faculty of Information and Communication Engineering from Anna University in 2015. Her research interests are wireless sensor networks, mobile computing, cryptography, network security, and machine learning. She has published a patent and several papers in reputed impact factor journals and international conferences. She was also invited to review papers from reputed international journals. She has been the organizing secretary for many national and international conferences conducted by her department. She has delivered keynote addresses and has been the session chair at national and international conferences conducted by other engineering colleges. She received the Cambridge International Certificate for Teachers and Trainers in the year 2011. She received the IETE-K S Krishnan Memorial Award in 2015 for the Best System Oriented Paper for her research paper titled "False Data Elimination in Heterogeneous Wireless Sensor Networks Using Location Based Selection of Aggregator Nodes" published in *IETE Journal of Research*, Vol. 60, No. 2, March–April 2014 issue. She has co-authored four books, namely, *Python Programming*, *Foundations of Data Science*, *Fundamentals of Data Science and Analytics*, and *Cryptography and Cyber Security*. She has delivered several guest lectures at Anna University and other engineering colleges. She is serving as a doctoral committee member for Ph.D. scholars. She is a lifetime member of the Indian Society for Technical Education and the Computer Society of India.

Dr. V. J. Subashini is a Professor of Computer Science and Engineering at Jerusalem College of Engineering, Chennai, India. She earned her Doctoral degree in the Faculty of Information and Communication Engineering and a Master's degree in Computer Science and Engineering from Anna University, Chennai. She received her Master's degree in Computer Applications and Bachelor's degree in Mathematics from Manonmaniam Sundaranar University, Tirunelveli, India. She has more than 24 years of professional experience and has made over 15 scholarly contributions to reputed peer-reviewed journals and national and international conference proceedings.

Mr. Prasidh Srikanth is the Director of Product Management at Palo Alto Networks, a leading American multinational cybersecurity company that specializes in providing advanced security solutions to organizations worldwide with headquarters in Santa Clara, California. He leads the product strategy and execution for the Data Security product line. Prior to joining Palo Alto Networks, Prasidh had a track record spanning startups like Securiti.ai, Bitglass, and global giants like Microsoft. Prasidh holds an M.S. in Computer Science from Stony Brook University, New York, and a B.E. in Computer Science from Anna University, Chennai, India.

Contributors

Kunal Abhishek
Cyber Security and Forensics
Centre for Development of Advanced
    Computing (C-DAC)
Patna, Bihar, India

S. Angel Deborah
Sri Sivasubramaniya Nadar College of
    Engineering
Anna University
Chennai, Tamil Nadu, India

A. Babiyola
Meenakshi Sundararajan Engineering
    College
Anna University
Chennai, Tamil Nadu, India

Madhumala R. Bagalatti
Dayananda Sagar Academy of
    Technology and Management
Visvesvaraya Technological University
Bengaluru, Karnataka, India

Bindu S.
BNM Institute of Technology
Visvesvaraya Technological University
Bengaluru, Karnataka, India

M. Chaitra
Credokey SoftTech Private Limited
Bangalore, Karnataka, India

Puspraj Singh Chauhan
Pranveer Singh Institute of Technology
APJ Abdul Kalam Technical University
Lucknow, Uttar Pradesh, India

Emmanvelraj M. Chirchi
Don Bosco Institute of Technology
Visvesvaraya Technological University
Bengaluru, Karnataka, India

Vanajaroselin Chirchi
The Oxford College of Engineering
Visvesvaraya Technological University
Bengaluru, Karnataka, India

C. Chitra
PSNA College of Engineering and
    Technology
Anna University
Chennai, Tamil Nadu, India

Prashant Dahiwale
Government Polytechnic
Gujarat Technological University
Daman UT, India

M. Deekshitha
Sri Venkateswara College of Engineering
Anna University
Chennai, Tamil Nadu, India

K. P. K. Devan
Easwari Engineering College
Anna University
Chennai, Tamil Nadu, India

N. Devi
Sri Venkateswara College of
    Engineering
Anna University
Chennai, Tamil Nadu, India

Chhaya Suryabhan Dule
KS School of Engineering and
    Management
Visvesvaraya Technological University
Bengaluru, Bengaluru, India

M. Esther Hannah
Women's Christian College
University of Madras
Chennai, Tamil Nadu, India

Sheeja V. Francis
Jerusalem College of Engineering
Anna University
Chennai, Tamil Nadu, India

V. Hari Santhosh
IoT Cloud Architect at trinamiX(BASF)

N. Hema
Vellore Institute of Technology
Deemed University
Vellore, Tamil Nadu, India

D. Indumathy
Rajalakshmi College of Engineering
Anna University
Chennai, Tamil Nadu, India

P. Indumathy
Easwari Engineering College
Anna University
Chennai, Tamil Nadu, India

Ankit Jain
Pranveer Singh Institute of Technology
APJ Abdul Kalam Technical
  University
Lucknow, Uttar Pradesh, India

A. Jemshia Mirriam
Sathyabama Institute of Science and
  Technology
Deemed to be University
Chennai, Tamil Nadu, India

S. L. Karthik Raj
The Oxford College of Engineering
Visvesvaraya Technological
  University
Bengaluru, Karnataka, India

Jaskirat Kaur
Punjab Engineering College
Deemed to be University
Chandigarh, India

Ramanpreet Kaur
Chandigarh Engineering College
Punjab Technology University
Mohali, Punjab, India

P. Kavitha
M.A.M School of Engineering
Anna University
Chennai, Tamil Nadu, India

T. Kavitha
New Horizon College of
  Engineering
Visvesvaraya Technological
  University
Bengaluru, Karnataka, India

E. C. Khushi
Dayanand Sagar University
Bengaluru, Karnataka, India

Prathibha Kiran
AMC Engineering College
Visvesvaraya Technological
  University
Bengaluru, Karnataka, India

K. Kiruthika Devi
Sri Venkateswara College of
  Engineering
Anna University
Chennai, Tamil Nadu, India

S. Koushik
Credokey SoftTech Private Limited
Bangalore, Karnataka, India

Balram Kumar
Chandigarh Engineering College
Punjab Technology University
Mohali, Punjab, India

Anitha Kumari
GITAM, Deemed to be University
Bengaluru, Karnataka, India

B. S. Liya
Easwari Engineering College
Anna University
Chennai, Tamil Nadu, India

R. Malathy
Jerusalem College of Engineering
Anna University
Chennai, Tamil Nadu, India

P. Manjula
Saveetha School of Engineering
Deemed to be University
Chennai, Tamil Nadu, India

Sanjay Mate
Sangam University and GTU
Atoon, Rajasthan, India

T. T. Mirnalinee
Sri Sivasubramaniya Nadar College of
    Engineering
Anna University
Chennai, Tamil Nadu, India

Jyoti R. Munavalli
BNM Institute of Technology
Visvesvaraya Technological University
Bengaluru, Karnataka, India

M. Nafees Muneera
Sathyabama Institute of Science and
    Technology
Deemed to be University
Chennai, Tamil Nadu, India

S. Nirmala
AMC Engineering College
Visvesvaraya Technological University
Bengaluru, Karnataka, India

K. Panimozhi
BMS College of Engineering
Deemed University
Bengaluru, Karnataka, India

A. Punitha
M.A.M School of Engineering
Anna University
Chennai, Tamil Nadu, India

M. M. Raghuwanshi
Symbiosis Institute of Technology
Deemed University
Maharashtra, India

S. Raja Shree
Sathyabama Institute of Science and
    Technology
Deemed to be University
Chennai, Tamil Nadu, India

S. Rajalakshmi
Sri Sivasubramaniya Nadar College
    of Engineering
Anna University
Chennai, Tamil Nadu, India

S. Rajarajeswari
Vellore Institute of Technology
Deemed University
Tamil Nadu, India

K. M. Rajasekharaiah
AMC Engineering College
Visvesvaraya Technological
    University
Bengaluru, Karnataka, India

V. Rajeswari
Karpagam College of Engineering
Anna University
Chennai, Tamil Nadu, India

T. R. Reshmi
Cryptology and Computing Research
    Group
Society for Electronic Transactions and
    Security (SETS)
Chennai, Tamil Nadu, India

K. Sakthipriya
Karpagam College of Engineering
Anna University
Chennai, Tamil Nadu, India

P. Sankar
Hindustan Institute of Technology and
    Science
Deemed to be University
Chennai, Tamil Nadu, India

V. Saranya
Sathyabama Institute of Science and
    Technology
Deemed to be University
Chennai, Tamil Nadu, India

S. Saraswathi
Sri Sivasubramaniya Nadar College of
    Engineering
Anna University
Chennai, Tamil Nadu, India

S. Saravanan
M.A.M School of Engineering
Anna University
Chennai, Tamil Nadu, India

G. Senbagavalli
AMC Engineering College
Visvesvaraya Technological University
Bengaluru, Karnataka, India

S. Shalini
Dayananda Sagar Academy of
    Technology and Management
Visvesvaraya Technological University
Bengaluru, Karnataka, India

S. Sheela
Global Academy of Technology
Visvesvaraya Technological University
Bengaluru, Karnataka, India

S. Shilpa
AMC Engineering College
Visvesvaraya Technological University
Bengaluru, Karnataka, India

Yasha Jyothi M. Shirur
BNM Institute of Technology
Visvesvaraya Technological
    University
Bengaluru, Karnataka, India

Anita Shukla
Pranveer Singh Institute of
    Technology
APJ Abdul Kalam Technical
    University
Lucknow, Uttar Pradesh, India

Parveen Singla
Chandigarh Engineering College
Punjab Technology University
Mohali, Punjab, India

Raghvendra Singh
Pranveer Singh Institute of
    Technology
APJ Abdul Kalam Technical University
Lucknow, Uttar Pradesh, India

V. M. Sivagami
Sri Venkateswara College of
    Engineering
Anna University
Chennai, Tamil Nadu, India

Vikas Somani
Sangam University
Atoon, Rajasthan, India

D. Bhuvana Suganthi
BNM Institute of Technology
Visvesvaraya Technological University
Bengaluru, Karnataka, India

S. Swarna Parvathi
Director Experiment Station
(BAEG)-Biological and Agricultural
    Engineering
University of Arkansas
Fayetteville, AR, USA

Shabeen Taj
Government Engineering College
Visvesvaraya Technological University
Bengaluru, Karnataka, India

N. Vanitha
Women's Christian College
University of Madras
Tamil Nadu, India

V. Vidhya
Sri Venkateswara College of
    Engineering
Anna University
Chennai, Tamil Nadu,
    India

1 Introduction to Internet of Things Security

Chhaya Suryabhan Dule
KS School of Engineering and Management, Visvesvaraya
Technological University, Bengaluru, India

K. M. Rajasekharaiah
AMC Engineering College, Visvesvaraya Technological
University, Bengaluru, India

1.1 INTRODUCTION

The tremendous growth in Internet of Things (IoT) devices and IoT technology facilitates various device interconnections with multiple technologies storing and processing data on a real-time basis bringing revolution to society. The Smart Internet of Things infrastructure basically consists of sensors, actuators, servers, and networks for communication, protocols, and layers [1]. The Traditional IoT architecture [2] consists of three basic layers: physical, network, and application layers used to access, monitor, and maintain system integrity [3]. The physical layer is responsible for sensing and collecting data using sensors and recognizing characteristics and identities of ambient smart objects. The network layer connects servers, network devices, and other smart things. The application layer provides application-specific services to users. The conventional architecture is extended by including the Business layer and the network layer is divided into processing and transport layers. The transport layer moves sensor data from the perception layer to the processing layer and vice versa using network technologies such as Wi-Fi, 3G, LAN, Bluetooth, RFID, and NFC. The huge amount of data received from the transport layer is stored, analyzed, and processed by the middleware processing layer. The processing layer is capable of managing and offering a wide range of services to the lower layers. It makes use of a variety of technologies, including big data processing modules, cloud computing, and databases. The IoT infrastructure comprises all applications, business models, and user privacy managed by the business layer. The enhanced architecture of IoT is influenced by human intelligence to respond to physical surroundings. The cloud computing-based IoT architecture consists of the cloud in the center, applications above it, and a network of smart things beneath it. This architecture is determined based on the type of data produced by IoT devices and data processing [4]. The cloud-computing-based IoT architecture is prioritized as it provides flexibility, scalability, and service platforms. Another IoT-based edge computing architecture incorporates network gateways and sensors for data processing and analytics, which has extra

DOI: 10.1201/9781003477327-1

layers for monitoring, pre-processing, storage, and security introduced between the physical and transport layers. The monitor layer is responsible for controlling power, resources, replies, and services. The filtering, processing, and analytics of sensor data are carried out by the preprocessing layer. Data replication, dissemination, and storage are storage capabilities offered by the temporary storage layer. Data integrity and privacy are ensured by the security layer performing encryption and decryption operations. In this section, we discuss the IoT architecture which consists of several layers comprising billions of heterogeneous interconnected IoT devices. Therefore, a standardization process is required to link these heterogeneous devices and exchange data. The IoT infrastructure is growing rapidly by adding smart devices which carry user's valuable and private information. Cyber-attacks on IoT devices and smart ecosystems affect people's lives, safety, and privacy. It raises the concern to establish security mechanisms and procedures to safeguard the user's safety and privacy. This book chapter discusses IoT regulation, standardization, and ethical, legal, and social activities to follow to maintain IoT ecosystem security.

1.2   IoT STANDARDIZATION

The IoT is a recent innovation that includes vehicles, smart cities, electronic health, home automation, energy management, industrial process control, and public safety. The IoT ecosystem collects data from IoT devices, transfers data from one device to another, processes and collaborates on that data, and takes action automatically. As per the survey carried out by Kinsey and Company, there is a slowdown in IoT growth which includes unreliable heterogeneous device connectivity and processing. The business organization and regulator should promote innovation and technology for safe interoperable personal information protection and property rights with established business models to facilitate and permit data sharing. IoT standards are the general knowledge among customers, vendors, organizations, and industries. IoT ecosystem can flourish multitude of standards, regulations, guidelines, procedures, and policies [5].

> **Standards**: Standards demonstrate how proposed techniques can be relied upon to prevent attacks, reduce risks, and mitigate them. Standardized techniques and procedures are needed in the IoT ecosystem to maintain the desired level of quality.
> **Regulations**: Regulations are common processes maintained by organizations and businesses to secure information and systems against cyber-attacks.
> **Guidelines**: Guidelines describe how to protect the system and its surroundings.
> **Policies**: Policies are the paper-recorded principles, processes, and procedures that meet the security requirements of an association unless the statement of objectives is approved by the organization's administration. Compliance and agreement are required in policy to assist businesses in securing information and assets, as well as gathering the set of rules, regulations, and allowed requests.
> **Procedures**: Procedures are sequences of instructions to achieve the goal and are used to implement standards and rules. During the early stages of IoT ecosystem design and implementation standards, laws, guidelines, policies,

and procedures were not mandatory despite the fact that they improve the system capacity with innovative technologies and procedures. The IoT-based infrastructure consists of standards given further.

Scalability: In IoT systems, medium-sized instances can be scaled up or down based on demand, security, storage, network, and analytics which need standards and policies.

Sustainability in Government Policies and Legislation: Standards are frequently utilized to protect consumers, commercial interests, and government directives.

Business Profit: Standards give the platform for establishment as well as assisting consumers in combining capital and operating expense reduction. It requires developing totally new IoT marketplaces, launching entirely new products, and broadening knowledge of IoT possibilities and techniques.

Alternative: Standards serve as a source for the most recent features and options, boosting daily life in the workplace, private spaces, and public spaces.

Protection: To reduce the attack surface, improve the visibility of security events, and safeguard, perceive, rectify, and offer details about the security environment, industry standards, and best practices must be used.

Standards also play a crucial role in ensuring the interoperability of "things" between various IoT devices. The security standard provides the guarantee of interoperability of things providing security functions targeted by hostile parties which address functionality (encryption techniques), application (encryption mode, like counter mode), and contextual use of functionality (as an application of encryption to the provision of confidentiality protection services). Security standards therefore need to handle simple mechanical interconnection, shared semantic and syntactic meaning, management of attributes, and effective management of the organization to respond to security breaches.

Organizational Interoperability: Organizational management security standards enforce a "need to know" process that specifies roles within organizations. The shared communication security (ComSec) framework allows organizations to communicate data securely without disclosing how data is handled before and after transfer. Hence, local IT security policies of the sending and receiving organizations are identical and supported by external measures.

Syntactic Interoperability: Syntactic interoperability is achieved in the IoT ecosystem consisting of heterogeneous components to understand symbol sets and their ordering.

Semantic Interoperability: Valid syntactic statements, pragmatics, and context bring clarity to semantics and help in semantic understanding. Semantic interoperability maximizes the availability of semantic content and the transfer of contextual knowledge. Authentication protocols provide context to message sets which are semantic containers containing syntactically valid information. This protocol is enhanced by incorporating a shared state used to identify the context. An authentication protocol goes through states that include "Identified", "Challenge issued", and "Response pending" prior to finalizing the state "Authenticated").

Electrical and Mechanical Interoperability: A device with a power connector IEC 60906-2 connection cannot simply accept power from other types of connectors. Similarly, USB type C cannot be simply connected to a serial port USB type A. It is also necessary to assure mechanical compatibility along with electrical interoperability which includes factors such as voltage level, ampere level, and frequency.

Radio Communication Interoperability: The communication within the IoT ecosystem is carried out using Radio (wireless) technology which necessitates a common understanding of the frequency band, modulation method, symbol rate, power, and other factors. Radio protocols must be designed to maximize link reliability accomplished by utilizing various types of forward error correction in the link [5, 6].

1.3   ETHICS – LEGAL, SOCIAL CONSIDERATION IN IoT SECURITY

The widespread use of IoT devices is beneficial, offering many solutions but it also raises extremely important ethical concerns. Many organizations, businesses, schools, and shopping centers utilize surveillance technologies which include GPS tracking, video monitoring, phone taps, surfing history, and many monitoring tools to maintain control over the business without workers' permission but it raises concerns about privacy invasion. It also raises concerns such as data loss or physical damage to systems due to cyber-attacks by hackers and viruses. Third-party organizations may masquerade as legitimate businesses in order to gather users' personal and financial data which can be used to carry out illegal transactions. Identity theft is another cybercrime that is increasing in the IoT environment. Bank and credit card information stolen by hackers can be used to impersonate their victims. All of these concerns are of utmost importance and need to be addressed and measures should be to avoid these ethical dilemmas. The appropriate data protection regulations and fundamental ethical principles must be compiled by all data collection devices used to gather personal information [7].

- The IoT infrastructure design must ensure to protection of an individual's personal rights which includes ownership of property, autonomy, and dignity. IoT applications must be carried out transparently and responsibly using established protocols.
- IoT applications must be designed to reduce adverse effects due to personal information gathering and processing avoiding any sort of distress physically or financially.

1.3.1   LEGAL ISSUES PERTAINING TO INTERNET OF THINGS

The legal concerns relating to an IoT service provider can be entirely resolved by establishing and executing agreements that include pertinent terms to protect both the service provider and IoT users. The major factors that should be taken into consideration are listed below:

Data Privacy and Protection: The Information Technology Act (ITA) of 2000 and the "Reasonable Practices and Procedures and Sensitive Personal Data or Information Regulations", 2011, enacted under section 43 A contain provisions relating to an individual's personal information. When the organization fails to establish and maintain "reasonable security methods and processes" to protect sensitive personal data and computer resources security, it is subjected to liability under Section 43A and such entity shall be responsible for paying the person impacted with damages. The violation of the confidentiality and privacy of the data acquired is punishable under section 72 of ITA. The service provider must orchestrate privacy policy, terms, and conditions, limitation of liability, duties of the service provider, consumer/user, indemnity, intellectual property rights, assignment/licensing, and dispute resolution to ensure privacy and protection of data and its sharing. The service provider can execute rigorously crafted nondisclosure agreements with the clients to guarantee compliance with Section 72 of the ITA.

Liability Issues: In the coming year, the data volume is increasing due to the increase in stakeholders. The service providers outsource the data collection, its storage, and processing to third parties which poses a threat to privacy disclosure. It is critical to take reasonable steps to ensure privacy and data protection clauses are protected from violation as per the Indian IT Act. There must be an assurance to maintain an appropriate balance in terms of "risk allocation" to limit the service provider's liability in the event of a breach of data privacy and non-disclosure requirements. The service provider can have software End User Licensing Agreements (EULA) that include the relevant provisions drafted and executed each time an IoT user agrees to use the service provider's software/services [7].

Data Ownership: In the IoT ecosystem, various stakeholders, IoT users, third parties, and a variety of data sources are involved resulting in valuable data being in the hands of various data processors. The IoT service provider as a data controller must determine the scope, extent, manner, and purpose of using personal data and third-party data processors to process data on the controller's behalf and under its control. The IoT service provider must ensure the line between the data controller and the data processor is not distorted. Furthermore, the Machine-Generated Information (MGI) and Machine-to-Machine Communication (M2M) generated in an IoT environment would raise ownership and liability concerns which need precise allocation of responsibilities which includes which user bears liability for any damage caused by IoT users and who owns the information generated by IoT applications.

E-Contracts Privity: The E-contracts in the IoT environment are EULA which represent terms and conditions of software and device use as per requirement of section 10 of the Indian Contract Act. There is no privity of contracts in an IoT environment which can cause complications in case of dispute. As a result, the agreement should include explicit provisions regarding third-party liability and dispute resolution.

Product Liability and Consumer Protection: The individual or business may suffer catastrophic losses due to IoT device failure, data loss, or software compromise. The device failures can be the result of device defects, network failures, or both and thus, it is crucial for IoT device manufacturers to obtain and maintain product liability insurance.

Intellectual Property Rights: An IoT environment facilitates data generation and content creation, including Machine Generated Data. The question that arises is, "Who claims the IP Rights in such content/data/process when original data is created by the interaction of various devices in an IoT environment"? The ownership of the title and claim to the IP Rights must be mentioned in the agreements executed between IoT service providers and device manufacturers/consumers.

1.4   SECURITY CONSIDERATION IN IoT STANDARDIZATIONS

A large amount of data is shared among various IoT devices which involves various security threats and risks without good security standards or security assessment mechanisms. These issues includes eavesdropping, hacking, disclosure of data and information, device security, system software exploitation, IoT device hijacking and ransom ware, insufficient IoT device testing and updates, lack of active device monitoring, lack of efficient and robust security protocols, impersonation, health and safety of users, denial of service (DOS/DDOS), invasions, trespassing, fake and rogue IoT devices, botnet attacks, physical attacks, unintentional damage or loss, disasters and outages, failure or malfunction, issues with dynamic systems, authentication, unsecured wireless networks, side channel attack, man-in-the-middle attack, identity theft, advanced persistent threat, jamming, function creep, buffer overflow, large-scale unauthorized data mining, surveillance, unauthorized access, deletion or modification of data, worms, viruses, and malicious codes, networked system openness, weak passwords, fixed firmware, resource constraints, tamper-resistant packages, or heterogeneous as well as other security threats.

1.5   SECURITY STANDARDS AND ASSESSMENT FRAMEWORKS FOR IoT-BASED SMART ENVIRONMENT

This section discusses different security standards and assessment frameworks that are identified for the security of the IoT ecosystem.

Cyber security Framework: The National Institute of Standard and Technology (NIST) has developed a cyber security framework to assist in managing security risk in IoT infrastructure based on industry standards and best practices. Industry standards and best practices consist of activities, outcomes, and informative references shared by infrastructure sectors providing detailed guidelines to develop organizational profiles. The cyber security framework consists of five key functions (identify, protect, detect, respond, and recover) to manage data and information security risks. The "identify" key function manages assets, business environment, and information technology governance through comprehensive risk assessment and management processes. The "protect" phase incorporates protective mechanisms to ensure critical service delivery and access control for data and information protection. The "detect" phase assists in developing and implementing appropriate activities to identify and detect anomalies in security, monitoring systems, and networks to uncover security incidents. The "respond" key functions develop and implement appropriate actions in response to detected security incidents. It includes suggestions to respond to security events, mitigation procedures, communication processes, and activities to improve security resiliency. The "recovery" phase develops and implements appropriate activities to maintain resilience plans and restore services compromised due to security incidents. The organization should also need to establish to recover from attacks [8].

NIST Risk Management Framework: The Risk Management Framework developed by NIST is a comprehensive measurable framework used by an organization to manage information security risk. This Phase consists of seven steps: prepare, categorize, select, implement, assess, authorize, and monitor. The "prepare" phase ensures necessary activities are carried out to manage security and privacy risks. The "categorize" phase categorizes the system and information processes, stored and transmitted based on impact analysis. The "select" phase assists in determining NIST SP 800-53 to protect systems based on risk assessment. The "implement" phase assists in implementing control and documenting to determine how controls are used. The "assess" phase determines whether controls are in place working normally as intended and producing the desired result. The "authorize" phase enables senior executives to make risk-based decisions to authorize the system and, finally, the "monitor" phase assists in continuously monitoring control implementation and system risks [8].

With growing security and privacy concerns in IoT-based smart environments, the framework discussed above has the potential to be adapted for specific IoT security and privacy risk management areas. The IoT ecosystem consists of a diverse massive number of devices that create security vulnerabilities owing to weak password protection, insecure interfaces, poor IoT device management, and lack of experienced personnel to handle heterogeneous systems. To improve such systems, standards are used that help in implementing new technologies which include communication protocol standards to employ encoding interfaces and provide automation platforms that minimize challenges imposed by multiple computing platforms, operating systems, and devices. A successful heterogeneous strategy is essential to enter the sectors of software-defined networking (SDN) and network functions virtualization (NFV), which shed light on system outcomes and resources. There are wide varieties of security standards. There are numerous new standards have emerged which are applicable to IoT technology The following standards and recommendations are the practices adapted to the IoT domain [9].

NIST-CPS PWG+: NIST's Cyber-Physical Systems Public Working Group (CPS PWG) focuses on IoT systems to increase efficiency and communication between systems and networks that improve life quality such as healthcare, traffic flow control, disaster recovery and response, electric power generation, and supply. Priority-based design principles for scheduling, reliability, security, and data interoperability are discussed in support of CPS [9].

NIST published its NISTIR 7628 guidelines in three volumes covering standards and guidelines for smart grid cyber security and analytical context to improve cyber security policies based on specific risks, characteristics, and vulnerabilities in September 2014.

In November 2016, NIST published SP 800-160 systems security engineering. It addresses IoT from an engineering viewpoint to develop more secure and long-lasting systems (including machine, human, and physical components).

NIST SP 800-53: This special publication on security issues provides security and privacy controls for information systems and organizations [9].

NIST SP 800-30: This special publication was developed to aid organizations in conducting risk assessments for information systems [9].

NIST Special Publication 800-37: This publication describes and provides guidelines for implementing the RMF in information systems and organizations [9].

NIST SP 800-39: This special publication helps to guide an integrated organizational-wide program to manage information security risk to organizational operations, assets, individuals, and other organizations.

NIST SP 800-12: The NIST SP 800-12 special publication is primarily intended for federal and governmental agencies, but it can also be used by others who focus on control and computer security within an organization.

NIST Special Publication 800-14: NIST SP 800-14 provides general descriptions of security principles to assist organizations in understanding cyber security policies.

NIST SP 800-53R1: NIST SP 800-53R1 provides general guidelines with the goal of protecting confidentiality, integrity, and availability.

Health Insurance Portability and Accountability Act (HIPAA): HIPAA provides guidelines to implement methods to secure employee or customer health information and prevent sensitive health information from being disclosed without his/her consent. It also provides plans for health issues and healthcare clearinghouses. These guidelines can be adapted for use in IoT-based smart health systems with the growing number of wearable IoT medical devices.[10]

Family Educational Rights and Privacy Act (FERPA): FERPA was implemented to secure students' private educational records and applied to all schools that receive funds from federal education programs.

Payment Card Industry Data Security Standards (PCI-DSS): This standard is developed with the goal of reducing credit card fraud providing guidelines to protect the safety of card information while storing and transmitting credit card data. This standard can be enforced in IoT devices such as smartphones used to process credit card transactions.

Cybersecurity Maturity Model Certification (CMMC): The CMMC standard is developed to assess defense contractors' cyber security capabilities, readiness, and sophistication. This standard is used by organizations to assess or measure the maturity of a security program regularly and provide guidance on how to progress to a high level of maturity.

Cybersecurity Capability Maturity Model (C2M2): C2M2 developed by the Department of Energy (DOE), USA, allows organizations to voluntarily measure the maturity levels of their cyber security capabilities on a consistent basis.

Federal Financial Institutions Examination Council (FFIEC) Cybersecurity Assessment Tool: The FFIEC Cybersecurity Assessment Tool (FFIEC-CAT) is intended to assist organizations in identifying cybersecurity risks and determining the maturity of cybersecurity programs. This tool provides a measurable process to assess cyber security readiness over time and risk levels across several categories such as delivery channels, connection types, external threats, and organizational characteristics.

North American Electric Reliability Corporation (NERC) 1300: This standard is intended to assist organizations in reducing risks to the reliability of bulk electric systems from the compromise of critical cyber assets.

North American Electric Reliability Corporation Critical Infrastructure Protection (NERC-CIP): The NERC-CIP standard is widely used to provide specific guidance on security mechanisms provided to secure power systems. NERC 1300 and NERC-CIP together enhance benefits to organizations in electricity distribution using smart inverters and IoT devices.

American National Standards Institute/International Society of Automation (ANSI/ISA 62443): The ANSI/ISA 62443 standard specifies requirements, processes, and techniques to provide automation and control systems for industry. It also helps in the development and maintenance of IoT products.

General Data Protection Regulation (GDPR): The GDPR was created for the European Union and imposes data privacy and security obligations on organizations that target or collect data about EU citizens using IoT devices and distributing personal data.

Systems and Organizations Controls (SOC2): SOC2 is a standard developed by the American Institute of CPAs to guide organizations to collect and store personal customer information using cloud services and maintain proper security to which vendors and third parties must adhere. SOC2 standards are designed to ensure the control at a service organization related to the security, availability, integrity, confidentiality, and privacy of the information processed by IoT-based smart systems.

Threat Assessment and Remediation Analysis (TARA): TARA is an international standard developed as part of MITRE portfolio engineering practices to achieve system security during the acquisition process. This standard helps in identifying and assessing vulnerabilities and selecting effective countermeasures to mitigate those vulnerabilities in an IoT-based smart environment [9].

Operationally Critical Threat, Asset, and Vulnerability Evaluation (OCTAVE): This standard was based on three fundamental components: creating asset-based threat profiles, identifying infrastructure vulnerabilities, and developing a security strategy and plans. It defines a comprehensive evaluation method to assist organizations in identifying critical information assets, threats to those assets, and vulnerabilities and determine what information is at risk.

Information Assurance for Small and Medium Enterprises (IASME) Governance: This standard was developed by the IASME consortium and used to certify a company's cyber security posture. It covers Risk assessment and management, monitoring, change management, incident response, and business continuity.

Health Information Trust (HITRUST): This Alliance created a framework that combines the CMMC and New York DOH Office of Health Insurance Programs concerned with organizational security and privacy.

Center for Internet Security V7 (CIS V7): This standard is developed to enhance security standards required by organizations which consist of actionable cyber security requirements.

Control Objectives for Information and Related Technologies (COBIT): The Information Systems Audit and Control Association (ISACA) created COBIT, which focuses on IT security, governance, and management in organizations to improve product quality by incorporating best practices to enhance security.

Nzism Protective Security Requirements (PSR) Framework: The New Zealand government created this framework as part of the Protective Security Requirements (PSR). It outlines the government's requirements to secure personnel, physical, and information security and establishes mandatory security standards for government departments and agencies.

COSO: The framework, namely, the committee of sponsoring organizations, is designed and implemented to guide organizations regarding risk management, internal control, and prevention of fraud.

Integrated Framework for Enterprise Risk Management: This framework is developed under the COSO umbrella and addresses the rise of enterprise risk and improves risk management approaches in response to changing business conditions.

Internal Control-Integrated Framework: This framework assists organizations in designing and implementing internal controls.

Australian Signals Directorate (ASD) Essential 8: The Essential 8 was developed by ASD in collaboration Australian Cyber Security Centre (ACSC) to assist organizations in protecting their systems against a variety of adversaries.

10 Steps to Cybersecurity: The National Cyber Security Centre in the United Kingdom has taken the initiative to provide ten general guidelines on how organizations can protect themselves in cyberspace.

Technical Committee on Cyber Security (TC CYBER) Framework: A framework developed by TC Cyber recommends a set of requirements to improve privacy awareness of individuals and organizations, Understanding telecommunication standards, cyber security ecosystem, protecting personal data and communication, consumer IoT security and privacy, cyber security for critical national infrastructures, network security, cyber security tools.

New Zealand Privacy Act 2020: The Privacy Act 2020, developed by New Zealand's Parliamentary Counsel Office, promotes and protects individual privacy.

Consortium for It Software Quality (CISQ): The CISQ security standard is designed, which is followed by software developers during application development. This standard is used by developers to assess the size and quality of programs and examines risks and vulnerabilities that exist in both complete and incomplete software applications.

Federal Risk and Authorization Management Program (FedRAMP): FedRAMP created a framework to provide a standardized approach to cloud service security authorizations used by the Government to assess cyber threats and risks to various infrastructure platforms, cloud-based services, and software solutions.

FISMA (Act for Federal Information Security): The standard developed by the Cybersecurity and Infrastructure Security Agency (CISA) is implemented with the goal for federal agencies to implement adequate measures for critical information protection from various types of attacks and develop highly effective cyber security programs.

Security Content Automation Protocol (SCAP): The SCAP standard was created by Open SCAP with an emphasis on automated configuration, vulnerability and patch checking, technical control compliance activities, and security measurement.

European Telecommunication Standard Institute: The ETSI is a non-profit organization that aims to create telecommunication standards for providing relevant information regarding various aspects related to cyber security and IoT. Prins et al. highlight several security challenges focusing on NFV, which include problem formulation (NFV-SEC 001), NFV-Related Management Software Security Features Cataloguing (NFV-SEC 002), Security and Trust Advice (NFV-SEC 003), and Privacy and Regulation: Report on the Implications of Lawful Interception (LI) (NFV-SEC 004).

IEEE Standard: P1915.1 is a customized standard that addresses security for virtualized systems. This standard provides a safe and secure prototyping model, structure, and analytics, which are used as a basis to secure SDN/ NFV sites.

Embedded Microprocessor Benchmark Consortium (EEMBC): This has evolved into an industry partnership that creates standards to assist system designers in selecting appropriate terminals and identifying the energy and performance characteristics of the systems. This standard belongs to a benchmark group of big data, cloud, networking, mobile devices (such as tablets and phones), IoT, ultra-low power microcontrollers, digital media, other application areas, automotive, and overall implement and perform analysis with FPMark (floating point), MultiBench (multicore), and CoreMark. The EEMBC SecureMark-TLS Benchmark simulates the cryptographic processes required by the protocol for a secure Internet connection.

GSMA (Global System for Mobile Communications): This is a very logical forum that connects everything in order to transition to a wireless world. It supports Wi-Fi standards, 3G/4G/5G networks, and additional wireless flow optimization. The GSMA has published guidelines, best practices, and self-assessment tools for IoT security.

The Foundation for IoT Security: This foundation was founded in September 2015 to secure IoT, accelerate adoption, and enhance potential benefits by educating and empowering those who identify, create, and use IoT systems and products.

The Open Web Application Security Project (OWASP): This standard was created to identify IoT security concerns and best security choices when developing, measuring, or deploying IoT tools and technologies. OWASP has ten IoT security projects: Firmware Analysis, IoT Vulnerabilities, IoT Attack Surface Areas, ICS/SCADA Software Weaknesses, IoT Testing Guides, IoT Security Principles, Community Information, IoT Security Guidance, IoT Framework Assessment, Design Principles; Consumer, Manufacturer Standards, and Developer.

Online Trust Alliance (OTA): OTA was founded in 2005 to educate users while developing and advancing best practices and tools to improve users' security, privacy, and identity. Trust and identity management are two critical aspects and OTA accomplishes this through data sharing and collaboration via working groups, committees, and training.

The Secure Technology Alliance (STA): In March 2017, STA became a multi-industry association known as the Smart Card Alliance. Its efforts revolve around the adoption, comprehension, and widespread application of embedded chip technology, smart cards, and connected hardware and software for the purpose of security. This includes secure tools and technologies for verification and validation, business, and IoT to ensure data privacy.

The Cloud Security Alliance (CSA): The CSA is dedicated to describing and raising awareness about cloud computing security and conducts specific research on cloud security, actions, certification, education, and products, with an emphasis on IoT.

I am the Cavalry: This organization was founded in 2013 with the goal of addressing computer security and public safety issues for specifically non-profit educational entities playing an important role in the potential impact on community protection and human life. The main focus of this standard is the IoT.

The International Electro-Technical Commission (IEC): The IEC 62443 is a cybersecurity standard for Industrial Control Systems released in 2009. IEC 62443 is the world's most widely used industrial security standard designed in 2007. This standard is constantly evolving, especially in areas such as IoT. The IEC 62351 standard, published in 1999, highlighted mainly end-to-end security, data and communications security in power system management, and associated information exchange [5].

The North American Electric Reliability Corporation (NERC): In 2007, NERC introduced the Critical Infrastructure Protection regulations for bulk power in North America. It is a regulated agency that requires legal compliance, including cyber security [5].

The IoT Global Council: It is a membership organization for new business leaders in the IoT industry founded in May 2014 and is concerned with the data and security of IoT.

The Internet Research Task Force (IRTF): The IRTF has established security-related works with IoT relevance and focus. This guideline includes working groups for DTLS in Constrained Environments (DICE), Authentication and Authorization for Constrained Environments (ACE), and Lightweight Implementation Guidance (LWIG).

The Industrial Internet Consortium (IIC): The IIC Is an umbrella standard with an industrial focus that has produced a specific cyber security framework for IoT.

IoT Security Foundation: The specific goal of this foundation is to enhance the benefits of secure IoT systems. It provides an appropriate mechanism to create, identify, and deploy IoT systems and products.

The US Federal Trade Commission (FTC): The US Federal Trade Commission (FTC) is the primary regulator responsible for IoT devices which includes governmental orders to allow for legitimate and legal liability considerations emphasizing IoT security which affects public safety and life-saving devices.

The International Standards Organization (ISO): The main objective of this standard is to cover cyber security with IoT. There are five sets of ISO standards that include cyber security features as given below.

- **ISO/IEC 27001**: This standard provides guidelines regarding the implementation of the best practices for providing security to management systems and protecting and maintaining data and information in terms of privacy, dependability, and accessibility
- **ISO/IEC 27032**: This standard focuses specifically on cyber security. However, the suggested controls are not as specific or rigid as ISO/IEC 27001. This standard identifies the vectors on which cyber-attacks rely which include cyberspace or the Internet.
- **ISO/IEC 27035**: This standard focuses on event management procedures, cyber security management schemes that are secure, and the support required to respond quickly and efficiently if something is incorrect. This standard also includes regulation in support of modernizing guidelines and procedures, reinforcing current controls.
- **IISO/IEC 27031**: This standard focuses on ICT continuity and employs logical step that begins with event management and prevents an uncontrolled event from becoming a risk to ICT continuity.
- **ISO/IEC 22301**: This standard supports business continuity for information management systems. It also focuses on retrieval and sustaining access to information in the critical event at the final stage of cyber resilience in the critical event occurrence allowing it to reappear and complete with protected functionality.

Furthermore, along with the established standard, there are new emerging standards with the aim to provide interoperability and standardization. Manufacturer Usage Description (MUD), introduces network capabilities to provide an additional layer of protection for devices connected to the IoT system and to improve the security of smart objects. The standard RFC process is currently being used to review MUD.

1.6   SECURITY CHALLENGES IN IoT SECURITY LAYERED MODEL

IoT infrastructure is comprised of various electronic devices, software, sensors, and actuators connected via wireless networks and responsible for collecting and exchanging data. The IoT device can generate, and analyze the behavior of people and take action. Even though IoT applications provide numerous benefits to human life, providing security and privacy services is still a major concern [11].

> **Security of the IoT Perception Layer**: The perception layer in IoT infrastructure consists of a variety of collecting and controlling modules using a variety of sensors such as temperature, sound, vibration, pressure, sensor networks, implantable medical devices (IMDs), Radiofrequency Identification (RFID), and the Global Positioning System. The detection of sensor nodes is affected by physical attacks or compromised by spoofing, node tempering, DoS, reply attacks, side-channel attacks [11].

> **Network Layer Security**: The network layer in IoT infrastructure consists of network and device interfaces, network and communication channels, and extraction of valuable information through intelligent processes In an IoT environment, to enable security to Internet protocol by extending IPv6 over low-power personal area network connected wirelessly to provide secure communication with WSN and secured end to end security without trustworthy gateways [11, 12].

> **Transport Layer Security**: Kothmayr et al. have proposed a two-way authentication scheme based on the DTLS protocol implemented in the transport layer. This security scheme is implemented by exchanging X.509 certificates using RSA keys. This X.509 communicates with 6LoWPANs using standardized communication stacks supported by UDP/IPv6 networking. For DTLS, paper [13] proposed Low personal area network header compression using standardized mechanisms to connect the compressed DTLS to the 6LoW-PAN standard which reduces the number of extra security bits significantly. In the subsequent work proposed by Lithe, the DLTS record header is embedded in DTLS packets integrating CoAP and DTLS. Lithe also proposed a novel scheme of header compression for DTLS to reduce energy consumption leveraging the 6LowPAN standard Because the border router used for 6LoWPAN is unable to assure authentication, the paper [8] proposed security protocols for the Transport Layer adopted in the Internet and do not necessarily imply that the same security levels can be achieved in Low-power as well as Lossy Network, which is still vulnerable to flooding, replay, and amplification attacks [11].

Application Layer Security: This layer in the IoT environment includes a wide range of service domains such as smart homes, healthcare, connected cars, and smart grids. The security threats in the application layer depend on the domain, hence each application area must consider its own security threats and devise countermeasures for each [14]. The majority of novel IoT devices include programmable systems running complex software and resemble general-purpose computers connected to the wireless network which can infected by computer viruses such as trojan [11].

IoT Authentication Scheme: IoT infrastructure connecting large IoT devices enabling M2M communication capabilities, traditional cryptography, legacy authorization, and authentication schemes are not suitable. IoT devices need to authenticate each other by exchanging information using M2M communication which poses a great challenge.

Chen et al. [9] proposed an access control model suitable for the distributed IoT environment; in this, capability-based access control data requestor communicates with any device in a group using a single token ensuring end-to-end security with IPSec using the identical local identifier to identify each device in the group and network prefix of the access group. The devices in the group can validate the token by using its unique identifier and prefix. It can also control access based on the requester's ULA in the token [13].

1.7 IoT SECURITY LIFE CYCLE

The security life cycle as shown in Figure 1.1, determines in-depth defense and security techniques and tools deployed in the various layers embedded in the IoT ecosystem. Security planning is a mandatory step to prevent and stop attacks, which ensures messages and data are encrypted. Access control and firewalls are deployed to ensure application security and that the IoT ecosystem is free of vulnerabilities. The IoT system contains a number of vulnerable components [15] which makes it critical to monitor the system continuously for attacks or anomalies that indicate an attack. An anomaly detection system is essential to respond to attacks due to a dynamically

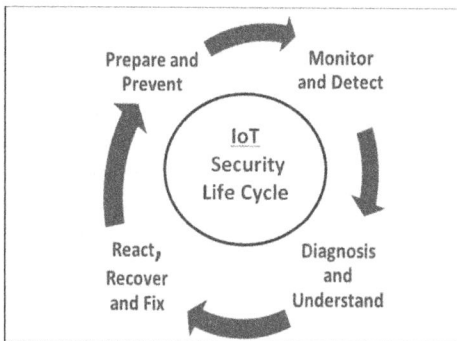

FIGURE 1.1 Security life cycle.

changing system [16]. To perform a precise analysis of the attack or anomaly, it is necessary to collect as much information as possible before the system changes. If the attack disrupts communication using interference, it is critical to determine the source of interference and do a real-time diagnosis of the attack or anomaly, which must respond quickly by executing various actions and activities [17]. Real-time response to an attack is an essential defense strategy. The IoT security life cycle consists of four phases: (1) prepare and prevent, (2) monitor and detect, (3) diagnose and understand, and (4) react, recover, and fix.

Prepare and Prevent: To prepare a system to defend against an attack several security blocks are required. The critical issue addressed in the planning process is to identify the location of security tools deployment which is a prerequisite for IoT systems.

Access Control: Sharing sensitive data among multiple parties using cryptographic mechanisms and access control mechanisms using access control lists. Access control models need to be designed to address various use cases in the context of IoT. To control which end-user applications access which IoT devices, what purpose or context, to control which IoT devices perform functions, collect information from the physical environment or receive data from other IoT devices, to control which applications are running on a gateway, in the cloud. To control which cloud users access data from a specific IoT device. Attribute-based access control model and context-based access control model need to be defined, implemented, and deployed for IoT to specify authorization, and for the architecture to manage and enforce control [18].

Application Security: Many attacks rely on errors in software and application code. There are a plethora of software security techniques which include ensuring memory safety and integrity of application execution control flow. Approaches range from compiler techniques that detect errors such as buffer overflow vulnerabilities in source code, to approach to instrument or randomize binaries to ensure vulnerabilities are not exploited by attackers. The other approach is monitoring application execution to ensure that the attacker has not altered the execution control flow. detecting and patching logical flaws in application code. Logical flaws include authentication bypass and insufficient authorization which cannot be detected by techniques used to detect buffer overflow or integrity of the execution control flow. To address the issue of logical flaws, it needs automatic tools to detect and patch these flaws. To maintain software security, it is difficult to determine specific techniques to handle a variety of IoT applications and scenarios [19].

Security Provisioning: The optimal allocation of security resources in a distributed system is referred to as the distributed security provisioning problem which needs to be addressed in comparison to traditional networks. The approach used to address security provisioning issues is based on Game Theory and Pareto Analysis. These approaches are not suitable in the context of IoT systems as the problem is much larger in scope. Furthermore,

IoT systems can change quickly and hence security provision allocation may need to change. The preliminary research on security provisioning for IoT systems has focused on static IoT scenarios and mobile IoT scenarios where mobility patterns are known in advance [1].

Monitor and Detect: The IoT ecosystem needs to provide a process for detecting attacks and monitoring the ecosystem to identify anomalies. Extensive research is being carried out on network intrusion detection and anomaly detection systems by various researchers. SNORT and Bro are open-source IADSes that rely on network data gathered by sniffers to detect attacks using signature matching over the data but these IADSes are not suitable for the IoT domain. The Kalis system is designed as a self-adapting, knowledge-driven intrusion detection system that employs a variety of communication protocols for real-time detection of an attack on an IoT. It collects data about network features and entities and uses data to dynamically configure the most effective set of detection techniques. Kalis employs an IoT attack taxonomy based on target types, system types, and components like IoT routers, hubs, and devices. It also employs a taxonomy of the relationships between monitored network/entity features and security incidents. By leveraging such knowledge, Kalis is suitable for detecting attacks as well as anomalies that indicate an attack and also suitable to determine the type of attack as compared to more traditional IDSes. Implementing IDS directly to IoT devices may not always be possible especially if it incorporates some complicated machine learning algorithms. One such solution employs edge computing by putting a system call logger on IoT devices, which is designed to make it difficult for malware to manipulate the log. The IDS engine is deployed on an edge server and analyses logs on a regular basis using a machine-learning approach. A promising open path is to develop community-based IDSes for IoT systems, analogous to SNORT. The second challenge is the automatic acquisition of knowledge about the IoT system to improve security. Relevant knowledge includes the understanding of IoT device types, software that runs on these devices, and specialized communication protocols to develop mechanisms for IDSes to automatically collect knowledge with the help of programs running on cloud and/or edge servers [19].

Diagnose and Understand: Understanding security-related events correctly is critical for a successful defense. If an IoT device is supposed to collect and deliver data on a regular basis, abnormal behavior should be detected using an anomaly detection system in order to determine whether the device or network is infiltrated, which is known as a jamming assault. It is crucial to perform such a diagnosis in a dynamic system. The major reason is mobility as IoT systems rapidly change which makes it critical to analyze the system while anomaly is being observed. Delaying such analysis may result in the loss of relevant information. Midi et al. [20] provided an early strategy for sensor networks that focuses on abnormalities in data capture and transmission across sensor networks. The method can determine whether data expected from a network sensor is not being transferred because the sensor [21] has been compromised or because the wireless network is jammed.

Link quality indicator (LQI) and received signal strength indicator (RSSI) are the two metrics that are being implemented to formulate the diagnosis in the IoT ecosystem. If it is determined that the data are not being delivered due to a jamming attack, the technique can identify the jamming source. Good diagnosis tools for IoT devices tackle various issues such as learning and characterizing network interference duration to distinguish between naturally occurring and attack-caused interference. A detailed diagnosis is conducted to identify whether the IoT device has been compromised and software components affected by malware. Such knowledge is crucial and helps to decide whether the IoT device can still run with limited capabilities. It also determines the purpose of the attack and the appropriate response action to take [1].

React, Recovery, and Fix: Protecting the system from attack detection and diagnosis are not sufficient mechanisms. The mechanism is to be designed to respond instantly to an attack by taking actions that allow the system to continue operation and block the attack. The existing procedures for making decisions on attack responses by human security analysts and administrators are not suitable for smart IoT systems. In smart systems, to support autonomous response, a policy-based response equivalent to database trigger design is developed. It is based on a security event or anomaly and provides actions for different security estimates based on incident conditions. Such policies are installed on sensors and are executed automatically whenever an incident or anomaly is detected. Many intriguing difficulties arise in the design of such an approach, such as how to deal with system partitions, which result in conflicting response actions by sensors in distinct partitions, and malicious sensors, which may attempt to undermine the responses of other benign sensors. Midi et al. discuss approaches to such challenges which include the effort to perform duplicate data gathering or transmission [1].

1.8   SUMMARY

This chapter presented the IoT ecosystem architecture and its basic components. It also covered the legal, ethical, and societal considerations that are taken into consideration for IoT security and steps taken to avoid ethical boundaries, and also talked about the security layered model and the security concerns it presents. Then, it addressed existing classification systems for IoT threats and security mechanisms, IoT authentication schemes and architectures, and security concerns and solutions in four layers: perception, network, transport, and application. This chapter also addresses IoT standardization, IoT security concerns, and a framework for security assessment. Finally, a review of the life cycle of IoT security is done.

REFERENCES

1. O. Said and M. Masud, "Towards Internet of Things: Survey and future vision," *International Journal of Computer Networks*, vol. 5, no. 1, pp. 1–17, 2013.

2. T. Kavitha, V. Ajantha Devi, S. Neelavathy Pari and Sakkaravarthi Ramanathan, *Internet of Everything: Smart Sensing Technologies*, Nova Science Publishers, Publication Date: June 17, 2022, doi: 10.52305/PNQM1088

3. J. Ranjith and M. V. Sarobin, "Security challenges prospective measures in the current status of Internet of Things (IoT)," *2022 International Conference on Connected Systems & Intelligence (CSI)*, Trivandrum, India, 2022, pp. 1–8, doi: 10.1109/CSI54720. 2022.9923984

4. Pallavi Sethi and Smruti R. Sarangi, "Internet of things: Architectures, protocols, and Applications" *Hindawi Journal of Electrical and Computer Engineering*, vol. 1, p. 25, 2017, doi: 10.1155/2017/9324035

5. Manju Lata and Vikas Kumar, "Standards and regulatory compliances for IoT security," *International Journal of Service, Science, Management, Engineering and Technology*, vol. 12, no. 5, Septembet–October 2021.

6. Hussain Alqarni, Wael Alnahari and Mohammad Tabrez Quasim, "Internet of things (IoT) security requirements: Issues related to sensors," *2021 National Computing Colleges Conference (NCCC)*, IEEE, 2021.

7. Bushra Siddiqua Oosman and Ravishankar Dudhe, "Review on the ethical and legal challenges with IoT," *International Conference on Computational Intelligence and Knowledge Economy (ICCIKE)*, Amity University Dubai, UAE, 17–18 March, 2021, 978-1-6654-2921-4/21/$31.00 ©2021.

8. E. Bertino, "IoT security a comprehensive life cycle framework," *2019 IEEE 5th International Conference on Collaboration and Internet Computing (CIC)*, Los Angeles, CA, USA, 2019, pp. 196–203, doi: 10.1109/CIC48465.2019.00033

9. Nickson M. Karie, et al., "A review of security standards and frameworks for IoT-based smart environments," *IEEE Access*, vol. 9, pp. 121975–121995, 2021.

10. Haider Ali Khan, et al., "IoT based on secure personal healthcare using RFID technology and steganography," *International Journal of Electrical and Computer Engineering*, vol. 11, no. 4, p. 3300, 2021.

11. Ankit R. Patel and Jignesh Kumar A. Chauhan, "IoT security a layered approach for attacks and defences," *International Journal of Innovative Research in Technology*, vol. 8, no. 7, 2021. ISSN: 2349-6002 IJIRT.

12. R. K. Kodali, S. C. Rajanarayanan, A. Koganti and L. Boppana, "IoT based security system," *TENCON 2019 - 2019 IEEE Region 10 Conference (TENCON)*, Kochi, India, 2019, pp. 1253–1257, doi: 10.1109/TENCON.2019.8929420

13. J. Zhang, H. Chen, L. Gong, J. Cao and Z. Gu, "The current research of IoT security," *2019 IEEE Fourth International Conference on Data Science in Cyberspace (DSC)*, Hangzhou, China, 2019, pp. 346–353, doi: 10.1109/DSC.2019.00059

14. A. Radovici, C. Rusu and R. Şerban, "A survey of IoT security threats and solutions," *2018 17th RoEduNet Conference: Networking in Education and Research (RoEduNet)*, Cluj-Napoca, Romania, 2018, pp. 1–5, doi: 10.1109/ROEDUNET.2018.8514146

15. T. Kavitha and D. Sridharan "Security vulnerabilities in wireless sensor networks: A survey" *International Journal on Information Assurance and Security (1554-1010) (JIAS)*, no. 5, pp. 031–044, 2010.

16. B. Al-Shargabi and O. Sabri, "Internet of Things: An exploration study of opportunities and challenges," *2017 International Conference on Engineering & MIS (ICEMIS)*, Monastir, Tunisia, 2017, pp. 1–4, doi: 10.1109/ICEMIS.2017.8273047

17. I. Mashal, O. Alsaryrah, T.-Y. Chung, C.-Z. Yang, W.-H. Kuo, and D. P. Agrawal, "Choices for interaction with things on Internet and underlying issues," *Ad Hoc Networks*, vol. 28, pp. 68–90, 2015.

18. R. Mahmoud, T. Yousuf, F. Aloul and I. Zualkernan, "Internet of Things (IoT) security: Current status, challenges and prospective measures," *2015 10th International Conference for Internet Technology and Secured Transactions (ICITST)*, London, UK, 2015, pp. 336–341, doi: 10.1109/ICITST.2015.7412116

19. K. Zhang, M. C. Y. Cho, C.-W. Wang, C.-W. Hsu, C. -K. Chen and S. Shieh, "IoT Security: Ongoing challenges and research opportunities," *2014 IEEE 7th International Conference on Service-Oriented Computing and Applications*, Matsue, Japan, 2014, pp. 230–234, doi: 10.1109/SOCA.2014.58

20. D. Midi and E. Bertino, "Node or Link? Fine-Grained Analysis of Packet-Loss Attacks in Wireless Sensor Networks," *ACM Transactions on Sensor Networks (TOSN)*, vol. 12, no. 2: 8:1–8:30, 2016.

21. T. Kavitha and S. Saraswathi, "New sensing technologies or/and devices for emergency response and disaster management" Book Chapter, IGI Global International Publisher, pp. 1–40, July 2017, doi: 10.4018/978-1-5225-2575-2, Ch001. ISBN: 9781522525752.

2 IoT Hardware, Platform, and Security Protocols

R. Malathy and Sheeja V. Francis
Jerusalem College of Engineering, Anna University,
Chennai, India

2.1 INTRODUCTION

The devices that are capable of connecting to a network via the internet are known as Internet of Things (IoT) devices. The IoT devices do not include traditional computers such as laptops and computers. There was a time when it was only possible to connect standard computers to the internet. But, today, any digital appliance capable of generating huge amounts of data and connected to the internet falls under IoT devices.

2.1.1 PROPERTIES OF IoT DEVICES [1]

i. **Sense**: An IoT device must be capable of sensing [2] its physical surroundings. It must be able to retrieve the data from external surroundings.
ii. **Sense and Receive Data**: The data and information that an IoT device collects are transferred to other devices via a network or connection.
iii. **Analyze**: Collecting data is meaningless if one cannot operate on it. Therefore, an IoT device must be able to analyze the data that it collects.
iv. **Controlled**: There must be a mechanism that ensures that an IoT device is under the direct control of the end user. Otherwise, it may lead to system failure or even hacking.

2.2 BUILDING BLOCKS OF IoT HARDWARE

The hardware utilized in IoT systems includes devices for a remote dashboard, devices for control, servers, a routing or bridge device, and sensors. These devices manage key tasks and functions such as system activation, action specifications, security, communication, and detection to support specific goals and actions. Figure 2.1 picturizes the building blocks.

Thing: The asset to be controlled or monitored can be standalone, or incorporated in the smart device. "Thing" in IoT is the asset that you want to control or monitor or measure, that is, observe closely. In many IoT products, the "Thing" gets fully incorporated into a smart device. For example, products like a smart refrigerator or an automatic vehicle. These products control and monitor themselves.

DOI: 10.1201/9781003477327-2

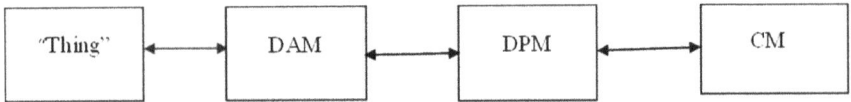

| "Thing" | ← → | DAM | ← → | DPM | ← → | CM |

FIGURE 2.1   IoT hardware: building blocks.

Data Acquisition Module (DAM): The Data Acquisition Module (DAM) focuses on acquiring physical signals from the thing which is being observed or monitored and converting them into digital signals that can be manipulated or interpreted by a computer. This is the hardware component of an IoT system that contains all the sensors that help in acquiring real-world signals such as temperature, pressure, density, motion, light, and vibration. The type and the number of sensors you need depend on your application [3]. This module also includes the necessary hardware to convert the incoming sensor signal into digital information for the computer to use. This includes conditioning of incoming signals, removing noise, analog-to-digital conversion, interpretation, and scaling.

Data Processing Module (DPM): The Data Processing Module (DPM) represents the unit that processes the data and performs operations on it, as well as storing it. It also requires storage capability. This happens because some data devices process the acquired data themselves, instead of transmitting them upstream. On the other hand, there are IoT devices which do not possess this capability and need intermediary entities in order to store and process the data. These entities are either gateway devices or cloud applications used for further aggregation.

Communication Module (CM): The last building block of IoT hardware is the communications module. This is the part that enables communications with your Cloud Platform and with third-party systems either locally or in the cloud.

2.3   KEY COMPONENTS OF IoT SYSTEMS

The hardware components for the IoT have arrived in a variety of shapes and sizes, depending on the tasks at hand. For instance, the core CPUs are used to operate the phones, the sensors are utilized to perceive data from the physical environment, and the data are processed and analyzed using edge devices. In the case of wired mode, the hardware aspects are the essential components of the IoT platform, and the capabilities of these devices have only grown in significance as IoT has progressed.

IoT Hardware components can vary from low-power boards to single-board processors like the Arduino Uno which are basically smaller boards that are plugged into mainboards to improve and increase their functionality by bringing out specific functions or features (such as GPS, light and heat sensors, or interactive displays). A programmer specifies a board's input and output, and then creates a circuit design to illustrate the interaction of these inputs and outputs.

Another well-known IoT platform is Raspberry Pi 2, which is a very affordable and tiny computer that can incorporate an entire web server. Often called "RasPi," it has enough processing power and memory to run Windows 10 on it as well as IoT Core. RasPi exhibits great processing capabilities, especially when using the Python programming language. BeagleBoard is a single-board computer with a Linux-based OS that uses an ARM processor, capable of more powerful processing than RasPi [4].

2.4 OTHER IoT HARDWARE DEVICES

Sensors: A sensor is an IoT device that senses physical changes in the environment and sends the data for manipulation via a network. Clouds store the data for future reference. Sensors monitor data and collect information constantly. These devices consist of a variety of modules [5] such as energy modules, RF modules, power management modules, and sensing modules.

Wearable Devices: Wearable devices are a benchmark revolution of the IoT industry. These are IoT devices that humans can wear on their bodies to regulate and perform a variety of tasks. These wearables are capable of tracking glucose levels, monitoring heart attack risks, coagulation, asthma, and daily steps, and tracking calorie consumption. The current smart wearable devices include those for the head (helmets, glasses), neck (jewelry, collars), arm (wristwatches, wristbands, rings), torso (clothing pieces, backpacks), and feet (shoes, socks).

Basic Devices: Traditional computers such as desktops, tablets, and cell phones are still an integral part of any IoT ecosystem. The desktop provides the user with a very high level of control over the system and its settings. The tablet acts as a remote and provides access to the key features of the system. A cell phone allows remote functionality and some essential settings modification. Other key connected devices include standard network devices like routers and switches.

2.5 IoT PLATFORMS

All IoT devices are connected to other IoT devices and applications [6, 7] to transmit and receive information using protocols. There is a gap between the IoT device and IoT application. An IoT platform bridges the gap between device sensors and data networks. It provides an insight into the data used in the backend application. An IoT platform is a set of components that allows developers to spread out the applications, remotely collect data, secure connectivity, and execute sensor management. It manages the connectivity of the devices and allows developers to build new mobile software applications. It facilitates the collection of data from devices and enables business transformation. It connects different components, ensuring an uninterrupted flow of communication between the devices. Some of them are listed below.

Amazon Web Services (AWS) IoT Platform: Amazon Web Service IoT platform offers a set of services that connect to several devices and maintain security as well. This platform collects data from connected devices and performs real-time actions.

Microsoft Azure IoT Platform: Microsoft Azure IoT platform offers strong security mechanisms, scalability, and easy integration with systems. It uses standard protocols that support bi-directional communication between connected devices and platforms. Azure IoT platform has Azure Stream Analytics that processes a large amount of information in real-time generated by sensors. Some common features provided by this platform are information monitoring, rules engine, device shadowing, and identity registry.

Google Cloud Platform: Google Cloud Platform is a global cloud platform that provides a solution for IoT devices and applications. It handles a large amount of data using Cloud IoT Core by connecting various devices. It allows applying Big Query analysis or Machine learning to the data. Some of the features provided by Google Cloud IoT Platform [8] are Cloud IoT Core, Speed Up IoT devices, Cloud Publisher-Subscriber, and Cloud Machine Learning Engine.

IBM Watson IoT Platform: The IBM Watson IoT platform enables the developer to deploy the application and build IoT solutions quickly. This platform provides the following services: real-time data exchange, device management, secure communication, Data sensing, and weather data services.

Artik Cloud IoT Platform: The Artik Cloud IoT platform was developed by Samsung to enable devices to connect to cloud services. It has a set of services that continuously connect devices to the cloud and start gathering data. It stores the incoming data from connected devices and combines this information. This platform contains a set of connectors that connect to third-party services.

2.5.1 COMPONENTS OF IoT PLATFORM

In a more sophisticated form, the platform consists of a variety of important building blocks: connectivity and normalization, device management, database, processing and action management, analytics, visualization, additional tools, and external interfaces.

1. **Connectivity and Normalization**

   Every IoT platform starts with a connectivity layer. It has the function of bringing different protocols and different data formats into one "software" interface. This is necessary in order to ensure all devices can be interacted with and data is read correctly. Having all device data in one place and in one format is the basic necessity to monitor, manage, and analyze IoT devices.

2. **Device Management**

   The device management module of an IoT platform ensures the connected objects are working properly and its software and applications are updated and running. Tasks performed in this module include device provisioning, remote configuration, management of firmware/software updates, and troubleshooting. As thousands or even millions of different devices become part of an IoT-enabled solution, bulk actions and automation are essential to control costs and reduce manual labor.

3. **Database**

Data storage is a central piece of an IoT platform. The management of device data brings the requirements for databases to a new level:

- **Volume**: The amount of data that needs to be stored can be massive. In many IoT solutions, only a minority of the generated data can be stored.
- **Variety**: Different devices and different sensor types produce very different forms of data.
- **Velocity**: Many IoT cases require the analysis of streaming data to make instant decisions.
- **Veracity**: In some instances, sensors produce ambiguous and inaccurate data. An IoT platform therefore usually comes with a cloud-based database solution that is distributed across different sensor nodes. It should be scalable for big data and should be able to store both structured (SQL) and unstructured data (NoSQL).

4. **Processing and Action Management**

The data that is captured in the connectivity and normalization module and that is stored in the database gets brought to life in this part of the IoT platform. A rule-based event–action–trigger allows the performance of "smart" actions based on specific sensor data. In a smart home, for instance, an event–action–trigger can be defined so that all lights get turned off when a person leaves the house. The technical realization often comes in the form of an If-this-then-that rule (IFTTT): If the GPS signal indicates Jason's smartphone is more than 5 yards away from his house, then turn off all the lights in his house.

5. **Analytics**

Many IoT use cases go beyond action management and require complex analytics to get the most out of the IoT data stream. In a smart home, for example, the analytics engine can provide the algorithms that allow the IoT platform to learn which combination of lights and heating is preferred by the user, at what time of the day, and in relation to the outside weather conditions. The analytics engine encompasses all dynamic calculations of sensor data, from basic data clustering to deep machine learning.

6. **Data Visualization**

Sometimes also referred to as "visual analytics," data visualization presents a much-underrated part of the IoT platform. The combination of the human eye and brain is still far superior to most analytic and rule-based engines. That is why data visualization is so important: it enables humans to see patterns and observe trends. Visualization comes in the form of line-, stacked-, or pie charts or 2D- or even 3D-models. The visualization dashboard that is available to the manager of the IoT platform is often also included in the prototyping tools that an advanced IoT platform provides.

7. **Additional Tools**

Advanced IoT platforms often offer an additional set of tools for the developer and the manager of the IoT solution. Development tools allow the IoT developer to prototype and test the IoT case. Management-focused tools support the daily operations of the IoT solution. An example is an "access management" tool that determines who has access to which device and to which data. Another tool is "reporting," which allows for data export (e.g., in a .csv or .json format) as well as data queries and other forms of structured output.

8. **External Interfaces**
   IoT-enabled businesses are rarely built standalone and on a green field. In established companies, it is crucial that the IoT integrates with existing ERP systems, management tools, manufacturing execution systems, and the rest of the wider IT ecosystem.

2.6 IoT PROTOCOLS IN OSI MODEL

The Open Systems Interconnection (OSI) model provides a map of the various layers that send and receive data. Each IoT protocol in the IoT system architecture enables device-to-device, device-to-gateway, gateway-to-data center, or gateway-to-cloud communication, as well as communication between data centers. The protocols in the layers of the OSI model are as follows.

I. **Application Layer**
   The application layer serves as the interface between the user and the device within a given IoT protocol.
   1. **Advanced Message Queuing Protocol (AMQP)**: A software layer that creates interoperability between messaging middleware. It helps a range of systems and applications work together, creating standardized messaging on an industrial scale.
   2. **Constrained Application Protocol (CoAP)**: A constrained-bandwidth and constrained-network protocol designed for devices with limited capacity to connect in M2M communication. CoAP is a document-transfer protocol that runs over (UDP).
   3. **Data Distribution Service (DDS)**: A versatile peer-to-peer communication protocol that does everything from running tiny devices to connecting high-performance networks. DDS streamlines deployment, increases reliability, and reduces complexity.
   4. **Message Queue Telemetry Transport (MQTT)**: A messaging protocol designed for lightweight M2M communication and primarily used for low-bandwidth connections to remote locations. MQTT uses a publisher-subscriber pattern and is ideal for small devices that require efficient bandwidth and battery use.

II. **Transport Layer**
   In any IoT protocol, the transport layer enables and safeguards the communication of the data as it travels between layers. The protocols of this layer are as follows:
   1. **Transmission Control Protocol (TCP)**: It offers host-to-host communication, breaking large sets of data into individual packets and resending and reassembling packets as needed.

2. **User Datagram Protocol (UDP)**: A communications protocol that enables process-to-process communication and runs on top of IP. UDP improves data transfer rates over TCP and best suits applications that require lossless data transmissions.

III. **Network Layer**

The network layer of an IoT protocol helps individual devices communicate with the router. The protocols in this layer are as follows:

1. **Internet Protocol (IP)**: Many IoT protocols utilize IPv4, while more recent executions use IPv6. This recent update to IP routes traffic across the internet and identifies and locates devices on the network.
2. **6LoWPAN**: This IoT protocol works best with low-power devices that have limited processing capabilities.

IV. **Data Link Layer**

The data layer is the part of an IoT protocol that transfers data within the system architecture, identifying and correcting errors found in the physical layer.

1. **IEEE 802.15.4**: A radio standard for low-powered wireless connection. It is used with ZigBee, 6LoWPAN, and other standards to build wireless embedded networks.
2. **LPWAN**: Low-power wide-area networks (LPWAN) networks enable communication across distances of 500 m to over 10 km in some places. LoRaWAN is an example of LPWAN that is optimized for low power consumption.

V. **Physical Layer**

The physical layer is the communication channel between devices within a specific environment.

1. **Bluetooth Low Energy (BLE)**: BLE works natively across mobile operating systems and is fast becoming a favorite for consumer electronics due to its low cost and long battery life.
2. **Ethernet**: This wired connection is a less expensive option that provides fast data connection and low latency.
3. **Long-Term Evolution (LTE)**: A wireless broadband communication standard for mobile devices and data terminals. LTE increases the capacity and speed of wireless networks and supports multicast and broadcast streams.
4. **Near Field Communication (NFC)**: A set of communication protocols using electromagnetic fields that allows two devices to communicate from within 4 cm of each other. NFC-enabled devices function as identity keycards and are commonly used for contactless mobile payments, ticketing, and smart cards.
5. **Wi-Fi/802.11**: Wi-Fi/802.11 is a standard in homes and offices. Although it is an inexpensive option, it may not suit all scenarios due to its limited range and 24/7 energy consumption.
6. **Z-Wave**: A mesh network using low-energy radio waves to communicate from appliance to appliance.

7. **ZigBee**: An IEEE 802.15.4-based specification for a suite of high-level communication protocols used to create personal area networks with small, low-power digital radios.

2.7 IoT SECURITY PROTOCOLS

IoT devices communicate using IoT protocols. IoT protocols ensure that information from one device or sensor gets read and understood by another device, a gateway, or a service. Different IoT protocols have been designed and optimized for different scenarios and usage [9]. The Top five IoT security protocols are seen in detail below.

MQTT – Message Queuing Telemetry Transport: MQTT is one of the most common security protocols used in IoT security. It was invented by Dr. Andy Stanford-Clark and Arlen Nipper in 1999. MQTT is a client-server communicating messaging transport protocol. The MQTT runs over TCP/IP or over other conventions that provide requested, lossless, two-way associations. The features of MQTT are that it's a simple and extremely lightweight protocol with easy and fast data transmission. It is designed for constrained devices as well as for low-bandwidth, unreliable, or high-latency networks. Minimum use of data packets ensures less network usage. Optimal power consumption saves the battery of the connected devices, making it perfect for mobile phones and wearables where battery consumption needs to be minimal.

CoAP – Constrained Application Protocol: CoAP is a web transfer protocol designed for constrained devices (like microcontrollers) and constrained networks such as low-power or lossy networks. The features of CoAP are similar to HTTP; CoAP is based on the REST model. Clients access the resources made available by servers under URLs using methods like GET, PUT, POST, and DELETE. CoAP is designed to work on microcontrollers, which makes it perfect for IoT as it requires millions of inexpensive nodes. It uses minimal resources, both on the device and on the network. Instead of a complex transport stack, it gets by with UDP on IP. It is one of the most secure protocols as its default choice of DTLS parameters is equivalent to 3,072-bit RSA keys. CoAP uses UDP (User Datagram Model) to transport information and therefore relies on UDP security aspects to protect the information. It uses Datagram TLS over UDP. CoAP has been designed to have a simple and user-friendly interface with HTTP for integration with the Web and supports functions such as multicast support and low overhead issues, thus contributing to security in the IoT.

DTLS – Datagram Transport Layer Security: The DTLS is an IoT security protocol designed to protect data communication among datagram-based applications. It is based on TLS (transport layer security) protocol and provides the same level of security. It is used in applications such as live video feeds, video streaming, gaming, VoIP, and instant messaging where loss of data is comparatively less important than latency. One of the features of DTLS is that it uses a retransmission timer to solve the issue of packet loss. If the timer terminates before the client receives the confirmation message from the server, then the client retransmits the data. The issue of reordering is solved by giving each message a specific sequence number. This helps in determining if the next message received is in sequence or not. If it is out of sequence, it is put in a queue and handled when the sequence number is reached. DTLS is unreliable and does not guarantee the delivery of data, even for payload information.

6LoWPAN: 6LoWPAN (IPv6 over Low Power Wireless Personal Area Networks) is a protocol for low-power networks like IoT systems and wireless sensor networks. 6LoWPAN plays a key role in domains like smart home automation, industrial monitoring, smart grids, and general automation. One of the features of 6LoWPAN is that it is used to carry data packets in the form of IPv6 over various networks. It Provides end-to-end IPv6 and hence provides direct connectivity to a wide variety of networks including direct connectivity to the internet. It is used for protecting the communications from the end-users to the sensor network.6LoWPAN security for IoT uses AES-128 link layer security which is defined in IEEE 802.15.4 for its security. Link authentication and encryption are used to provide security and additional security is provided to TLS mechanisms, which runs over TCP.

ZigBee: It provides efficient M2M communication from 10–100 m away in low-powered embedded devices like radio systems. It is a cost-effective, open-source wireless technology. ZigBee provides standardization at all layers, which enables compatibility between products from different manufacturers. Due to its mesh architecture, devices tend to connect with every device in the vicinity. This helps in expanding the network and making it more flexible. ZigBee uses "Green Power" which facilitates lower energy consumption and cost. ZigBee helps in the scalability of networks as it supports a high number (about 6,550) of devices. ZigBee supports two security models:

a. **The Centralized Security Network**: This provides higher security and is also more complicated as it uses a third device called Trust Centres which are applications that run on the device trusted by other devices within the ZigBee network. The Trust Centre forms a centralized network and configures and authenticates each device to join the network by giving it a unique TCLK (TC Link Key). The TC also determines the network key. To join the network, each device must be configured with the link key which is used to encrypt the network when passing it from the TC to a newly joined entity.

b. **The Distributed Security Network**: In DSN, there is no Central Node or Trust Center; this makes it simpler but less secure than the CSN. Every router can start distributed networks on its own. When a node joins the network, it only receives the network key.

2.8   IoT DEVICE SECURITY AND CHALLENGES

The IoT paves the way for technical revolution. However, it is easier to conduct attacks on IoT devices. Some dynamic threats and challenges for IoT devices are listed below:

i. Due to the lack of security mechanisms and impotent encryption processes, transmitted data can be attacked or misused by unauthorized interference during data transmission across the network. Maintaining data integrity and confidentiality must be needed during data transmission.

ii. Personal information or messages from the sender to the receiver can be affected by the attacker and wrong information can be sent.

iii. IoT devices are resource-constrained, which means they have limited computational and storage capabilities. It is not possible for the developers to add the required features to defend against security threats.

iv. Some security vulnerabilities in network services like Exploitable UDP Services, Denial-of-Service (DoS) via Network Device Fuzzing, Buffer Overflow, and DoS might allow an attacker to acquire unauthorized access to the IoT device.

v. An insecure cloud interface or insecure mobile interface due to poor authentication and unencrypted data transmission might allow unauthorized access to the device and the collected data.

vi. In some IoT applications, device life is shorter, and some applications have longer device life. Security techniques should be designed based on device lifetime and application. If the device's life is longer, then it should have the option to update security features to defend it from new attacks and security threats.

To protect devices from the above-mentioned security threats it is important not only to provide reliable but also realistic (i.e., efficient) solutions. There are some limitations to applying traditional security schemes directly in the IoT devices based on hardware, software, and network which are discussed below.

Most IoT devices are battery-dependent. Thus, using a low-power CPU that has the lowest computation power is considered of paramount importance. However, using such a processor makes it difficult to execute computationally heavy cryptographic algorithms. IoT devices have limited memory than other traditional devices and use a lightweight version of the operating system. Therefore, traditional security algorithms cannot be used in these devices with small memory as the process might not get enough space after booting up the operating system and some software. However, traditional security algorithms are not designed with memory efficiency consideration. Sometimes, IoT devices are needed to be placed in remote areas, and it is not possible to monitor them continuously. If a single IoT device from a network is attacked by an unauthorized person, an attacker might change the programs of the device or replace the device with another malicious node.

2.9 SUMMARY

The Internet of Things (IoT) is a new revolution in the capabilities of the endpoints that are connected to the internet, and is being driven by the advancements in capabilities in sensor networks, mobile devices, wireless communications, networking, and cloud technologies. The first part of the chapter covers the building blocks of IoT and its characteristics. A taxonomy of IoT systems is proposed, comprising various IoT levels with increasing levels of complexity. Domain-specific IoT and their real-world applications are described. The readers can learn about functional blocks of IoT systems including device communication, services, management, security, and application blocks, and learn about IoT protocols for link, network, transport, and application layers. Link layer protocols determine how the data is physically sent over the network.

REFERENCES

1. T. Kavitha, V. Ajantha Devi, S. Neelavathy Pari and Sakkaravarthi Ramanathan, *Internet of Everything: Smart Sensing Technologies*, Nova Science Publishers, Publication Date: June 17, 2022, doi: 10.52305/PNQM1088.
2. T. Kavitha and S. Saraswathi, "New sensing technologies or/and devices for emergency response and disaster management," Book Chapter, IGI Global International Publisher, pp. 1–40, July 2017, doi: 10.4018/978-1-5225-2575-2, Ch001. ISBN:9781522525752.
3. Iulia-Antonia Bîrlog, Dumitru-Marius Borcan, and George-Manuel Covrig, Bucharest University of Economic Studies, Romania, "Internet of Things Hardware and Software", *Informatica Economică*, vol. 24, no. 2, 2020.
4. Arshdeep Bahga and Vijay Madisetti, *Internet of Things: A Hands-on Approach*, Universities Press (India) Private Limited, 2014.
5. Alessandro Bassi, Martin Bauer, Martin Fiedler, Thorsten Kramp, Rob van Kranenburg, Sebastian Lange, and Stefan Meissner, *Enabling Things to Talk – Designing IoT Solutions with the IoT Architecture Reference Model*, Springer Open, 2016.
6. T. Kavitha and G. Senbagavalli, "The future of travel in public bus service: How a mobile bus ticketing system is revolutionizing the public travel," *2023 International Conference on Applied Intelligence and Sustainable Computing (ICAISC)*, Dharwad, India, 2023, pp. 1–7, doi: 10.1109/ICAISC58445.2023.10200016
7. T. Kavitha, S. Kumari, S. A. Kamble, D. N. Rachana and C. K. DhruvaKuma, "Design of IoT based Smart Coin Classifier using OpenCV and Arduino," *2022 IEEE 2nd International Conference on Mobile Networks and Wireless Communications (ICMNWC)*, Tumkur, Karnataka, India, 2022, pp. 1–5, doi: 10.1109/ICMNWC56175.2022.10031997. E-ISBN: 978-1-6654-9111-2.
8. Ronald L. Krutz and Russell Dean Vines, *Cloud Security: A Comprehensive Guide to Secure Cloud Computing*, Wiley-India, 2010.
9. Jan Holler, Vlasios Tsiatsis, Catherine Mulligan, Stamatis Karnouskos, Stefan Avesand and David Boyle, *From Machine to Machine to Internet of Things*, Elsevier Publications, 2014.

3 Secure Device Management and Device Attestation in IoT

V. Hari Santhosh
IoT Cloud Architect at trinamiX(BASF)

A. Babiyola
Meenakshi Sundararajan Engineering College,
Anna University, Chennai, India

C. Chitra
PSNA College of Engineering and Technology,
Anna University, Chennai, India

3.1 INTRODUCTION

The Internet of Things (IoT) poses a lot of security problems because there are so many gadgets and they all have different hardware and software configurations.

Secure device management is critical for ensuring the security and privacy of IoT devices and data and key management is an important aspect of secure device management in the IoT, as it allows for secure authentication, secure communication, and secure storage of data [2]. Device firmware updates are essential for patching vulnerabilities in IoT devices, but they also pose security risks if not implemented properly. Blockchain technology can be used to enhance the security of IoT device management by providing tamper-proof audit trails and secure data sharing. The proposed architecture uses a combination of encryption, digital signatures, secure bootstrapping, and blockchain technology to provide end-to-end security for IoT devices. Device attestation is the process of verifying the identity and integrity of IoT devices. It involves ensuring that IoT devices are genuine, have not been tampered with, and are running authorized firmware. Device attestation is essential for preventing attacks that involve impersonation or modification of IoT devices [4]. The device attestation techniques can be categorized into three categories: hardware-based, software-based, and hybrid. Hardware-based device attestation involves using hardware features of IoT devices [11], such as secure enclaves, to attest to their identity and integrity. Software-based device attestation involves using software features of IoT devices, such as digital certificates and cryptographic keys, to attest to their identity and integrity. Hybrid device attestation involves combining

DOI: 10.1201/9781003477327-3

hardware-based and software-based techniques to achieve stronger security guarantees like using blockchain technology for device management.

3.2 SECURE DEVICE MANAGEMENT

IoT device management is a critical aspect of the IoT ecosystem, as it ensures that IoT devices are secure, reliable, and perform optimally. IoT device management involves various processes, such as device provisioning, configuration management, monitoring, firmware and software updates, and maintenance [1].

This architecture diagram in Figure 3.1 shows the key steps involved in IoT device management, including device provisioning, configuration management, monitoring, firmware and software updates, and maintenance. Figure 3.1 also highlights the use of machine learning algorithms, blockchain technology, and cloud computing for secure and efficient IoT device management.

At the center of the architecture is the IoT device management platform, which provides a centralized management interface for managing IoT devices. The platform communicates with the IoT devices over the Internet or a local network and uses machine learning algorithms to analyze device data and detect potential issues. Anomaly detection and predictive analytics are used to generate alerts and notifications for administrators to take action.

Device provisioning is the process of registering and enrolling IoT devices on a management platform. During this process, the devices are authenticated and authorized before being allowed to enroll on the platform. Once enrolled, the devices are assigned unique identifiers and credentials that allow them to communicate securely with the platform. Device provisioning can be performed manually or automated, depending on the number of devices and the complexity of the system.

Configuration management is the process of configuring the settings and parameters of IoT devices to ensure their secure and optimal operation. Configuration management may include setting network parameters, security policies, device encryption, and firewall rules. Configuration management can be performed manually or automated, and it is essential to ensure that IoT devices are configured correctly to mitigate security risks [1].

Monitoring and alerts are critical processes in IoT device management, as they enable administrators to monitor the health and performance of IoT devices and detect anomalies or security breaches promptly. Monitoring and alerts can be performed manually or automated, and they can use various techniques such as machine learning algorithms, anomaly detection, and predictive analytics to detect potential issues [2].

Keeping IoT devices up-to-date with the most recent security patches, bug fixes, and feature additions requires regularly updating their firmware and software. Firmware and software updates can be deployed manually or automated, and it is essential to verify the authenticity of the updates and ensure that the devices are not compromised during the update process.

Maintenance is the process of managing and maintaining IoT devices throughout their lifecycle, including performing regular maintenance tasks such as updating firmware, replacing batteries, and performing hardware repairs [2]. Maintenance can

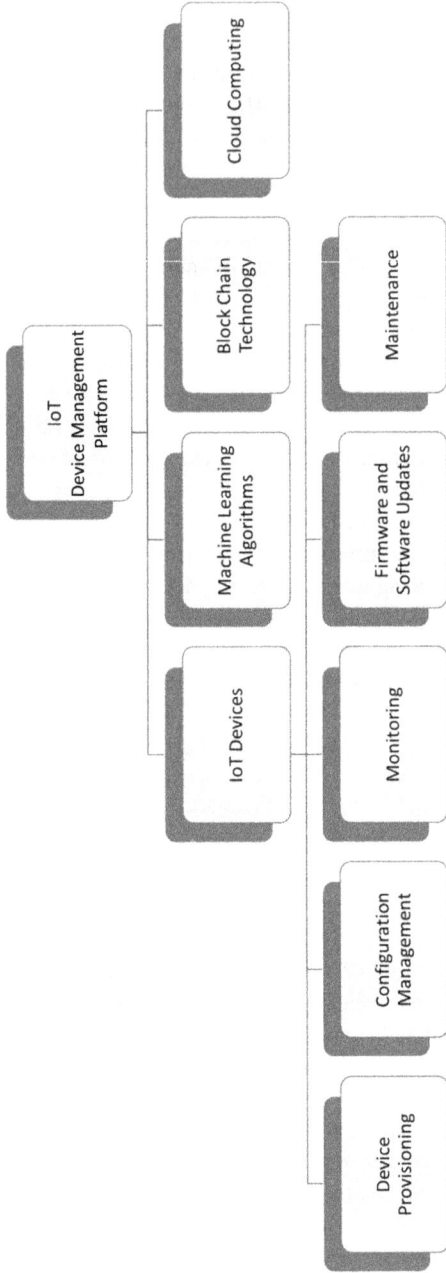

FIGURE 3.1 IoT device management techniques.

be performed manually or automated, and it is essential to ensure that IoT devices are maintained correctly to ensure their reliability and longevity.

Blockchain technology is used for secure and decentralized management of device identities, authentication credentials, and transaction data. This provides a tamper-proof and secure way to manage IoT devices and enables secure peer-to-peer communication between IoT devices.

Cloud computing is used for scalable and efficient IoT device management, providing on-demand computing resources and storage for managing large numbers of devices. Cloud computing also enables real-time data processing and analysis, allowing administrators to monitor and manage devices in real-time.

3.3   APPROACHES TO DEVICE ATTESTATION

To put it another way, remote attestation occurs when a trusted verifier performs a reliable check on the state of an untrusted prover while being physically separated from both parties. Since the prover in this situation is (a process executing on) an IoT device, it is untrusted due to the risk of malware infection [7]. The verifier is a safe, off-site server run by the system administrator. To demonstrate to the system operator that the IoT devices are malware-free, we employ remote attestation. This can be done remotely, eliminating the requirement for the operator to have physical access to the devices, and enabling fully automated verification of every component of an IoT system on a massive scale [7]. The comparison of IoT Device Attestation Techniques is shown in Table 3.1.

3.3.1   Hardware-Based Device Attestation

Hardware-based device attestation involves using hardware features of IoT devices to attest to their identity and integrity. Hardware-based device attestation workflow is presented in Figure 3.2. Trusted Platform Modules (TPMs), Secure Enclaves, and Physically Unclonable Functions (PUFs) are commonly used hardware-based techniques.

TABLE 3.1
Comparison of IoT Device Attestation Techniques

Technique	Hardware-Based	Software-Based	Hybrid
Trusted Platform Modules (TPMs)	✔	✘	✔
Secure Enclaves	✔	✘	✔
Physically Unclonable Functions (PUFs)	✔	✘	✔
Digital Certificates	✘	✔	✔
Cryptographic Keys	✘	✔	✔
Code Signatures	✘	✔	✔

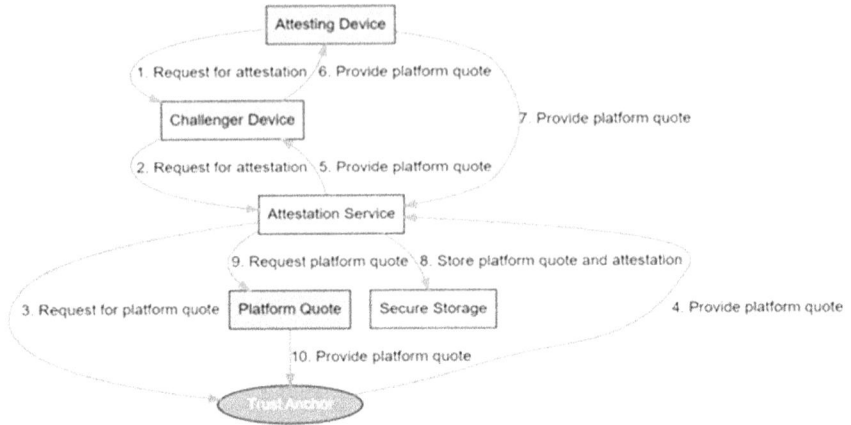

FIGURE 3.2   Hardware-based device attestation workflow.

Trusted Platform Modules (TPMs) are hardware security chips that are integrated into many modern computer systems, including IoT devices. TPMs provide a root of trust for device attestation by storing cryptographic keys and performing secure operations such as key generation, signing, and verification [9]. TPMs are tamper-resistant, and their security features are built into the hardware, making them difficult to manipulate or compromise.

Secure Enclaves are another hardware-based technique used for device attestation. Secure Enclaves are hardware-isolated execution environments that can be used to securely run sensitive code and data. By executing code within a Secure Enclave, the code can be protected from external tampering or observation. Secure Enclaves can be used to provide a root of trust for device attestation by securely generating, storing, and using cryptographic keys [5].

Physically Unclonable Functions (PUFs) are hardware-based techniques that leverage the physical characteristics of the underlying hardware to generate unique and unclonable identifiers [9]. PUFs exploit the randomness and variability of physical phenomena such as voltage fluctuations, temperature variations, and manufacturing defects to generate device-specific keys or signatures. PUFs can provide a root of trust for device attestation by generating unique and unclonable keys or signatures that can be used to attest to the identity and integrity of a device.

The above steps highlight the interactions between the Attesting Device, Challenger Device, Attestation Service, Trust Anchor, Platform Quote, and Secure Storage during the hardware-based attestation process.

Trust Anchor: The trust anchor represents a trusted entity or authority, such as a hardware manufacturer or a certification authority. In the context of hardware-based attestation, the trust anchor establishes the foundation of trust by providing a set of trusted measurements or cryptographic keys [3]. For example, the Intel Trusted

Platform Module (TPM) serves as a trust anchor by securely storing cryptographic keys and measurements that attest to the integrity of a computing platform.

Attesting Device: The attesting device is the target device or system that undergoes the attestation process to provide evidence of its integrity, security features, or compliance with specific standards [3]. For example, a server in a data center can serve as the attesting device, where it generates a platform quote containing measurements of its hardware components, firmware, and software configurations.

Challenger Device: The challenger device is responsible for requesting and receiving the attestation evidence from the attesting device. It verifies the integrity and security properties of the attesting device based on the provided evidence [6]. For example, a cloud service provider's infrastructure acts as the challenger device, which requests attestation evidence from servers in order to validate their security posture and ensure compliance with security policies.

In this scenario, the trust anchor, such as the Intel TPM, establishes trustworthiness by providing secure storage for keys and measurements. The attesting device, such as a data center server, generates a platform quote with measurements that demonstrate its integrity. The challenger device, represented by the cloud service provider's infrastructure, requests and verifies the platform quote to ensure the server's security and compliance. The hardware-based attestation process allows the challenger device to gain confidence in the trustworthiness and security of the attesting device, enabling secure and trusted interactions within a larger system or network [6].

3.3.2 SOFTWARE-BASED DEVICE ATTESTATION

Software-based device attestation involves using software features of IoT devices to attest to their identity and integrity. Digital certificates, cryptographic keys, and code signatures are commonly used software-based techniques [10]. The steps of the Software-based device attestation workflow are discussed in Figure 3.3.

   Digital certificates are a widely used software-based technique for device attestation. Digital certificates are electronic documents that bind an identity to a public key. They can be used to verify the identity and authenticity of a device or entity. Digital certificates can be issued and verified by trusted third-party certificate authorities, providing a chain of trust that can be used to establish the identity and integrity of a device [10].

   Cryptographic keys are another software-based technique used for device attestation. Cryptographic keys can be used to sign and verify digital messages, ensuring their authenticity and integrity. Public-key cryptography is often used in device attestation to provide a secure communication channel between devices or between a device and a remote server [10].

   Code signatures are a software-based technique used to attest to the authenticity and integrity of software running on a device. Code signatures are digital signatures that are applied to software code to verify its origin and integrity [10]. Code signatures can be used to ensure that software updates and patches come from a trusted source and have not been tampered with.

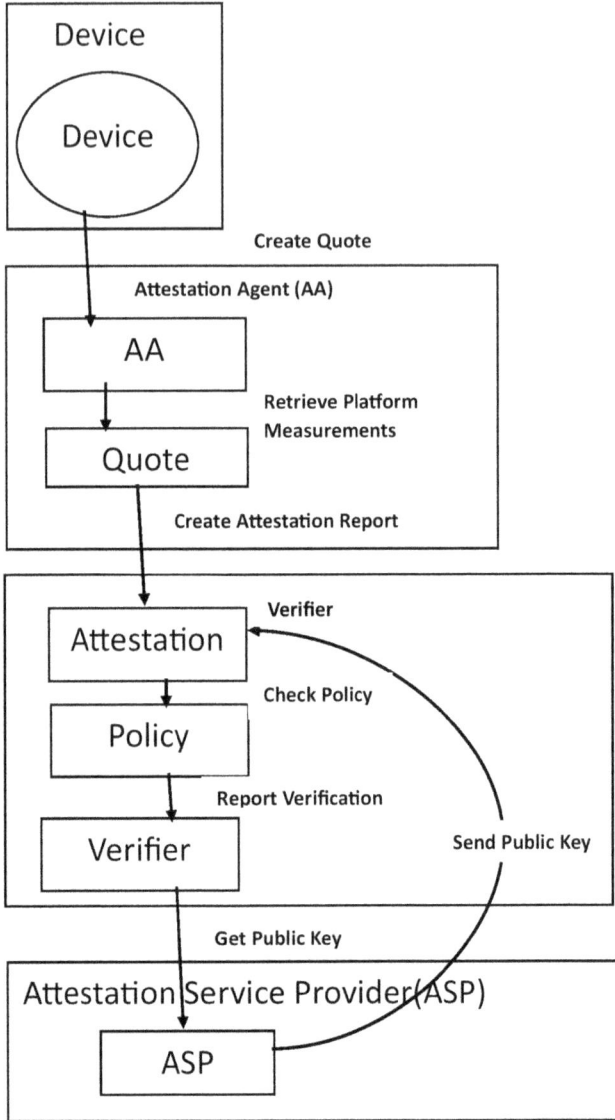

FIGURE 3.3 Software-based device attestation workflows.

The flow of information and interactions between the Device, Attestation Agent (AA), Verifier, and Attestation Service Provider (ASP) is described in the software-based device attestation process.

The Device is the device being attested. The Attestation Agent (AA) is responsible for facilitating the attestation process. The Verifier verifies the attestation report and performs policy checks. The ASP is the service that provides the necessary support for attestation.

The Attestation Agent (AA) generates an attestation report based on the quote and platform measurements. The Attestation report is sent to the PolicyCheck component for policy validation. The PolicyCheck component checks the attestation report against predefined policies. The Verifier receives the validated attestation report from the PolicyCheck component [9].

The Verifier contacts the ASP to obtain the public key. The ASP sends the public key to the Verifier. The Verifier verifies the attestation report using the received public key. The workflow is completed, and the result of the attestation process is determined.

3.3.3 Hybrid Device Attestation

Hybrid device attestation involves combining hardware-based and software-based techniques to achieve stronger security guarantees [9]. A device may use hardware-based attestation to verify its identity and integrity, and software-based attestation to verify the authenticity of firmware and software updates. The steps are shown in Figure 3.4.

For example, a device may use a hardware-based technique such as a TPM or Secure Enclave to generate and store a unique device identifier and cryptographic keys [9]. The device may then use a software-based technique such as digital certificates or code signatures to verify the authenticity and integrity of firmware and software updates [4].

Hybrid attestation can provide a balance between security and flexibility. Hardware-based attestation provides stronger security guarantees but may be less flexible than software-based attestation. Hybrid attestation can provide the benefits of both hardware-based and software-based attestation while mitigating their limitations.

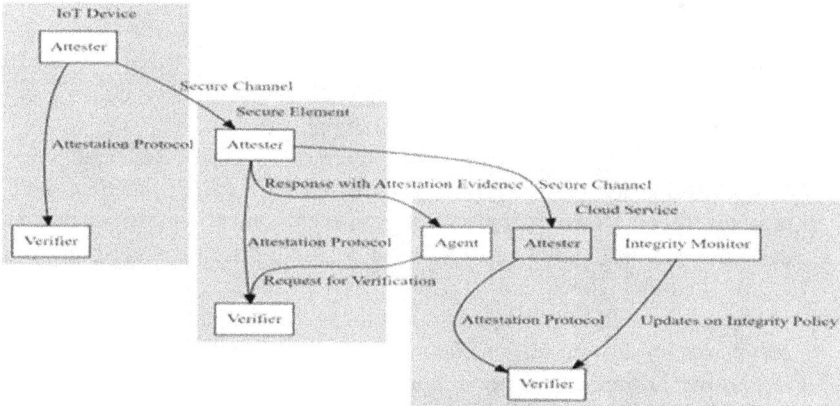

FIGURE 3.4 Hybrid-based device attestation workflow.

3.4   THREATS AND WEAKNESSES OF DIFFERENT AUTHORIZATION SCHEMES

Device attestation and management involve various authorization schemes to ensure secure communication and access control. However, these authorization schemes can also have certain threats and weaknesses that may compromise the security of the system. Here are some of the threats and weaknesses of different authorization schemes used in device attestation and management

Password-Based Authorization: Passwords are the most commonly used authorization scheme in device attestation and management. However, passwords can be easily compromised through various methods, such as brute-force attacks, dictionary attacks, and phishing attacks. Users may also use weak passwords, which can be easily guessed, and reuse passwords across different systems, which can lead to credential-stuffing attacks.

Token-Based Authorization: Token-based authorization involves the use of tokens or access keys to authenticate devices and provide access control. However, if tokens are not properly managed, they can be easily stolen or leaked, allowing unauthorized access to the system. Tokens may also be vulnerable to replay attacks, where an attacker intercepts a token and reuses it to gain access to the system.

Certificate-Based Authorization: Certificate-based authorization involves the use of digital certificates to authenticate devices and provide access control. However, if certificates are not properly managed, they can be easily stolen or compromised, allowing unauthorized access to the system. Certificates may also be vulnerable to man-in-the-middle attacks, where an attacker intercepts a certificate exchange and replaces the legitimate certificate with a fake one.

Biometric-Based Authorization: Biometric-based authorization involves the use of biometric data, such as fingerprints, voice recognition, and facial recognition, to authenticate devices and provide access control. However, biometric data can be easily stolen or compromised, and may not be reliable in all situations. For example, fingerprints can be lifted from surfaces and used to create fake fingerprints, and facial recognition may not work properly in low-light conditions or if the user is wearing glasses or a hat.

Role-Based Authorization: Role-based authorization involves assigning different roles and permissions to users based on their job functions and responsibilities. However, if roles and permissions are not properly defined and managed, users may have access to sensitive information or perform actions that they should not be able to do. Role-based authorization may also be vulnerable to privilege escalation attacks, where an attacker gains access to a lower-level role and then elevates their privileges to gain access to sensitive information or perform unauthorized actions.

In summary, while different authorization schemes have their own strengths and weaknesses, it is important to properly manage and secure them to ensure the overall security of the device attestation and management system.

3.5   SUMMARY

In conclusion, Secure Device Management and Device Attestation are essential components of IoT security. Device Attestation plays a vital role in verifying the identity and authenticity of IoT devices, while Secure Device Management helps ensure that devices are securely managed throughout their lifecycle.

There are three primary methods that can be utilized when it comes to the process of device attestation. These methods are known as hardware-based device attestation, software-based device attestation, and hybrid device attestation. Each method offers a number of benefits as well as a number of drawbacks; the method that is selected is dependent on the particular use case as well as the security needs. Even though device attestation adds a robust layer of protection to the IoT, there are still a number of vulnerabilities and dangers that must be addressed. The possibility of device manipulation is one of these, along with the possibility of replay attacks and the vulnerability of the device's root of trust. Moreover, the compromise of a single device can lead to a significant security breach in the entire IoT ecosystem.

To mitigate these threats and weaknesses, continuous monitoring and evaluation of the device's security posture are essential. This includes implementing secure boot, runtime integrity checking, and secure firmware updates. Additionally, the use of secure communication protocols, such as Transport Layer Security (TLS) and Secure Socket Layer (SSL), can further enhance the security of the device. Looking ahead, the adoption of IoT devices is expected to continue to grow, and with it, the need for secure device management and device attestation. As such, there is a need for ongoing research and development of more robust and secure IoT security mechanisms to ensure the protection of sensitive data and critical infrastructure.

REFERENCES

1. A. Khan, A. Ahmad, M. Ahmed et al., "Authorization Schemes for Internet of Things: Requirements, Weaknesses, Future Challenges and Trends," *Complex & Intelligent Systems*, vol. 8, pp. 3919–3941, 2022.
2. M. Conti, S. Kumar, and C. Lal, "Secure Device Management in the Internet of Things," *IEEE Communications Magazine*, vol. 55, no. 10, pp. 132–139, Oct. 2017.
3. D. Ma, Y. Yang, Y. Xu, and J. Chen, "Device Attestation for the Internet of Things: A Survey," *IEEE Internet of Things Journal*, vol. 5, no. 3, pp. 1733–1743, June 2018.
4. A. Rostami, S. P. Mohanty, and M. M. Tentzeris, "Secure Device Attestation and Management for the Internet of Things," *IEEE Transactions on Information Forensics and Security*, vol. 12, no. 6, pp. 1339–1350, June 2017.
5. A. Azab, M. Zhang, Y. Wang, and Y. Li, "Device Attestation: A Survey," *IEEE Communications Surveys & Tutorials*, vol. 21, no. 4, pp. 3764–3794, Fourthquarter 2019.
6. T. K. Das, S. Misra, and S. K. Ray, "Device Attestation: An Overview," *Journal of Network and Computer Applications*, vol. 127, pp. 1–13, June 2019.
7. R. Nagy, M. Bak, D. Papp, and L. Buttyán, "T-RAID: TEE-Based Remote Attestation for IoT Devices," in Gelenbe, E., Jankovic, M., Kehagias, D., Marton, A., and Vilmos, A. (eds.), *Security in Computer and Information Sciences*. EuroCybersec 2021. Communications in Computer and Information Science, vol. 1596, 2022.
8. A. Ibrahim, A.-R. Sadeghi, and G. Tsudik, "US-AID: Unattended Scalable Attestation of IoT Devices," *2018 IEEE 37th Symposium on Reliable Distributed Systems (SRDS)*, Salvador, Brazil, 2018, pp. 21–30, doi: 10.1109/SRDS.2018.00013

9. J. Julku, J. Suomalainen, and M. Kylänpää, "Delegated Device Attestation for IoT," In Lauret, J. M., Abdel-Maguid, M., Ararweh, Y. and Benkhelifa, E. (eds.), *2021 8th International Conference on Internet of Things: Systems, Management and Security, IOTSMS 2021*, IEEE Institute of Electrical and Electronic Engineers, pp. 1–8. [9704959], 2021.

10. A. E. Braten, F. A. Kraemer, and D. Palma, "Autonomous IoT Device Management Systems: Structured Review and Generalized Cognitive Model," *IEEE Internet of Things Journal*, vol. 8, no. 6, pp. 4275–4290, 15 March 2021, doi: 10.1109/JIOT.2020.3035389

11. T. Kavitha, V. Ajantha Devi, S. Neelavathy Pari, and Sakkaravarthi Ramanathan, *Internet of Everything: Smart Sensing Technologies*, Nova Science Publishers, Publication Date: June 17, 2022, doi: 10.52305/PNQM1088

4 Secure Operating Systems and Software

S. Koushik and M. Chaitra
Credokey SoftTech Private Limited, Bangalore, India

4.1 INTRODUCTION

A computer system has hardware, operating system (OS), and user applications as its components. The hardware consists of input/output devices, memory, and processing units, which are combined and provide essential computing resources for the system. The application programs are software users use to solve their computing needs. The user uses and interacts with many such programs to increase work efficiency [1].

The operating system forms a middle layer between the hardware and user applications. The main task of the operating system is to control and coordinate between the hardware and applications. The operating system covers many roles and functions.

Operating systems require greater computing power and more resources. Also, the architecture is complex. Currently, operating systems follow microkernel architecture to modularise the operating system's design [2].

Figure 4.1 shows an abstract view of Computer Systems. This approach is used in modern operating systems where external hardware components assist in input–output operations.

Microprocessors are used as processing units, have more functionality, and are primarily used on personal computers. With the introduction of microcontrollers, many hardware components like processors, memory, and input/output are embedded into a single chip. Smartphones, digital cameras, and handheld devices are a few examples of embedded systems. The devices mentioned above have microcontrollers as their processing units and transceivers as their input–output devices. As a result, these devices require significantly less power, and the cost of the entire system is meagre [3].

The Internet of Things (IoT) is the network of electronic devices, instruments, or anything attached to sensors that enable data collection and exchange between objects over a network. Any device equipped with microcontroller units and transceivers is called a "smart object".

With continuous advancements in computing and experimentation, IoT has become a global phenomenon in the current computing context. IoT is continuously evolving, and many challenges open up for researchers to experiment and bring advancements to IoT platforms.

Smart devices and smartphones are increasing daily with increased connectivity with the help of Wi-Fi and a 4G network. Moreover, especially with the 5G network, the data collected from the devices increases as more devices get added [4].

DOI: 10.1201/9781003477327-4

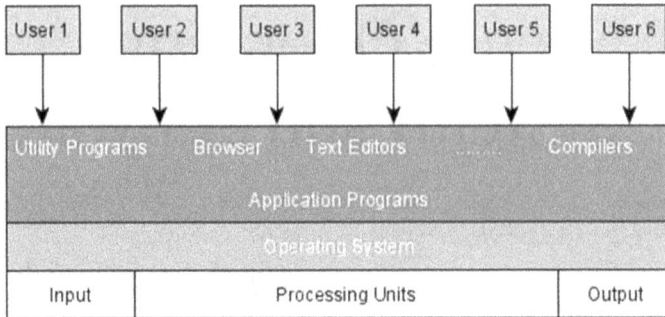

FIGURE 4.1   Abstract view of computer system.

The smart devices currently have micro software that enables input–output transceivers to function on smart devices. However, with growing data collection, it is crucial to have a fully functional operating system to manage resource allocation on the device instead of connecting to a server for decision-making [5–8].

Many operating systems exist for IoT devices. In the following section, we discuss the list of operating systems and their features and functionalities [9].

1. **Contiki**: Contiki is a flexible and portable operating system for networked, memory-constrained systems that uses a hybrid Protothread approach. It primarily concerns low-power wireless IoT devices. IoT Contiki is an open-source software project to develop a lightweight operating system for large-scale deployment. Contiki also offers a rich set of network features, including IPv6, CoAP, and RPL.
2. **RIOT**: RIOT is an open-source embedded operating system designed for network- and memory-constrained systems and IoT devices. RIOT intends to have hardware-independent and hardware-dependent code with a well-defined interface [10]. RIOT supports programming languages like C and C++ with multi-threading support.
3. **TinyOS**: TinyOS is an embedded, component-based operating system and platform for low-power wireless devices. TinyOS presents hardware abstraction, provides a wrapper around the abstraction, and provides an interface for standard functionalities like communication and sensing. TinyOS provides a concurrency model in which tasks are non-preemptive and run in first-in-first-out order and pre-emptive events.
4. **LiteOS**: LiteOS is a Unix-like, interactive, open-source OS designed for wireless sensor networks. It has a Unix-based programming environment consisting of three components: LiteFS, LiteShell, and the kernel. An interface enables the users to interact with the device using UNIX commands executed by kernel and LiteShell. LiteFS provide filesystem support. LiteOS also supports a plug-and-play routing stack, as it lacks built-in networking protocols.
5. **FreeRTOS**: FreeRTOS is an open-source, real-time operating system for embedded devices which is developed by Amazon. It enables small,

low-power edge devices to be easily programmed, deployed, connected, and managed. It uses Nabto to communicate with peers and supports kernel and user modes. In addition, it uses C programming language based on tasks and routines supporting dynamic memory allocation. The operating system is equipped with AWS SDKs and libraries. The operating system is used to connect with AWS cloud services.

6. **Mantis OS**: Mantis OS is an open-source embedded multithreaded operating system for wireless microsensor platforms written in C programming language. Implementing MANTIS in a lightweight RAM footprint fits in less than 500 bytes of memory, including the kernel, scheduler, and network stack. The user-level threads are executed based on the Round-Robin scheduling approach based on some priority levels assigned. The operating system is designed to be flexible and provide cross-platform support.

7. **Nano-RK**: RK stands for Resource Kernel. Nano RK is a fully pre-emptive reservation-based, wireless sensor networking real-time operating system (RTOS) from Carnegie Mellon University written in C programming language. The design of the operating system has higher priorities for shorter jobs. It supports features such as a watchdog timer and deep sleep mode to run the power down when no tasks are needed. It disallows dynamic task creation. The operating system is a reservation-based operating system which is structured to multi-task.

8. **SOS**: It has modular components which are loosely connected at compile time and are dynamically loaded. These modular components perform specific tasks that follow flexible priority scheduling and communicate via messages.

9. **NutOS**: This is an open-source modular real-time operating system that gets pre-empted by hardware interrupt and unblocked by the event. The queuing mechanism of events is in decreasing order of their priority levels.

10. **uC/OS-III**: It is highly portable, scalable, and pre-emptive in real-time. It is a multi-tasking kernel that supports semaphores and message queues and has built-in performance measures. In addition, the scheduling mechanism of tasks is of Round-Robin fashion as there are several priority levels.

4.2 ANATOMY OF IoT OS AND ITS DESIGN

Currently, the computing capability is meagre on IoT devices, which require many nodes. These nodes are, in turn, connected to gateways. The gateways are then further connected to the cloud. Most of the computing, decision-making, and data storing happens on the cloud. The desirable features of the operating system are a small memory footprint, real-time decision-making, energy efficiency, and reliability [11–13].

Desirable features of IoT operating system:

1. **Architecture**: Modern Operating systems use a microkernel approach as they are modular and smaller. But power and memory management become essential to the software on smaller devices. A monolithic kernel helps

manage the power and memory efficiently as module interaction is significantly less. In addition, the layered approach of an operating system for IoT devices makes the software more manageable and less complex.

2. **Programming language support**: The operating systems must support native programming languages like C or C++ for the users to build applications for the devices. For example, TinyOS uses an extension of C programming. In addition, the nesC programming language focuses on component-based and event-driven approaches and is challenging to learn. Therefore, support for programming language. API development should leverage the underlying capabilities of the operating system rather than adding overhead. Due to less memory and device constraints, enabling high-level programming language on the device is challenging.

3. **Resource Management**: An abstraction layer of the operating system for the hardware is known to provide resource management on IoT devices. As it is a fact that IoT performs in a constrained environment, IoT-based operating systems should have higher management capabilities when compared to the general-purpose operating system [14].

   a. **Process Management**: The operating system performs process management activities fairly, sharing essential resources. Multi-threading support efficiently enables memory management by allocating stack memory to each thread. In addition, operating systems like Contiki support a lightweight mechanism called protothread, which enhances the capabilities of the event-driven programming model.

   b. **Memory Management**: Various process threads allocate and deallocate memory. Any operating system manages memory in two methods, i.e., static allocation and dynamic allocation. The static allocation method allocates the memory and cannot be altered during the runtime, whereas dynamic allocation frees and allocates memory during runtime. There are drawbacks in dynamic memory, like memory leaks and page misses, to name a few. Due to significantly less memory on the device, they were compacting.

   c. **Energy Management**: Input–output operations are the most energy-consuming operations. Therefore the operating system should provide energy-efficient solutions like sleep/wake and duty-cycle modes. Energy-efficient approaches are achieved by adding a piece of hardware or with the help of software. However, a hardware-based method to reduce energy is not cost-efficient as additional hardware requires additional cost. On the other hand, the software-based technique is more practical but can introduce overhead. However, a software-based approach helps to reduce energy consumption and different layers [15].

   d. **Communication Management**: IoT devices provide a continuous, seamless and ubiquitous mechanism to communicate between IoT devices. Many researchers have investigated communication protocol design specifically for IoT. Protocols defined at different layers, namely, MAC, network, and transport layer, play an essential role in managing network performance positively.

e. **File Management**: Due to the lack of hardware support on IoT devices, file management and file system design take precedence while the device stores the data. Usually, such devices have flash-based drives and use efficient data storage and retrieval mechanisms. Coffee is one such file management system on flash-based storage devices.

4. **Schedulers**: The scheduling strategy depends on the algorithm employed in the device. As IoT operates in real-time, even the design of the scheduler should be real-time. Throughput, latency, and wait time decide the scheduler's performance, thereby, the IoT system's performance. Schedulers also should enable multi-tasking on IoT systems.

5. **Portable**: Major hardware platforms like Arduino, Raspberry, or Zolertia are built on the principles of ARM. These hardware devices have similar features and capabilities. The operating system design should leverage the features of the underlying hardware.

6. **Security**: The usage of IoT devices is in almost every industry for daily activities. IoT devices record various attacks, and researchers have extensively researched the issue. Therefore, security should be an integral part of operating systems built for IoT [16].

4.3 SECURITY CHALLENGES AND OPPORTUNITIES IN IoT OS

Most IoT devices work on the basis of standard operating systems, which currently cannot meet specialised security requirements. Furthermore, the IoT concept encompasses a wide range of appliances, gadgets, technologies, software, and communication protocols. As a result, numerous security vulnerabilities get generated by this heterogeneous ecosystem, which could severely impair every element of our lives connected to the IoT. Therefore, IoT OS security is crucial and should provide security and privacy services [17].

A challenge at the physical layer level involves developing an adequate operating system that supports various features and simultaneously adheres to security policies. The difficulty lies in creating standardised, secure operating systems that are less vulnerable and can deliver all the security and privacy functions for restricted devices [18]. Also, security patching improves security, but it is limited due to battery power and can expose IoT systems to various security risks. It is also necessary to ensure that the operating system communicates with other operating systems on multiple platforms capable of supporting the same policies to satisfy the shared security policy enforcement on both machines [19].

Several protocols and standards at the information layer empower IoT devices and applications. Unfortunately, deploying the most commonly used protocols may cause insecurity, resulting in sensitive data leakages like user credentials, configuration information, and software updates that make them vulnerable to attacks [20].

New security requirements must be followed since IoT devices are exposed to the Internet at the application layer. There is a risk that attackers can install malicious modules on nodes and can steal user credentials or inject malicious code [21].

There are various threats identified at each layer: Fabricating Attack, Data Alteration and Modification, Man-In-The-Middle Attack, Collision, Selective

Forwarding Attack, Timing, Denial of Service (DOS) Attack, Internet Smurf Attack, Black Hole Attack Detection, Homing Attack, Wormhole Attack Detection, Sybil and Clone ID Attack Detection, Sinkhole Attack Detection, Resource exhausting attack and RPL Attack, and Malware and Ransomware attack. Unfortunately, IoT is being developed quickly without considering the severe security risks involved and any potential legislative adjustments [22].

4.4   DESIGN OF SECURE OS IN IoT

Various authors discuss multiple types of models for designing IoT. The First IoT reference model is given by ITU-T IoT reference model ITU-T Y.2060. Following this are three-layer, four-layer, five-layer, and six-layer architecture models. In common, six layers in IoT are considered [10, 17].

The details of the six layers are in Figure 4.2.

1. **Application Layer**: This layer collects and categorises the information and provides applications based on the client's requirement for all IoT. Some examples of IoT-based applications are Smart Home, Smart Mobility, and many more.
2. **Network Layer**: The network layer assists in processing the information and is also responsible for transporting data over the Internet and ensuring connectivity among various devices.
3. **Security Layer**: As the information gets transferred over the Internet, there is a possibility for various attacks. Hence security is necessary to ensure a secure connection and information transmission to the intended recipient. This layer provides security over IoT devices. It ensures the information is encrypted, sent across the communication or network layer, and decrypted at the recipient.
4. **Middleware Layer**: This layer processes the secured information after security checks by the security layer received from various sensor devices. It also has direct access to the database to store all the processed information.

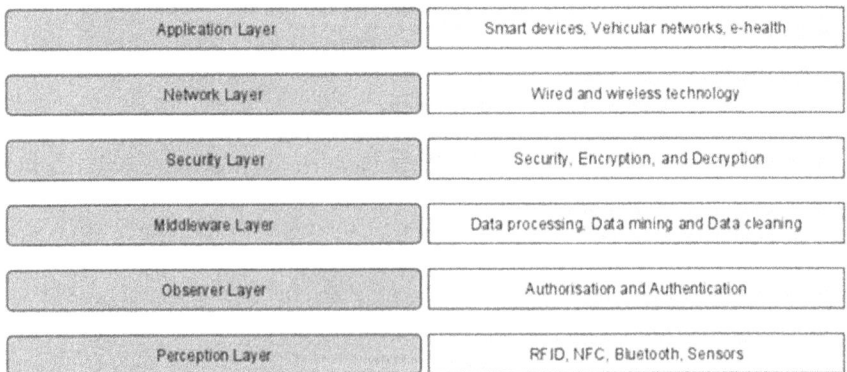

Application Layer	Smart devices, Vehicular networks, e-health
Network Layer	Wired and wireless technology
Security Layer	Security, Encryption, and Decryption
Middleware Layer	Data processing, Data mining and Data cleaning
Observer Layer	Authorisation and Authentication
Perception Layer	RFID, NFC, Bluetooth, Sensors

FIGURE 4.2   Six-layer architecture of IoT.

In addition, this layer can distinguish related and unrelated data, in which associated data gets processed, and unrelated data gets removed.

5. **Observer Layer**: This layer acts like a monitoring agent which monitors the processed data and performs high-level security checks, and if there is no threat observed, the received data from the Perception layer gets passed to the middleware layer for further processing. User authentication is done at this layer.

6. **Perception Layer**: This layer is similar to the Physical layer of the Open Standard Interface (OSI) model. It has data sensors that sense the physical environment based on parameters like temperature and speed. It is known to gather the data from different sensor devices, convert it into digital signals, and pass the collected information to the Observer layer for further processing.

4.4.1 THE ARCHITECTURE OF THE PROPOSED SECURE IoT OPERATING SYSTEM

The proposed architecture diagram for the Secure IoT operating system is shown in Figure 4.3. The devices module has various sensors and actuators. A sensor keeps track of any changes in the environment and alerts the users. The sensors are programmed to sense the change based on the user's requirements. An actuator responds to a signal by taking appropriate action.

The display component involves display devices like a monitor, which provides a user interface enabling the users to interact and displays the relevant message to the user based on the action taken. It also shows error messages and alerts to users. The computing devices in IoT can be any computer devices used to compute or decide whether the collected data has to proceed with the communication layer for further processing or display an alert message, basically selecting the next course of action.

The communication layer ensures the connectivity of various devices via various protocols and Application Programming Interfaces (APIs). It serves as an information bridge, facilitating a connection between the sensing and service layers to transmit data to the application layer with all the necessary checks. The major functionalities at the communication layer involve routing, flow control, and reliability control. While the routing component handles the transmission of packets from source to destination, flow control ensures the transmission of packets from sender to receiver. Reliability control ensures the transmission of error-free and reliable data frames from sender to receiver.

Channel-Aware Routing Protocol (CARP) is one of the routing protocols used for IoT as it would need lightweight packets. IoT devices typically connect to the Internet via an Internet Protocol (IP) network. The communication API is crucial in IoT since it enables seamless information transmission between IoT devices or other devices connected via the same network or over the Internet. APIs expose the required data allowing the devices to transmit data to IoT-enabled applications, serving as a data computing interface.

The devices have low power and fewer resources, and the management layer manages different tasks. The components are put together to ensure the smooth functioning of the IoT device.

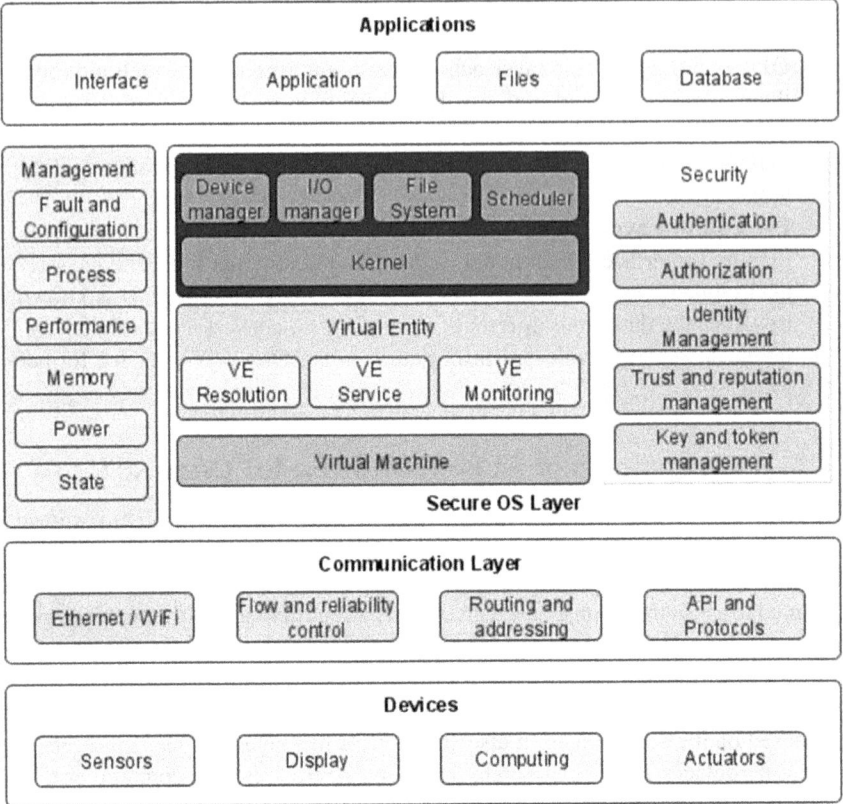

FIGURE 4.3 Architecture of proposed secure IoT operating system.

IoT devices are prone to fault and failure due to multiple devices interacting with each other. In addition, a complex, dense network of IoT edge devices requires the unconditional function of the device to perform complex tasks. The fault and configuration manager helps apply a dynamic and intelligent solution to design fault-tolerant systems.

Process management involves processing tasks in a multi-processor environment on IoT. Unlike older devices, the design of multi-core systems helps to process and communicate between tasks. Processing and communicating with functions complement each other. The processor manager manages the communication performance and increases the computational workload.

The performance of the IoT devices depends on the computational tasks and the amount of wake state the device is performing. Many researchers [23–25] have discussed and performed experimentation to achieve more remarkable performance. The performance manager layer understands the data to be processed at the device level. The layer also understands the data to be pushed to the cloud for a higher level of computation.

Memory is integral to IoT devices, and flash memory helps store information. Various principles of Direct Memory Access (DMA) and Dynamic Memory Management (DMM) are used in this layer to manage memory efficiently.

Power and state directly influence the performance of the device. Power management is essential as most IoT-enabled embedded devices have a low power supply (battery) operated. A sensor's sleep/wake state decides the power used on the device. The management component's power and state layer efficiently manages the device's power and state.

The operating system should be able to run on multiple devices, and the proposed architecture is per virtualisation technology, where the interaction to lower levels happens through virtual layers. Containerisation enables easy portability and exploits the possibility of using technologies built for more significant devices easily adaptable to IoT devices. Virtual entity resolution explores the option of resolving the different input–output components, while virtual entity service provides the required service and scheduled tasks on the hardware. Virtual entity monitor helps in task resolution within time.

Many approaches are tried and tested for building kernels for IoT devices. Standard techniques include microkernel, monolithic kernel, and hybrid kernel. Since IoT devices run on tight schedules and resources, yet certain features assist the device to function smoothly, the architecture proposes a hybrid kernel approach. Linux-based kernels are usually microkernels compatible with Linux/Unix-based operating systems. However, swapping modules to the controller from memory requires more power and resources. Hence hybrid kernel would keep the necessary kernel elements on the controller while the required device software is paged in when the sensors are in a "wake" state for the controller to function.

The device manager manages all the devices, changes in hardware configuration, and enables hardware on request by other components. It also communicates with device drivers to identify conflicts between devices.

I/O Manager comprises a set of standard routines which support heterogeneous devices. It also controls the data exchange between heterogeneous devices and processors, which might involve dynamic switching.

The File System is an integral part of any OS, mainly responsible for managing the information in the file on secondary storage with the help of indexing for easier access. The file system includes disk, database, transactional, network, and special types. File security ensures that only authenticated and authorised users can access the files. Therefore, file security is a vital part of IoT application security.

Schedulers in OS are the special system software that handles the scheduling of jobs and tasks, thus maintaining the processor's efficiency. From the ready queue, it chooses a job and sends it to the CPU for execution. Amin et al. [26] have proposed a secure and fast hardware scheduler, an online security-aware hardware scheduler that avoids attacks and ensures the jobs are completed well within the deadlines in real-time.

Security module comprises of following components.

1. **Authorisation**: Authorisation verifies whether the user has the right to access the resources. Authorisation is a security mechanism to determine user privileges and access levels to system resources such as data, application files, and services. Only authorised users will be able to proceed with using IoT applications. Authorisation can happen through API keys, Basic Auth, OAuth, and Hash-based Message Authentication Code HMAC.

2. **Authentication**: Authentication is a security mechanism that verifies the user's identity. Using an account password, One-Time Password (OTP), Tokens, or Personal Identification Number (PIN) ensures authentication. Enterprises must keep their networks secure and permit only authorised users to access the resources and applications. Secure IoT device authentication can be achieved by providing Secure digital identities, cryptographic keys, and Public Key Infrastructure (PKI), which are used as secure identities required for data exchanges, and it creates trust concerning the origin of the data source.

3. **Identity Management**: As many IoT devices are in different networks and might be external to home or organisation's security network policies, they might be vulnerable to various attacks. Simple certificates cannot address the multiple levels of authorisations, roles, and information these complex environments require. Identity and access management governs devices to be allowed inside the network and the interconnectivity of the devices.

4. **Trust and Reputation Management**: IoT devices also employ traditional aspects of trust in Information Systems. The trust-based application depends on System security and user safety. Important concepts involved in trust management are behaviour trust, reputation, honesty, and accuracy [27].

   A device is trustworthy if there is a firm belief in the competence of the device to act as expected such that this firm belief is not a fixed value associated with the device, but rather it is subject to the device's behaviour and applies only within a specific context at a given time. The reputation of an entity is an expectation of its behaviour based on other entities' observations or the collective information about the entity's past behaviour within a specific context at a given time. A recommender is said to be honest if the information about a particular entity within a specific context at a given time received from the entity is the same information the entity believes in. A recommender is said to be accurate if the deviation between the information received about the trustworthiness of a given entity "y" in a specific context at a given time and the actual reliability of "y" within the same context and time is within a precision threshold [27].

5. **Key and Token Management**: IoT's dynamic and heterogeneous nature demands key and token management. In such cases, a significant challenge is to ensure secure communication. Secure communication is achieved by Group Key Management (GKM) in which secure links are provided between group members, which would be revoked when a node leaves or joins the group, thus maintaining secrecy. Also, an approach is proposed in [28] where a new master token is generated for managing essential dissemination across a group of subscribers. The forward and the backward secrecy can be achieved using rekeying operations.

At the topmost layer, the application layer involves various applications and interfaces implemented by standard protocols. IoT is essentially the network of interconnected devices that interact and exchange information via interfacing, share a standard communication protocol, and present the details per user's requirements in a simple, understandable manner.

4.4.2 FEATURES OF PROPOSED SECURE OS DESIGN FOR IoT

The features of the proposed secure OS include the following:

1. The operating systems' main responsibility is to handle I/O operations and I/O device access control.
2. The proposed OS ensures a protected and secure mode with authentication and authorisation features.
3. Secure IoT architecture manages *File Systems* and secure file sharing with access control.
4. *Device* and *Power Management* is also critical for IoT devices.
5. *Error Handling and Fault Tolerance* mechanisms are built-in features of this architecture, thereby making the operating systems more robust.
6. Program or application specific security features are crucial as IoT devices are deployed and used in heterogeneous environments. In addition, the data transfer between devices requires optimum security to prevent external threats.
7. Secure communication among various parties involved in the application logistics and adoption of intelligent devices using IoT devices demonstrates accountability and traceability.

4.4.3 APPLICATIONS OF PROPOSED SECURE OS DESIGN

1. **Time-Critical Applications**: In health care systems, IoT gets equipped with sensors which locate various health-related critical requirements in real-time.
2. **Smart Mobility (Vehicular Ad-Hoc Networks)**: Automation in vehicles makes it more secure, efficient, consistent, and controlled access via applications. It also has sensors assisting drivers, thereby reducing accidents.
3. **Agro-Industrial Applications**: Alerting system to farmers regarding production estimation, risks and environmental conditions, assistance to farmers regarding seed selection and full tracking and tracing system from farm to fork. It can also monitor Air, Soil, Irrigation, and Fertilizers.
4. **Advanced Metering and Monitoring Infrastructure**: With advancements in home automation systems, there is a need for intelligent and advanced metering agents that constantly monitor and meter the usage of various smart equipment.
5. **Smart Supply Chain and Logistics**: The adoption of secure OS in IoT devices helps supply chain management from contracts and planning to complete monitoring of asset conditions, thereby providing features such as traceability, accountability, liability, and trust in smart logistics.

4.5 SUMMARY

IoT is an integral and essential part of computing nowadays. The IoT platform has seen many transitions, and further improvements are underway. This chapter discusses the IoT and its uses. Smart devices are used almost daily, from industrial to home use. The abstract view of the IoT and its architecture discusses different layers of IoT and how smart devices work. Various operating systems are in the market, which is discussed in detail. The anatomy of the operating system and the security aspects of the operating system are discussed in detail further. In contrast, the firmware and middleware provide a strong, robust computing environment, a lack of security coupled with scheduling. Different layers are responsible for tasks performed in real-time on any IoT-related devices. Many operating systems operate on hard-deadline task completion, an essential feature of smart devices. An architecture proposed in designing a secure operating system for IoT devices discusses in detail the different layers and the tasks of the operating system. The architecture is coupled with various complex elements that interact with each other, thereby enabling multiple security parameters at different levels. The architecture also concludes management layers, responsible for different management actions. This theoretical architecture paves the way for combining the security for an operating system designed for IoT.

AUTHORS' CREDIT

Koushik S., Director at CredoKey SoftTech Pvt. Ltd., prepared the initial draft of the manuscript. Chaitra M. co-authored, reviewed, and revised fundamental portions of the manuscript. All authors read and approved the final manuscript.

COMPETING INTERESTS

The authors declare that they have no competing interests.

REFERENCES

[1] M. U. Farooq, M. Waseem, S. Mazhar, A. Khairi and T. Kamal, "A review on internet of things (IoT)," *International Journal of Computer Applications*, vol. 113, pp. 1–7, 2015.
[2] E. Baccelli, O. Hahm, M. Gunes, M. Wählisch and T. C. Schmidt, "OS for the IoT-goals, challenges, and solutions," in *Workshop Interdisciplinaire sur la Sécurité Globale (WISG2013)*, 2013.
[3] P. Gaur and M. P. Tahiliani, "Operating systems for IoT devices: A critical survey," in *2015 IEEE Region 10 Symposium*, 2015.
[4] F. Javed, M. K. Afzal, M. Sharif and B.-S. Kim, "Internet of Things (IoT) operating systems support, networking technologies, applications, and challenges: A comparative review," *IEEE Communications Surveys & Tutorials*, vol. 20, pp. 2062–2100, 2018.
[5] J. Alawadhi, A. M. AlJanabi, M. A. Khder, B. J. A. Ali and R. F. Al-Shalabi, "Internet of Things (IoT) security risks: Challenges for business," in *2022 ASU International Conference in Emerging Technologies for Sustainability and Intelligent Systems (ICETSIS)*, 2022.
[6] J. Gubbi, R. Buyya, S. Marusic and M. Palaniswami, "Internet of Things (IoT): A vision, architectural elements, and future directions," *Future Generation Computer Systems*, vol. 29, pp. 1645–1660, 2013.
[7] A.-C. G. Anadiotis, L. Galluccio, S. Milardo, G. Morabito and S. Palazzo, "Towards a software-defined network operating system for the IoT," in *2015 IEEE 2nd World Forum on Internet of Things (WF-IoT)*, 2015.

[8] F. Wang, H. Xing and J. Xu, "Real-time resource allocation for wireless powered multiuser mobile edge computing with energy and task causality," *IEEE Transactions on Communications*, vol. 68, pp. 7140–7155, 2020.

[9] S. Jain and A. Kajal, "Effective analysis of risks and vulnerabilities in Internet of Things," *International Journal of Computing and Corporate Research*, vol. 5, 2015. https://www.mdpi.com/1424-8220/19/8/1793

[10] E. Baccelli, C. Gündoğan, O. Hahm, P. Kietzmann, M. S. Lenders, H. Petersen, K. Schleiser, T. C. Schmidt and M. Wählisch, "RIOT: An open source operating system for low-end embedded devices in the IoT," *IEEE Internet of Things Journal*, vol. 5, pp. 4428–4440, 2018.

[11] Y. B. Zikria, H. Yu, M. K. Afzal, M. H. Rehmani and O. Hahm, *Internet of Things (IoT): Operating System, Applications and Protocols Design, and Validation Techniques*, vol. 88, Elsevier, 2018, pp. 699–706.

[12] H. Luan and J. Leng, "Design of energy monitoring system based on IOT," in *2016 Chinese Control and Decision Conference (CCDC)*, 2016.

[13] N. M. Kumar and P. K. Mallick, "The Internet of Things: Insights into the building blocks, component interactions, and architecture layers," *Procedia Computer Science*, vol. 132, pp. 109–117, 2018.

[14] A. Musaddiq, Y. B. Zikria, O. Hahm, H. Yu, A. K. Bashir and S. W. Kim, "A survey on resource management in IoT operating systems," *IEEE Access*, vol. 6, pp. 8459–8482, 2018.

[15] F. Flammini, D. Dobrilović, A. Gaglione and D. Tokody, "LoRaWAN Technology Mapping to Layered IoT Architecture". https://www.researchgate.net/profile/Francesco-Flammini/publication/344501334_LoRaWAN_Technology_Mapping_to_Layered_IoT_Architecture/links/600fe4f645851553a06ff244/LoRaWAN-Technology-Mapping-to-Layered-IoT-Architecture.pdf

[16] A. Raoof, A. Matrawy and C.-H. Lung, "Secure routing in IoT: Evaluation of RPL's secure mode under attacks," in *2019 IEEE Global Communications Conference (GLOBECOM)*, 2019.

[17] S. A. Hamad, Q. Z. Sheng, W. E. Zhang and S. Nepal, "Realizing an internet of secure things: A survey on issues and enabling technologies," *IEEE Communications Surveys & Tutorials*, vol. 22, pp. 1372–1391, 2020.

[18] D. Li, Z. Zhang, W. Liao and Z. Xu, "KLRA: A Kernel level resource auditing tool for IoT operating system security," in *2018 IEEE/ACM Symposium on Edge Computing (SEC)*, 2018.

[19] D. Vogt, B. Döbel and A. Lackorzynski, "Stay strong, stay safe: Enhancing reliability of a secure operating system," in *Proceedings of the Workshop on Isolation and Integration for Dependable Systems (IIDS 2010)*, Paris, France, April 2010.

[20] Z.-K. Zhang, M. C. Y. Cho, C.-W. Wang, C.-W. Hsu, C.-K. Chen and S. Shieh, "IoT security: Ongoing challenges and research opportunities," in *2014 IEEE 7th International Conference on Service-Oriented Computing and Applications*, 2014.

[21] M. Asim and W. Iqbal, "Iot operating systems and security challenges," *International Journal of Computer Science and Information Security*, vol. 14, p. 314, 2016.

[22] S.-K. Choi, C.-H. Yang and J. Kwak, "System hardening and security monitoring for IoT devices to mitigate IoT security vulnerabilities and threats," *KSII Transactions on Internet and Information Systems*, vol. 12, pp. 906–918, February 2018.

[23] M. A. Husnoo, A. Anwar, R. K. Chakrabortty, R. Doss and M. J. Ryan, "Differential privacy for IoT-enabled critical infrastructure: A comprehensive survey," *IEEE Access*, vol. 9, pp. 153276–153304, 2021.

[24] B. S. Dhak and P. L. Ramteke, "Evaluation of Kernel-level IoT security and QoS aware models from an empirical perspective," in *Proceedings of the 3rd International Conference on Communication, Devices and Computing*, 2022.

[25] T. Kulik, B. Dongol, P. G. Larsen, H. D. Macedo, S. Schneider, P. W. V. Tran-Jørgensen and J. Woodcock, "A survey of practical formal methods for security," *Formal Aspects of Computing*, vol. 34, pp. 1–39, 2022.

[26] A. Norollah, H. Beitollahi, Z. Kazemi and M. Fazeli, "A security-aware hardware scheduler for modern multi-core systems with hard real-time constraints," *Microprocessors and Microsystems*, vol. 95, p. 104716, 2022.

[27] G. Fortino, L. Fotia, F. Messina, D. Rosaci and G. M. L. Sarné, "Trust and reputation in the internet of things: State-of-the-art and research challenges," *IEEE Access*, vol. 8, pp. 60117–60125, 2020.

[28] M. Dammak, S.-M. Senouci, M. A. Messous, M. H. Elhdhili and C. Gransart, "Decentralized lightweight group key management for dynamic access control in IoT environments," *IEEE Transactions on Network and Service Management*, vol. 17, pp. 1742–1757, 2020.

5 Wireless Physical Layer Security

V. Saranya
Sathyabama Institute of Science and Technology,
Deemed to be University, Chennai, India

P. Manjula
Saveetha School of Engineering, Deemed to be University,
Chennai, India

*S. Raja Shree, A. Jemshia Mirriam, and
M. Nafees Muneera*
Sathyabama Institute of Science and Technology, Deemed
to be University, Chennai, India

5.1 INTRODUCTION

An extensive range of applications like environmental monitoring, social networks, banking, and other financial operations are increasingly using wireless networks. The security of wireless networks is therefore of vital societal relevance. Rather than providing security at the physical transmission layer, security is typically provided over the upper layers, such as the logical layers of networks in communication. The principal means of maintaining data confidentiality is encryption, which is a technology that is effective in the majority of current circumstances. However, difficulties such as the management of keys or the complexities of computation make the adoption of data encryption challenging in several emerging networking topologies. In the planning and use of wireless networks, security is a crucial concern. Because of the constrained computational capacities found in certain emerging wireless networks (e.g., radio-frequency identification (RFID) tags, specialized sensors), along with their vast scale and decentralized organizational structure, conventional methods of ensuring security in such networks are unfeasible. By utilizing the inherent capability of radio propagation physics to enable certain types of protection, physical layer security (PLS) has the potential to overcome these issues.

The physical layer (layer 1) is the first and bottommost layer in the seven-layer Open System Interconnect (OSI) paradigm of networking [1]. The name "physical layer" can be a bit confusing. A lot of people who study computer networks get the idea that the physical layer is all about the actual hardware of the network. Actually, the physical layer deals with data transmission and reception, encoding, topology

DOI: 10.1201/9781003477327-5

and physical network design, cabling among devices and mediums, and hardware specifications.

Jiang et al. [2] discussed the general concepts of PLS (eavesdropping scenario) as it is shown in Figure 5.1.

When the two terminals T2 and T3 are not adjacent, radio-frequency signals are seen toward the output of the main channel, and the eavesdropper's channels are typically dissimilar. The most noticeable effects of physical processes that produce natural disparities for wireless communications are fading and route loss. For example, if terminal T1 transmits a stream of video, the signal received at terminal T3 may be noticeably worse than the one received by terminal T2, and this worsening may even make it hard for terminal T3 to interpret the content of the video stream. The physical reality of communication channels is not taken into consideration by the standard secure communication architecture. In particular, it ignores the signal degradation brought on by noise or fading. This discovery easily prompts the development of a more ideal model of communication, commonly known as the wiretap channel (Figure 5.2), and explicitly introduces noise and distortion in both the main and eavesdropper's channels.

All communication systems use various cryptosystem keys like private keys and public keys to address the challenges of authentication, privacy, and secrecy in the

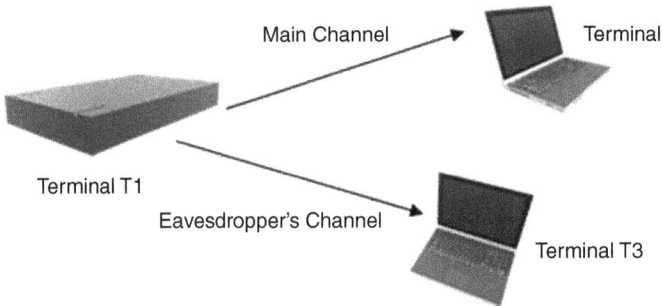

FIGURE 5.1 General concepts of physical layer security.

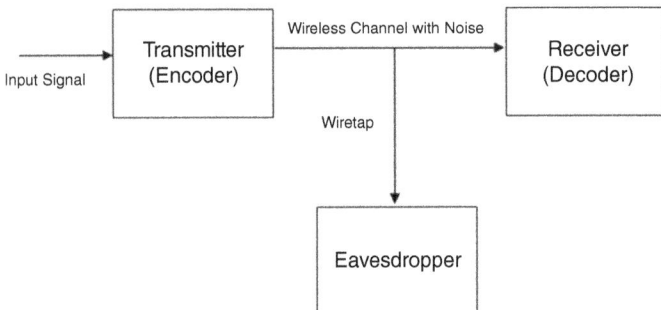

FIGURE 5.2 Wiretap channel model.

upper layers of networking protocol [3]. Today, a lot of findings in information theory, signal processing, and cryptography point to the security benefits of designing secure systems that take into account the constraints of the physical layer. The goal of lower layer (PHY) security is to improve the efficiency of confidentiality which in turn is described as the data rate of private messages.

In wireless networks, confidentiality and secure data transmission are predominant issues as wireless network systems expand continuously. Hence, the need to provide secure data communication with the presence of attackers (jammers and eavesdroppers) cannot be over-accentuated. The conventional cryptographic approaches are currently focused on the application layer. However, these approaches are associated with key generation and management, including computational complexity and cost. PLS ensures secure communication without the use of a cryptographic approach and also achieves a higher rate of secrecy.

Cryptography and PLS are distinct approaches to safeguarding information, each serving a unique purpose. While both share the goal of data protection, they employ fundamentally different methods and are suited to different contexts. Below, comparative analyses are provided.

1. **Purpose**

   Cryptography: Cryptography primarily focuses on securing data during transmission or storage. It achieves this by employing mathematical algorithms and encryption keys to encode and decode data, ensuring that only authorized individuals can access the information.

   Physical Layer Security: In contrast, PLS is oriented toward safeguarding data by leveraging the inherent characteristics of the physical transmission medium itself. Its objective is to create obstacles that make it challenging for unauthorized parties to intercept or comprehend the transmitted data.

2. **Techniques**

   Cryptography: Cryptography relies on algorithms like encryption and decryption to transform plain text data into cipher text, rendering it meaningless without the appropriate decryption key. It provides data confidentiality, integrity, and authenticity.

   Physical Layer Security: PLS harnesses the distinctive properties of the communication channel, such as signal attenuation, fading, and noise. It employs methods such as beamforming, interference management, and signal manipulation to impede eavesdropping attempts.

3. **Vulnerabilities**

   Cryptography: Cryptography may be susceptible to attacks if adversaries can breach the encryption algorithm, steal encryption keys, or execute man-in-the-middle attacks. Ongoing advances in computing power and cryptographic attack techniques can undermine the security of encrypted data.

   Physical Layer Security: PLS is less prone to cryptographic attacks but relies on the assumption that the physical channel's properties are not easily manipulated by eavesdroppers. Nevertheless, it can be vulnerable to specific types of physical layer attacks, such as jamming or signal interception.

4. **Applications**

Cryptography: Cryptography finds extensive use in securing data across computer networks, internet communications, mobile devices, and storage systems. Additionally, it plays a crucial role in protecting sensitive information such as passwords and financial transactions.

Physical Layer Security: PLS is predominantly employed in wireless communication systems, where ensuring the security of the wireless channel is imperative. It is commonly applied in scenarios like military communications, wireless sensor networks, and the security of Internet of Things (IoT) devices.

5. **Trade-offs**

Cryptography: Cryptography provides robust security for data in transit or at rest, contingent upon the strength of the encryption algorithm and the security of encryption keys. However, it may introduce computational overhead.

Physical Layer Security: PLS offers an additional layer of protection by capitalizing on the physics of the transmission medium. Nevertheless, its effectiveness may not match cryptography in all scenarios.

In summary, cryptography and PLS are complementary strategies for securing information. Rather than being in competition, they can be employed in tandem to bolster overall security, particularly in situations where safeguarding both the transmission medium and the data is imperative. Each approach possesses its unique strengths and weaknesses, with their efficacy dependent on specific use cases and threat scenarios.

5.2   PHYSICAL LAYER ATTACKS

The physical layer is the lowest layer in wireless communication systems and is in charge of carrier frequency production, frequency division, modulation, and signal detection. In order to demodulate the information signals, wireless communication systems send the data stream to the top layers after receiving them through the physical layer. The characteristics of transmission in the physical layer are impaired by various sorts of physical attacks. Attacks are divided into two types such as active assaults and passive assaults and this is shown in Figure 5.3.

5.2.1   ACTIVE ATTACKS

Interference and jamming are common tactics in active threats. The broadcasting signal was affected by these two sorts of attacks in a few particular frequency bands. When a transmitter malfunctions, a jamming attack takes place, and when there is interference at the receiver, the signal cannot be received. Jamming attacks are also referred to as malicious attacks, and interference is brought on by users who are using the same channel as hostile attackers. In an earlier study, the authors discussed various types of jamming attacks [4, 5] which are given below.

```
                        ┌──────────────────┐
                        │ Security Attacks │
                        └──────────────────┘
                    ┌───────────┴────────────┐
                    ▼                        ▼
          ┌──────────────────┐     ┌──────────────────┐
          │ Passive Attacks  │     │  Active Attacks  │
          └──────────────────┘     └──────────────────┘
                  │                        │
                  ├──▶ ┌────────────────┐  ├──▶ ┌───────────────────────┐
                  │    │ Traffic Analysis│  │    │ Denial of service Attack│
                  │    └────────────────┘  │    └───────────────────────┘
                  └──▶ ┌────────────────┐  ├──▶ ┌───────────────────────┐
                       │ Eavesdropping   │  │    │ Masquerade Attack      │
                       └────────────────┘  │    └───────────────────────┘
                                           ├──▶ ┌───────────────────────┐
                                           │    │ Replay Attack          │
                                           │    └───────────────────────┘
                                           ├──▶ ┌───────────────────────┐
                                           │    │ Information Disclosure │
                                           │    └───────────────────────┘
                                           ├──▶ ┌───────────────────────┐
                                           │    │ Message Modification   │
                                           │    └───────────────────────┘
                                           └──▶ ┌───────────────────────┐
                                                │ Resource Consumption   │
                                                └───────────────────────┘
```

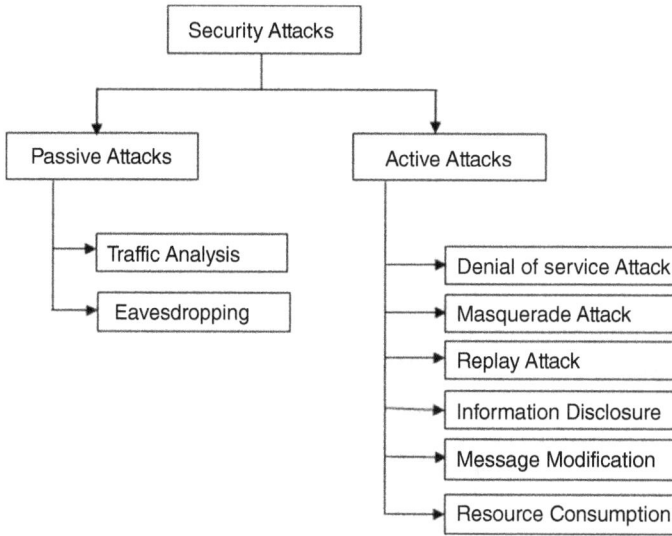

FIGURE 5.3 Security attacks in wireless networks.

1. **Deceptive Jamming**: In this method, the attackers deliver affected data packets to users across the network while ensuring that the packets are perceived as regular datagrams by the consumers. This type of interference is extremely damaging and difficult to detect.
2. **Barrage Jamming**: Several frequencies are simultaneously attacked, disrupting any communication between users within the barrage jamming coverage area. The vast range of frequencies has an impact on the transmission power limit factor. The attackers jam weak frequencies while attacking a wide variety of frequencies.
3. **Spot Jamming**: Spot jamming is a fairly straightforward and popular device that uses high-strength signals to block the original signal. Individual frequency jamming is the major emphasis of spot jamming.
4. **Sweep Jamming**: In this jamming technique, multi-hop technology is employed by the attackers as it is simple to cover a wide range of frequencies and it directly targets the frequency-hopping technology.

5.2.2 PASSIVE ATTACKS

Passive assaults are categorized into two methods: traffic analysis and eavesdropping. These assaults are brought on by several fundamental problems with the wireless medium, broadcast messages, and names [6]. Due to the broadcasting nature of wireless communication systems, it is impossible to keep out unauthorized intruders because any user within the service area can access, use, and analyze the wireless signal [2]. With open access to wireless communication networks in information transmission systems, intruders can easily access user information. Attackers are monitoring network changes and traffic flow in order to gather information from ongoing conversations. For instance, in a wireless sensor network, a hacker might simply determine the location of the base station and make adjustments [7].

5.3 TECHNIQUES IN PHYSICAL LAYER SECURITY

The research community should pay more attention to the prospects and difficulties of achieving high levels of Physical Layer (PHY) security for mobile communication systems. Various existing studies discussed a number of ways for PHY security, including Pre-processing Schemes [8, 9], Coding Techniques [10], Key Generation and Exchange [11–13], Artificial Noise Schemes [14], and Game Theoretic Schemes, Signal Processing, Cooperative Communications, etc. Figure 5.4 shows the techniques in PLS and are described below.

1. **Secure Key Generation**: An important component of cryptosystems is key generation. The two authentic parties coming to an agreement on a mutual key have to accomplish privacy, authentication, and integrity services for symmetric key cryptosystems. Yet, the distribution of keys is one of the complex issues in this system. Furthermore, the conventional key generation technique demands more work in terms of complexity and permissible secrecy. Key generation methods have two broad categories: Keyless Security and Secret Key-based security that are utilizing the Physical Layer (PHY). The Keyless Security method known as the wiretap channel model sends data without encryption or the creation of a shared key.
2. **Directional Modulation (DM)**: The DM approach is utilized such that the information is directed toward the destination which leads to the baseband modulated signal equal to the constellation diagram of the receiver.
3. **Spatial Modulation (SM)**: It is a novel MIMO (Multiple Input, Multiple Output) [15] antenna transmission method that strikes a good compromise between transmission rate and hardware overhead. The system's spectral efficiency is increased without using more expensive RF lines because it fully utilizes the channel index information to send a large number of bit streams. In spatial networks, both the aerial index and modulation symbols

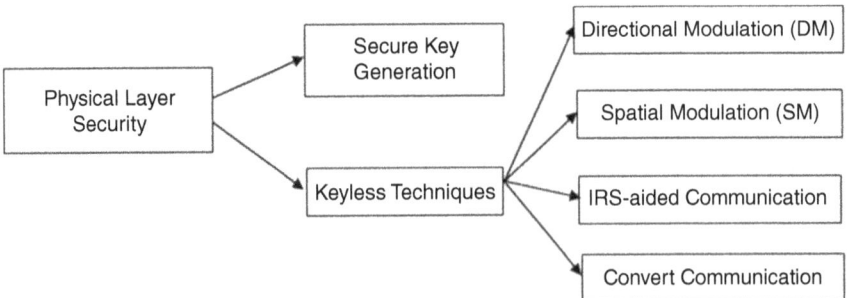

FIGURE 5.4 Physical layer security techniques.

convey private information, so intercepting one of them could result in the disclosure of sensitive data. The SM network has significant communication security flaws in contrast with the MIMO Network. Hence, it is crucial from a strategic standpoint to use PHY security technologies to protect SM systems.

4. **Covert Communication**: With covert communication, it is ensured that there is a low enough chance that an unlawful observer will notice a message being transmitted to the receiver. An essential secure transmission technique called covert transmission seeks to conceal the transmitter's transmission behavior. PHY security measures, which work to prevent Eves from listening in on the transmitted data, cannot provide the same level of communication security as covert communication techniques.

5. **Intelligent Reflecting Surface (IRS) Aided Communication**: This approach helps to improve the performance of data transmission in wireless networks. An artificial electromagnetic surface known as an IRS is made up of several passive reflective units. Fine-grained Three-Dimensional Beamforming (3DBF) is utilized to enhance received power, increase channel quality, and increase communication range. IRS actively engages in the signal transmission process by transforming the traditional unforeseen as well as unmanageable wireless communication environment into a configurable and largely immutable transmission space.

5.4   PHYSICAL LAYER SECURITY SCHEMES FOR IoT

There have already been a large number of PLS strategies created in the literature, the following schemes are used primarily in the IoT field.

Artificial Noise (AN) Injection: The idea behind this strategy is to send both the information-carrying signal and the AN at the same time in order to hinder the performance of the attacker. The range space and null space of the channel matrix of the authorized user are where the information-bearing signal and the AN are respectively injected. In this way, the AN has little negative effect on the authorized receiver and only worsens the eavesdropper [16]. An efficient way to give the authorized transmission connection a channel quality edge is through AN injection. The majority of AN-based PLS methods, however, require numerous antennas to be deployed at the transmitter [17, 18], which goes against the IoT [19] devices' need for low cost and small size. The cooperative AN injection emerges as a possible remedy to resolve this problem and guarantee the security of IoT communication.

Bit Flipping: The bit flipping method is commonly employed to secure communications between a large number of sensor nodes and a Legitimate Fusion Centre (LFC). In this approach, the sensor nodes are divided into two distinct groups: the strong group and the weak group, based on the strength of their channel gains to the LFC. The strong group consists of sensors with superior channel qualities, which are responsible for transmitting genuine information-bearing data. Conversely, the weak group comprises sensors with poorer channel qualities and their role is to transmit manipulated or false data, thereby disrupting the Eavesdropping Fusion Centre (EFC). By categorizing the sensor nodes into these two groups, taking into account the variation in their channel gains to the LFC, the bit-flipping method effectively leverages the statistical independence between the legitimate channel and the wiretap channel. As a result, the received signal-to-noise ratio (SNR) at the EFC is significantly lower compared to that at the LFC. This significant degradation of performance at the EFC hampers its ability to gather accurate information, ensuring enhanced security in the communication system.

Compressive Sensing (CS): When compared to the Nyquist sampling rate, CS may compress sparse signals at a significantly reduced rate. PLS has recently been achieved using the CS approach [20]. Multiplying by a measurement matrix results in the linear transformation of the sparse information-bearing signal in CS. To ensure the confidentiality of transmission, conceal the measurement matrix from the eavesdropper using the m sequence as suggested by the author [20]. The Received Signal Strength Indicator (RSSI) values of packets exchanged between authorized users are used to generate a random seed, which in turn creates an m-sequence. By exploiting the fact that the channel coefficients of the genuine link and the eavesdropping link are unrelated, it becomes possible to detect the presence of a listener. This method ensures the confidentiality of information since the eavesdropper is unable to calculate the same measurement matrix as the legitimate nodes due to the lack of correlation between the channel coefficients of the valid link and the eavesdropping link.

Cooperative Secrecy: IoT often consists of large physical items like sensors, actuators, and controllers. Utilizing cooperation between these kinds of low-power devices can nevertheless meet the consumers' needs for secrecy even while each individual device's computing power is constrained. In cooperative secrecy, the fundamental idea is to employ friendly nodes as jammers, generating artificial interference to hinder the eavesdropper's signal reception. This approach combines secure beamforming and cooperative jamming to enhance security at the physical layer [21].

Despite the limited computing power of each individual low-power device, their collaboration can effectively fulfill consumers' requirements for confidentiality. Cooperative secrecy involves leveraging friendly nodes to act as jammers, emitting intentional interference to disrupt the eavesdropper's ability to receive the signal. The author [22] introduced cooperative jamming (CJ) strategies in the context of amplify-and-forward (AF) and decode-and-forward (DF) systems. These tactics involve relay nodes transmitting weighted artificial noises to degrade the eavesdropper's channel. To enhance the security of the physical layer, the author [21] combined secure beamforming with cooperative jamming. They formulated an optimization problem in their study to minimize the risk of secrecy outage while still meeting the secrecy rate criterion.

Advantages and Disadvantages of Current Physical Layer Security Techniques: One limitation of artificial noise (AN) injection as a PLS method is its vulnerability to existing algorithms due to its construction using a pseudo-random number generator, which can be easily exploited. Moreover, the additional energy required to transmit the fake noise signal reduces the secrecy gain provided by AN injection. In comparison, the CS-based secure transmission method is more energy-efficient as it doesn't require additional power. However, it relies on the exchange and secrecy of a measurement matrix among the legitimate transceivers, leading to protocol design complexities and non-negligible overheads. These drawbacks of CS-based approaches can be addressed by employing the bit-flipping technique, which reduces implementation complexity. Nevertheless, the bit-flipping technique requires a weak group of sensors to transmit bogus data, resulting in power and bandwidth wastage.

The strategy of cooperative secrecy is commonly embraced as a PLS approach, primarily due to its ability to empower low-power devices in countering powerful eavesdroppers while efficiently managing resources. However, the complexity of protocol design increases due to the need for additional signaling to coordinate multiple network devices.

Physical layer encryption is a cross-layer approach that merges physical layer secret key generation with application layer encryption. Its notable advantage is its smooth integration with established network security protocols utilizing application-layer cryptographic techniques. However, the success of physical layer encryption greatly relies on the ability of communicating parties to reach a consensus on the generated keys, which presents challenges, particularly in wireless scenarios [23].

5.5 SECURITY LIMITATIONS AND POSSIBLE ATTACKS

There may be numerous restrictions with PHY security solutions. Also, it is susceptible to a number of common security threats, including those that target integrity, privacy, and authentication. Denial of Service (DoS) is a popular security attack that targets nodes and causes malicious behavior or intentional failure. It can be done by utilizing the victim system's resources, sending large volumes of unnecessary information, and preventing the victim from using the communication system's resources and services. These types of assaults can majorly affect PHY security in wireless communication networks.

In PHY security systems, a brute force attack is also a possibility because some channel measures that are a component of the security system, like the key exchange, can be anticipated and retrieved using a forceful assault. The majority of PLS solutions are appropriate for communication systems with inadequate computational resources, which is good news. This indicates that the probability of a brute force assault in these systems is quite low.

The jamming attack ranks among the most frequent assaults on wireless communication networks. The amount of flow that the wireless system could jam is substantially increased by the jamming devices' location, making a Flow Jamming Attack (FJA) which is known as an intelligent attack. FJA are the cleverest attacks because they maximize the quantity of wireless system traffic that may be blocked while using the least amount of power [3, 24]. With the application of PHY security solutions, this attack has a high likelihood. The key generation process' probing phase is significantly influenced by the FJA. FJA is regarded as a linear programming, together with its model. Finding scheduling algorithms for gathering real-time information about the data flow in the communication system during an FJA is one of the important areas of research that addresses the jamming attack.

5.6   SUMMARY

Wireless Sensor networks are widely employed in real-world applications and the military to accumulate event-driven and real-time data as it becomes a crucial part of day-to-day life. These applications depend on wireless network protocol for providing secure transmission and data integrity. However, the nature of wireless networks is insecure and attackers may attempt illegal access to modify the information and interrupt the flow of data transmission.

Future networks face new security challenges due to emerging wireless technologies such as Secure Key Generation, Directional or Spatial Modulation, and covert and IRS-aided communication. Extensive research efforts have recently focused on developing efficient and secure transmission techniques for wireless communications by leveraging the radio channel propagation characteristics at the physical layer (PHY). This research evaluates the existing PHY secure communication methods, considering both theoretical and technological aspects. It addresses privacy concerns related to PHY security technologies, including secure key generation, covert communication, DM, and IRS. Additionally, it identifies the difficulties associated with PHY security and suggests potential avenues for future research.

REFERENCES

1. W. Stallings, *Cryptography and Network Security Principles and Practices*, Prentice Hall PTR, 2006.
2. W. Jiang, Y. Zhang, J. Wu, W. Feng and Y. Jin, "Intelligent Reflecting Surface Assisted Secure Wireless Communications with Multiple-Transmit and Multiple-Receive Antennas", *IEEE Access*, vol. 8, pp. 86659–86673, 2020, doi: 10.1109/ACCESS.2020. 2992613
3. D. Wyner, "The Wiretap Channel", *Bell System Technical Journal*, vol. 54, pp. 1355– 1387, 1975.
4. A. Mpitziopoulos, D. Gavalas, C. Konstantopoulos and G. Pantziou, "A Survey on Jamming Attacks and Countermeasures in WSNs", *IEEE Communications Surveys & Tutorials*, vol. 11, no. 4, pp. 42–56, 2009.
5. W. Fang, F. Li, Y. Sun, L. Shan, S. Chen, C. Chen et al., "Information Security of PHY Layer in Wireless Networks", *Journal of Sensors*, Hindawi Publishing Corporation, 2016, doi: 10.1155/2016/1230387

6. W. Xu, W. Trappe, Y. Zhang, and T. Wood, "The Feasibility of Launching and Detecting Jamming Attacks in Wireless Networks", in *Proceedings of the 6th ACM International Symposium on Mobile Ad Hoc Networking and Computing*, ACM, pp. 46–57, 2005.

7. S. Hong, C. Pan, H. Ren, K. Wang, A. Nallanathan, "Artificial-Noise-Aided Secure MIMO Wireless Communications via Intelligent Reflecting Surface", *IEEE Transactions Communication*, vol. 68, pp. 7851–7866, 2020.

8. X. Yu, D. Xu, Y. Sun, D. W. K. Ng and R. Schober, "Robust and Secure Wireless Communications via Intelligent Reflecting Surfaces", *IEEE Journal on Selected Areas in Communications*, vol. 38, no. 11, pp. 2637–2652, 2020, doi: 10.1109/JSAC.2020.3007043

9. S. Hong, C. Pan, H. Ren, K. Wang, A. Nallanathan and H. Li, "Robust Transmission Design for Intelligent Reflecting Surface Aided Secure Communications", in *GLOBECOM 2020 - 2020 IEEE Global Communications Conference*, Taipei, Taiwan, pp. 1–6, 2020, doi: 10.1109/GLOBECOM42002.2020.9322565

10. B.-J. Kwak, N.-O. Song, B. Park, D. Klinc, and S. W. McLaughlin, "Physical layer security with yarg code", in *Proceedings of the 1st International Conference on Emerging Network Intelligence*, pp. 43–48, 2009.

11. H. Shen, W. Xu, S. Gong, Z. He and C. Zhao, "Secrecy Rate Maximization for Intelligent Reflecting Surface Assisted Multi-Antenna Communications", *IEEE Communications Letters*, vol. 23, no. 9, pp. 1488–1492, 2019, doi: 10.1109/LCOMM.2019.2924214

12. X. Yu, D. Xu and R. Schober, "Enabling Secure Wireless Communications via Intelligent Reflecting Surfaces", in *Proceedings of IEEE Global Communication Conference (GLOBECOM)*, pp. 1–6, 2019.

13. J. Chen, Y.-C. Liang, Y. Pei and H. Guo, "Intelligent Reflecting Surface: A Programmable Wireless Environment for Physical Layer Security", *IEEE Access*, vol. 7, pp. 82599–82612, 2019, doi: 10.1109/ACCESS.2019.2924034

14. X. Guan, Q. Wu and R. Zhang, "Intelligent Reflecting Surface Assisted Secrecy Communication: Is Artificial Noise Helpful or Not?", *IEEE Wireless Communication Letter*, vol. 9, pp. 778–782, 2020.

15. T. Kavitha, M. Satish Kumar, G. Srihari, L. Umasankar and N. V. Babu, "Scaled and Nonlinear Multi-Objective Model for Downlink and Uplink Exposure in Massive MIMO", *Physical Communication*, vol. 57, 102004, 2023. ISSN: 1874-4907.

16. S. Goel and R. Negi, "Guaranteeing Secrecy Using Artificial Noise", *IEEE Wireless Communication*, vol. 7, pp. 2180–2189, 2008.

17. X. Zhang, M. R. McKay, X. Zhou and R.W. Heath, "Artificial-Noise-Aided Secure Multi-Antenna Transmission with Limited Feedback", *IEEE Wireless Communication*, vol. 14, pp. 2742–2754, 2015.

18. G. Wang, C. Meng, W. Heng and X. Chen, "Secrecy Energy Efficiency Optimization in AN-Aided Distributed Antenna Systems with Energy Harvesting", *IEEE Access*, vol. 6, pp. 32830–32838, 2018.

19. T. Kavitha, V. Ajantha Devi, S. Neelavathy Pari and Sakkaravarthi Ramanathan, *Internet of Everything: Smart Sensing Technologies*, Nova Science Publishers, Publication Date: June 17, 2022, doi: 10.52305/PNQM1088

20. A. Mukherjee, "Physical-Layer Security in the Internet of Things: Sensing and Communication Confidentiality under Resource Constraints", *Proceedings of IEEE*, vol. 103, pp. 1747–1761, 2015.

21. L. Hu, H. Wen, B. Wu, F. Pan, R. F. Liao, H. Song, J. Tang and X. Wang, "Cooperative Jamming for Physical Layer Security Enhancement in Internet of Things", *IEEE Journal of Internet Things*, vol. 5, pp. 219–228, 2018.

22. L. Dong, Z. Han, A. P. Petropulu and H. V. Poor, "Improving Wireless Physical Layer Security via Cooperative Relays", *IEEE Transaction, Signal Process*, vol. *58*, pp. 1875–1888, 2010.

23. T. Kavitha and D. Sridharan, "Security Vulnerabilities in Wireless Sensor Networks: A Survey", *International Journal on Information Assurance and Security (1554-1010) (JIAS)*, no. 5, pp. 031–044, 2010.

24. X. Lu, W. Yang, X. Guan, Q. Wu and Y. Cai, "Robust and Secure Beamforming for Intelligent Reflecting Surface Aided mmWave MISO Systems", *IEEE Wireless Communications Letters*, vol. 9, no. 12, pp. 2068–2072, 2020, doi: 10.1109/LWC.2020. 3012664

6 Lightweight Crypto Mechanisms and Key Management in IoT Scenario

Yasha Jyothi M. Shirur, Bindu S., and Jyoti R. Munavalli
BNM Institute of Technology, Visvesvaraya Technological University, Bengaluru, India

6.1 INTRODUCTION

The Internet of Things (IoT) makes use of networks and devices with software, sensors, and other technologies built into it for the exchange of data. The data transfer takes place online between two devices and is commonly known as *machine-to-machine communication*. These devices can be easily linked to the cloud since the internet uses a set of protocols for efficient data transport. Also, with the aid of a cloud computing platform, the infrastructure may be scaled up benefiting both the manufacturers and customers. It is widely used across many industries due to its benefits and simplicity in significant sectors such as medical care, smart cities, medicine, agriculture, business, and automotive industries, in automation and control for carrying out the operation automatically, and in making smarter judgments with more information, which is the goal of the Continuous Information Tracker [1]. It is used in continuous fault monitoring at the earlier stages of occurrences saving both time and money. It also has a few drawbacks such as security vulnerabilities and dependency on networks [2].

The main concern is making IoT devices secure. Attacks happen in a variety of forms as they are based on network connectivity. The biggest issue is the transmission of sensitive data. The businesses adhere to a set of network security standards such as integrating certain antivirus software, protecting the firewall, and enabling encrypted email communication [3]. Data protection can be provided using physical security, logical forms of protection, organizational administration panels, and other methods that allow access restriction for unauthorized users or processes [4]. The most crucial ones are data generation, compilation, storing, and sharing by any company or individual user. It is important to keep the data safe from tampering and unlawful access. There are various types of attacks such as the following:

Take Control: Unidentified individuals hack open the house door and automobile doors remotely.

DOI: 10.1201/9781003477327-6

69

Information Theft: Unauthorized individuals access the information stored on various devices owned by others.

Disrupt Services: Receive spam information such as incorrect information on pacemakers. There are numerous data security measures and controls in place to guard against unauthorized access to the data such as access management, where users who have permission from role-based access control can access unique information. With the use of automatic access control, network administrators govern all data.

Authentication: Authentication improves data security and guards against data misuse. Authentication verifies passwords, access tokens, bio-metrics, or swipe cards, are required for authorized users.

Tokenization: Both random and sensitive data are handled using tokenization. Instead of employing any mathematical procedures, it keeps the data in a secure database. It offers a few token values to which the data in the survey table is referred. Therefore, the original data is kept in a secure network.

Encryption: Using an encryption key and an algorithm, normal text is transformed into encrypted text. Cipher text is the name for the altered text. Data access by unauthorized users is prohibited. Users who have access to the authorized key can only decrypt data. All password formats, facial recognition software, and login credentials use this technique. Handling the encryption keys carefully is a requirement.

The development of cryptographic technology is paramount as observed in the substantial research being done on new attack, design, and implementation methods. Modern methods include Lightweight Cryptography (LWC). A cryptographic technique or protocol called "lightweight cryptography" is designed for use in limited situations, such as those posed by RFID tags, sensors, contact-less smart cards, and medical equipment [5]. LWC has a tiny computational complexity or footprint. Its international standardization and guidelines compilation are now underway, and it aims to increase the applicability of cryptography on restricted devices.

The goal of LWC is to offer security solutions that can operate on devices with limited resources by consuming less memory, less computational power, and less energy. In comparison to traditional cryptography, LWC is anticipated to be quicker and simpler [6].

6.1.1 Different Lightweight Crypto Mechanisms

Embedded systems are frequently constrained by processing speed, memory capacity, storage capacity, and energy consumption. The system is directly impacted by the cryptographic technology used in Embedded systems to provide tamper-resistant hardware and software security measures. A sizeable portion of the module's surface is made up of memory components. The cost is dependent on the component's surface. The optimized code produces outcomes more quickly and the faster a set of instructions is carried out, the less the power consumed.

Traditional cryptography approaches prioritize delivering high-security levels while ignoring the needs of constrained devices. Recently, LWC has evolved and it focuses on creating cryptographic protocols for devices with limited connectivity, power supply, hardware, and software capabilities.

Recent schemes include software and hybrid implementations for lightweight devices and hardware designs that are often thought to be better suited for ultra-constrained devices. Without using redundant parts, hardware designs accomplish the exact functionality. The primary design objective is in gate-level implementation which minimizes the number of logic gates needed to implement the cipher [7]. A tiny GE indicates that the circuit will be inexpensive and power-efficient. While implementations of 1000 GE are being researched for even smaller devices, including 4-bit micro-controllers, an implementation of 3000 GE can be regarded as adequate for restricted devices. Other important concerns are power limitations and energy consumption. Power limitations affect active devices, such as passive RFID tags, which depend on a host device to function, while energy consumption is significant when a gadget is powered by batteries. Hardware designs also consider attacks and pertinent defenses linked to power analyses.

Typically, all that is needed for software implementations to function is a CPU. The key design objectives are to lower the cipher's memory and processing demands. Throughput and power efficiency are prioritized in implementations. They provide a significant portability benefit over hardware implementations.

Hybrid approaches combine two strategies for utilizing their finest qualities. The fundamental cipher capability is implemented by hardware, and data and communication manipulation are handled by software. The creation of cryptographic co-processors is a widespread procedure. The communication bandwidth between hardware and software components has the biggest impact on throughput. Specific communication applications, such as RFID tags, portable electronics, and internet servers, are the focus of hybrid solutions.

6.1.2 SYMMETRIC CRYPTOGRAPHY

For lightweight and ultra-lightweight ciphers, the common security range is 80–128 bits. Although 128-bit security is customary for mainstream applications, 80-bit security is judged adequate for constrained devices like 4-bit micro-controllers and RFID tags. For one-way authentication, security of 64–80 bits would be sufficient [8].

Three alternative techniques are used to implement lightweight ciphers. First, researchers try to improve the performance of well-known and in-depth ciphers like AES and DES. Modern AES hardware uses 2400 GE and acts as a benchmark for more contemporary ciphers.

In the second case, researchers develop and employ completely original ciphers. In the third case, researchers blend elements of numerous well-studied ciphers with well-known personal traits. The lack of decryption is another factor that can reduce the need for such ciphers, particularly for ultra-lightweight cryptography. This technique can be used with devices that just need one-way authentication. The key should also be hard-wired to the device in order to further limit the GE brought on by the lack of key generation operations.

6.1.3 BLOCK CIPHERS

A cloud-based tool called Block Cipher was created to assist companies in creating blockchain apps using web APIs and callbacks. Users can deploy contracts and generate, decode, and create transactions using its APIs.

A conventional block cipher with a modest level of security that can be utilized in limited devices is DES [9]. The DESXL variant combines the two types whereas the DESL version of the cipher uses key whitening to boost security while achieving a 20% size reduction.

AES, Camellia, CLEFIA, and IDEA are among other conventional ciphers that are being researched in this area. The ISO/IEC, NESSIE, and CRYPTREC projects have been approved for camellia usage. While the software implementation is quick, the hardware implementation exceeds the 3000 GE bound. CLEFIA employs 128-, 192-, and 256-bit keys and has a block size of 128 bits. It was created by Sony, and both the hardware and software are quite effective. The ISO 29192-2 standard specifies it. PGP v2.0 uses IDEA, which works well in embedded software.

A significant point in LWC and the benchmark for lightweight ciphers is PRESENT. It is a 64-bit block cipher developed in two versions PRESENT-80 and PRESENT-128. It uses 80-bit and 128-bit keys and has a 128-bit block size. The novelties of PRESENT include a fully wired diffusion layer devoid of any algebraic unit and the substitution of eight different S-Boxes with a single, carefully chosen one. A promising ultra-lightweight cipher with a hybrid structure of block and stream cipher is Hummingbird-2. For each message handled, it has the option to generate a Message Authentication Code (MAC), creating a one-way authentication protocol. It performs better than PRESENT and encrypts data at high rates.

Lightweight stream ciphers continue to fall short of lightweight block ciphers despite efforts to improve them. The lengthy startup procedure before initial use is their main flaw. Additionally, some communication protocols cannot make use of stream ciphers. Due to the simplicity and speed of their hardware, they are still in the foreground.

6.1.4 ASYMMETRIC CRYPTOGRAPHY

To function on devices with the aforementioned resource restrictions, asymmetric algorithms and protocols must also be modified. This is a challenging undertaking since asymmetric ciphers require much more computing power than their symmetric equivalents and are frequently utilized with sophisticated hardware. With limited devices like 8-bit microcontrollers, the performance disparity is larger. Even an optimized asymmetric technique, such as elliptic-curve cryptography (ECC), runs 100–1000 times slower than a conventional symmetric algorithm, such as AES, which results in two or three orders of magnitude more power consumption.

Typical asymmetries in cryptosystems. One-way trapdoor functions are the foundation of conventional public key cryptography. These functions are based on a number of challenging mathematical puzzles. Three reliable cryptosystems are available. The foundation of conventional public key cryptography is one-way trapdoor functions. These operations are built upon a number of challenging mathematical puzzles. Three trustworthy cryptosystems exist: (1) based on the Integer Factorization

Problem, RSA, Rabin (FP), (2) Elliptic Curve Discrete Logarithm Problem-Based ECC/HECC (ECDLP), and (3) the Discrete Logarithm Problem-Based ElGamal algorithm mathematical issues [10].

6.2   LIGHTWEIGHT CRYPTOGRAPHY APPLICABLE TO VARIOUS IoT DEVICES

Nowadays, embedded systems are widely used in a variety of applications, including smart cards, cars, telecommunications, home automation systems, computer networking, digital consumer electronics, defense, and aerospace. IoT is the technology that enables the interconnection of these embedded devices such as sensors and actuators through the internet in order to exchange data, improve processes, and monitor the equipment to benefit businesses. These procedures frequently involve sensitive or important information that must be shielded from the outer world. Therefore, the main concern is safety [11]. However, there are some limitations in terms of computational power, memory capacity, chip size, and power consumption while ensuring security for these devices.

In LWC, cryptographic protocols or algorithms are created specifically for use in limited contexts, such as RFID tags, contactless smart cards, sensors, embedded systems, and medical equipment.

Future solutions, from industrial to applications in daily life, will all include the IoT. This new technology appeals to everyone by giving things a higher level of intelligence and automating judgments. Applications that work with data, however, are more susceptible to other kinds of attacks. In order to protect communications between IoT edge nodes, researchers are continually investigating new solutions. IoT nodes should be inexpensive and power-efficient, which results in decreased computing performance. On the other hand, a secure communication layer consumes a lot of power and needs strong hardware. A viable method for reducing computing complexity while preserving the appropriate level of security is to use LWC methods.

6.3   KEY MANAGEMENT

Key management in cryptography plays an important role. It is even more important in connected devices like IoT networks. The IoT has heterogeneous connected devices sharing different kinds of data in the network. Security at different layers must be provided for these IoT devices. Management of keys in an effective and efficient way is required in current times. LWC should also concentrate on effective and efficient key management [28–30] to secure the information collected and shared through IoT devices [12].

Key management follows a life cycle that includes the following stages:

- Generation
- Distribution
- Exchange and usage
- Storage
- Destruction of the secret keys.

Key generation is of three types: Symmetric, Asymmetric, and Hybrid. In symmetric cryptography, both encryption and decryption of data use the same key. This method is secure and comparatively faster. If the attacker gets access to this key, then data can be decrypted. Symmetric encryption does not provide authentication. In asymmetric cryptography, two types of keys are used; public keys and private keys. Public keys are distributed widely whereas private keys are known only to the owner. Public key generation requires a lot of computations and these depend on the type of cryptographic algorithms used. So, the public key is openly distributed keeping the private key private. In such systems, any sender encrypts a message using the receiver's public key, but the encrypted message can only be decrypted with the receiver's private key [13]. Key generation utilizes four algorithms: AES, RSA, 3DES, and ECC [14].

Key distribution is done in four ways: public announcements, publicly available directories, ublic-key authority, and public-key certificates. In public announcements, the public key is broadcast to everyone. In publicly available directories, the public key is stored in a trusted public directory. Public-key authority is like a directory but with more tightened security. In public certification, the authority provides a certificate to allow key exchange. Key distribution is performed via a secure TLS or SSL connection.

The key used for encryption needs to be stored and this is done through Hardware Security Module or CloudHSM. The key is usable only for a certain time period, after which it expires. The deletion instruction deletes the key from the key manager database permanently. NIST standards require that deactivated keys be kept in an archive, to allow reconstruction.

IoT consists of different sensors that work in synchronization to provide different services such as smart cities, smart agriculture, home automation, healthcare, military, safety, and personal management systems [15]. In traditional systems, a single cryptographic key is used for security whereas it is not sufficient in IoT systems. When data is dealt with in real-time, authentication and security should also be provided in real-time [16]. Asymmetric or a combination of symmetric and asymmetric cryptography is used for IoT key management. Some of the techniques used in IoT key management are mutual key management, group key management, XOR-based key management, Elliptic-Curve-Cryptography-based key management, Smart-Object-based reliable key management, and CA-less key management. Evolutionary algorithms, Artificial Intelligence, Machine learning, and Genetic Algorithms provide promising outcomes in IoT security [17].

Cluster management in a network is a challenging task as sensor nodes move around with changes in topology randomly. Key management in cluster networks is carried out in three phases. In phase 1, key generation for each node is done. A master key is generated using a hash function. Then pairwise key is generated to communicate to the destination node. The master key is activated before sensor nodes are configured in the network. In phase 2, key distribution is performed by monitoring the security aspects. Finally, the secured link is created in phase 3. With this type of key management, the performance parameters such as energy efficiency and packet loss rate are improved. As the hashing function is used, encryption is not required for the whole of the data but only for the hash value [18]. As IoT networks are resource-constrained key management that uses symmetric cryptography, partial key per distribution is suitable. Storing partial keys also reduces memory requirements in IoT devices. This model provides security from eavesdropping, replay attacks, node capture attacks, Sybil attacks, and man-in-the-middle attack [19].

As IoT networks support lightweight devices, the key management system must be efficient. For that, the Hashed Advanced Encryption Standard (HAES) algorithm is implemented. This study implements the key management in two phases: data owner and data user. The data owner phase authenticates the data and uploads it securely. Authentication involves key generation, signature creation, and data access policy. In the data user phase, a user sends an access request for data to the blockchain. Further, the data owner, based on policies, accepts or rejects the request. HAES algorithm is compared with other algorithms like DES, AES, RC4, and blowfish, and HAES provided the highest security level of 97.89% [20]. Matrix-based key management allows any node to compute pairwise keys along with its neighbors. This model allows scalability and extensibility for IoT networks. Computation cost is less [10]. A chaotic cryptography-based privacy model improves the security. The key generation process is optimized by optimal pairs of keys through a self-adaptive sailfish optimization algorithm [22].

6.4   KEY MANAGEMENT CHALLENGES IN IoT [23]

i. **Device Security**: Devices used in the network should be secured through encryption. They must not be controlled by external intrusion. The sensors send data to the cloud to reduce the computation burden on network resources [21].

ii. **Communication Security**: The protocols used in communication should be highly secured and efficient. The protocols used must be compatible with heterogeneous networks.

iii. **Cloud/Data Center Security**: Data packets should not only be secured during transmission but also secured in the database they are stored in. Denial of service is a common attack in the cloud environment.

iv. **Software Vulnerabilities**: Such kinds of vulnerabilities give control of IoT devices to adversaries. The third-party apps used could compromise security. Malicious software, worms, Trojan viruses, etc., attack and data is leaked from target devices. The applications used on IoT devices must be updated and tested for bugs [24].

v. **Chip Vulnerabilities**: Hardware Trojan is inserted in the payload that is injected into the device, causing information leakage and further manipulation of the information [24].

vi. **Authentication Security**: IoT authentication is done through one-time password authentication, ECC-based mutual authentication, ID and password-based authentication, and certificate-based authentication [25].

vii. **Data Privacy**: As more devices get added to the IoT network, security needs to be provided at various layers for device security, network, and applications [26].

viii. **Machine Phishing**: Hackers infiltrate IoT devices and send false signals [27].

In order to address the security challenges in IoT applications the security block accomplished with light cipher algorithms can be implemented in the IoT architecture [28]. The cryptographic algorithm is designed and coded in Verilog, verified for functionality, and synthesized to develop intellectual property (IP) cores. These developed IP cores can be integrated at various levels of IoT chip fabrication. One such optimized light cipher algorithm suitable for IoT application is developed and discussed in detail in the form of a case study [28, 29]. This case study will help the researchers to understand the advantages associated with integrating the security algorithm with hardware while developing the IoT architecture.

6.5 CASE STUDY: DESIGN AND IMPLEMENTATION OF SYNTHESIZABLE TWO-LEVEL CRYPTOSYSTEM FOR HIGH-SECURITY ENABLED IoT APPLICATIONS

The cryptographic algorithms are simple to implement using software; they are typically too slow for real-time applications like IoT, network routers, embedded systems, and storage devices [30]. The features of IoT and VLSI can be combined with modern technology to develop a smart efficient system to meet industry requirements [31]. In order to address this, the optimized security Intellectual Property (IP) can be implemented in Very Large-Scale Integration (VLSI) which plays an important role in data security in IoT. The IoT can be configured from two different perspectives:

1. The entire IoT application can be developed as a System on Chip (SoC).
2. Each layer of the IoT is embedded with the VLSI capability and can be used based on the application requirement [32].

It is crucial to implement cryptographic algorithms in hardware. In this case study, an optimized and efficient synthesizable security IP is developed which if integrated with any of the IoT chips enhances the security of the data transmission and reception.

6.5.1 LIGHT CIPHER SECURITY BLOCK DESIGN IMPLEMENTATION

The basic security block includes a transmitter and a receiver block on either side, as depicted in Figure 6.1. Figure 6.2(a) and (b) illustrate the internal low-level modules of the encryption and decryption systems.

Domain Parameters Register Block: The use of various domain parameters in different applications is a useful rule of thumb when choosing domain parameters. Another helpful rule to follow when choosing domain parameters is to randomly choose the curve parameters a & b, the cofactor h, the generator point G in the chosen curve, the order of the curve n, and the modulus p. The SEC, or Standards for Efficient Cryptography, has established several common domain parameters out of which a suitable one can be picked.

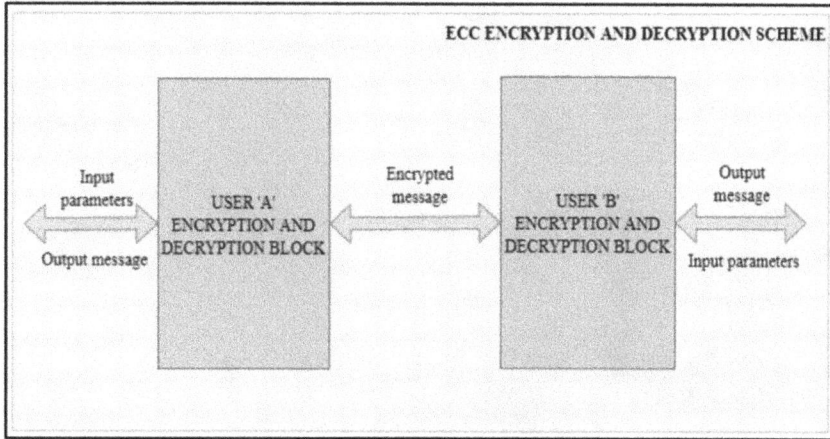

FIGURE 6.1   Encryption and decryption scheme.

Private Key Register Block: This register stores the sender's private key on the recipient side and the sender's private key on the sender side.

Plain Text Register Block: The message to be encoded is stored in this register.

ECC Arithmetic Operations: To create public keys and encrypt the plain text, general operations such as point addition, multiplication, and doubling are carried out.

Public Key Register Block: This register saves the recipient's generated public key. Encryption of the original content is done using this key.

Encrypted Message Register: This register stores the encrypted message which later needs to be transmitted.

Encrypted Cipher Text in point coordinate Form: Here, ordinary text is encrypted and transformed into point form so that it can be transmitted without fear of being attacked.

The Decryption block receives as input the encrypted content PC and the domain parameters used for encryption and saved in the appropriate Register Blocks. The ECC arithmetic process blocks are common on both sides and transmit Plain Text (M) to the recipient after decrypting the encrypted text.

6.5.2   ENCRYPTION AND DECRYPTION PROCESS

A detailed flow chart of encrypting and decrypting schemes is shown in Figure 6.3(a) and (b). The process can be described as follows:

- After determining the curve's parameters (a, b), choose the generating polynomial G based on the proper values for "a & b".
- A private key is selected from the range (0, n) which is now chosen as "ka" after a suitable "n" has been chosen. The public key is then calculated using the formula $Pa = ka*G$.

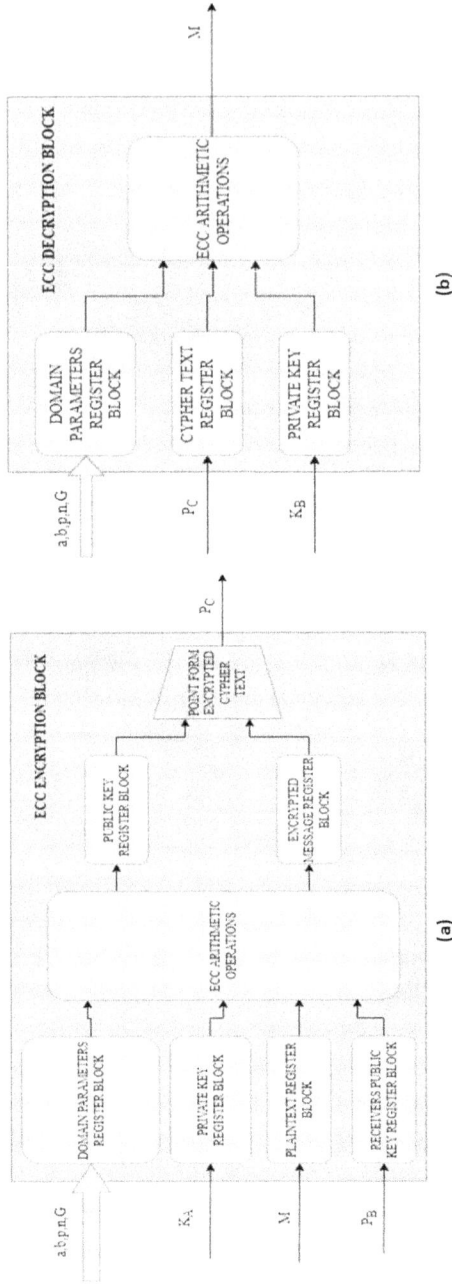

FIGURE 6.2 Internal low-level details of (a) encryption block and (b) decryption block.

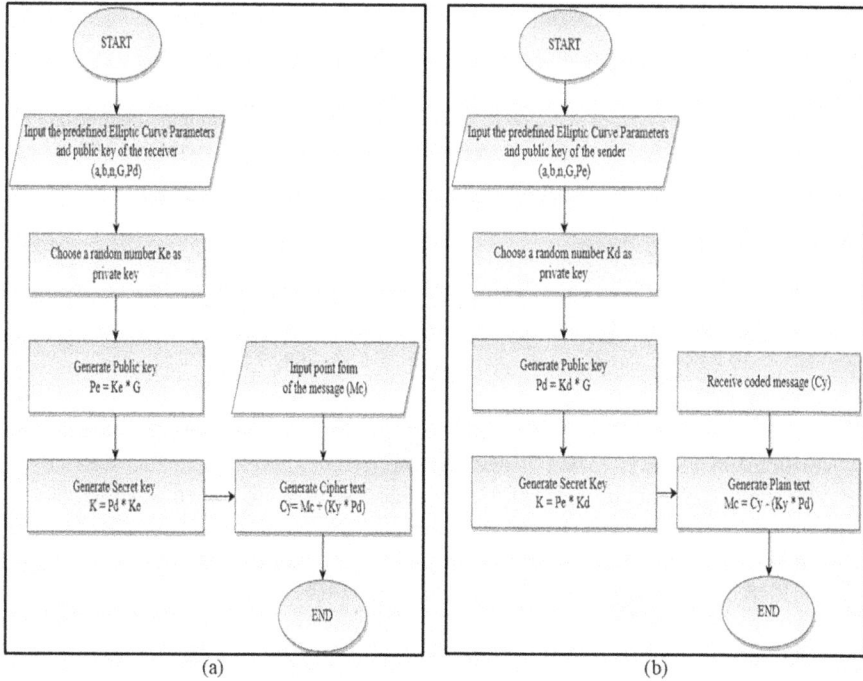

FIGURE 6.3   Flowchart of (a) encryption process and (b) decryption process.

- To encrypt plain text, a secret key is calculated by multiplying the sender's private key by the recipient's public key.
- The Plain Text message is transformed into the "Pm" curve's point coordinates, where the point is encrypted.
- Cm = (kG, Pm + kPb) is the formula used to create cipher text.
- The coordinates of the ECC are used to decrypt the message. The y-coordinate of the encrypted test contains the original content of the message.
- To obtain the original content text message, kb*G*k is calculated and the "y" coordinate is subtracted from it.
- The received original content text message must be transformed back into a readable message because it is in point form.

6.5.3   RESULTS AND DISCUSSION

The improved light cipher ECC approach employs stream cipher rather than block cipher because stream ciphertext attacks need the same amount of effort for every encrypted character as they do for each block of message in block ciphertext. It is therefore extremely difficult to decipher the ciphertext in a stream cipher.

For example, consider the following scenario: A message "A1" is encrypted at the sender's end, and at the receiver's end, it is decoded for the retrieved and received

TABLE 6.1

Comparison of Point Doubling and Point Addition

Method	Total Number of Point Doubling Required	Total Number of Point Addition Required
Scalar multiplication (Standard method)	–	20
Scalar multiplication (Implemented method)	4	1

ciphertext. At first, these alphanumeric are mapped to ECC points based on mapping values given in Table 6.1.

Verification of Encryption Block: As shown in Figure 6.4(a), the alphanumeric in the first character of the message that is "A" is coded as [0, 1]; this is given as input to the encryption module. The point message is encrypted to another point of the same size in the hexadecimal format, that is, [1f, 23] which is shown in decimal format in Figure 6.4(b), that is, [31, 35]. Take the next character of the original message, which is "1" is coded as [12, 26], this is given as input to the encryption module. The point message is encrypted to another point of the same size as shown in Figure 6.4(c) in the hexadecimal format that is [f, 33] which is shown in decimal format in Figure 6.4(d), that is, [15, 51].

Verification of Decryption Block: Take the first two position numbers "[31, 35]" given as input to the Decryption module as shown in Figure 6.5(a). The received ciphertext is decrypted and the result which is the coded message is (0, 1) in decimal format as shown in Figure 6.5(b) and the hexadecimal

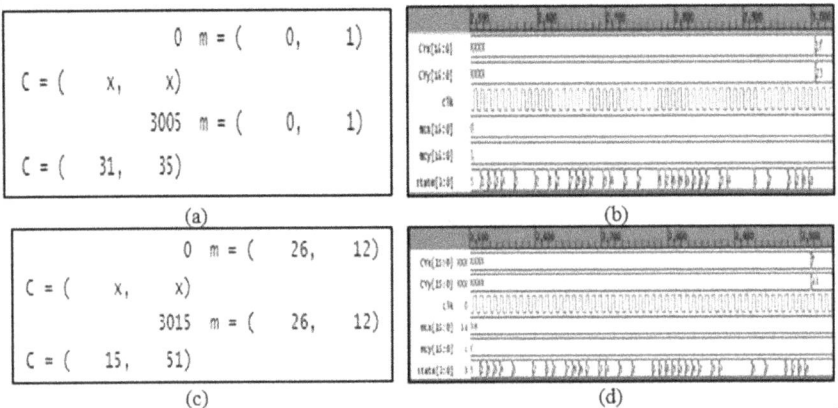

FIGURE 6.4 (a) Encrypting the content "A" of the original message, (b) waveform of "A" encryption, (c) encrypting the content "1" of the original message, and (d) waveform of "1" encryption.

(a)

(b)

(c)

(d)

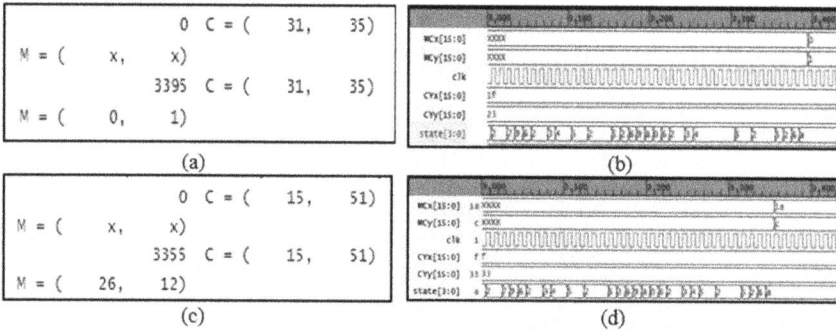

FIGURE 6.5 (a) Decrypting the character "A", (b) waveform for decryption of "A", (c) decrypting the message "1", and (d) waveform for decryption of "1".

format obtained during the simulation as shown in Figure 6.5(c). Take the next two position numbers "[15, 51]". This is given as input to the Decryption module as shown in Figure 6.5(d). The received ciphertext is decrypted and the result is a coded message as shown in Figure 6.5(c), which is (26, 12) in decimal format and the hexadecimal format is (1a, c) as shown in Figure 6.5(d).

For each design, synthesis reports of power, area, and timing are generated using Cadence XC Simulator for both before and after optimization. As depicted in Figure 6.6(a), the design's total area before optimization was 29666.448 square units, and after optimization, it was reduced to 29608.992 square units. Figure 6.6(b) illustrates how the timing of the clock was 500R before optimization and 490R after optimization. It was observed that there is a slight increase in the power consumption.

6.5.4 VIRTUAL IMPLEMENTATION

For the suggested design's FPGA virtual implementation, the Vivado Xilinx tool is employed. ZedBoards [15] is the board that is utilized for this virtual implementation.

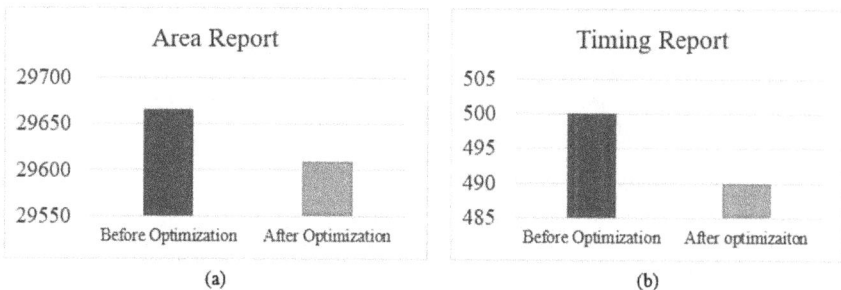

(a)

(b)

FIGURE 6.6 Analysis of (a) area based on synthesis reports generated and (b) timing based on synthesis reports generated.

The point multiplication algorithm's implementation has been revised; the new way requires fewer computations to obtain the same results. Point multiplication takes up most of the computation time when ECC is implemented. As indicated in Table 6.1, the improvements made to the implemented method result in output with fewer computations, which cuts down on computational time and lowers channel consumption.

The scalar multiplication is carried out using the left-to-right procedure. The approach utilized here and conventional point multiplication are compared in Table 6.1, and it is found that the latter requires less computations, which reduces computational time and channel utilization.

6.6 SUMMARY

The lightweight architecture is the best choice for resource-constrained IoT devices. It uses a symmetric cryptography algorithm and provides high security. This algorithm is used by many applications and networks. As of now, ECDH and ECES have been put into use; ECDH handles key exchange using ECC and DH, and ECES handles solely encryption using ECC. Here, the two current approaches are integrated, and certain modifications are also made to improve their effectiveness. In this approach, a modified form of ECES is utilized for encryption together with ECDH for key exchange. When compared to the conventional version, the encrypted text produced by the improved ECES algorithm is now optimized and takes almost 50% less area compared to the conventional approach.

ACKNOWLEDGMENTS

The authors thank the management of BNM Institute of Technology and Visvesvaraya Technological University for their support and encouragement rendered to submit this book chapter.

REFERENCES

1. Ghazal, T. M., Hasan, M. K., Alshurideh, M. T., Alzoubi, H. M., Ahmad, M., Akbar, S. S., Al Kurdi, B., & Akour, I. A., "IoT for Smart Cities: Machine Learning Approaches in Smart Healthcare—A Review", *Future Internet*, 13(8), 2021. doi:10.3390/fi13080218
2. Aljanabi, S., & Chalechale, A., "Improving IoT Services Using a Hybrid Fog-Cloud Offloading", *IEEE Access*, 9, 13775–13788, 2021. doi:10.1109/ACCESS.2021.3052458
3. Eichelberg, M., Kleber, K., & Kämmerer, M., "Cybersecurity Protection for PACS and Medical Imaging: Deployment Considerations and Practical Problems", *Academic Radiology*, 28(12), 1761–1774, 2021. doi: 10.1016/j.acra.2020.09.001
4. Ogonji, M., Okeyo, G., & Wafula, J., "A Survey on Privacy and Security of Internet of Things", *Computer Science Review*, 38, 100312, 2020. doi: 10.1016/j.cosrev.2020.100312
5. Bojjagani, S., Sastry, V. N., Chen, C. M., Kumari, S., & Khan, M. K., "Systematic Survey of Mobile Payments, Protocols, and Security Infrastructure", *Journal of Ambient Intelligence and Humanized Computing*, 14(1), 609–654, 2023.

6. Shamala, L. M., Zayaraz, D. G., Vivekanandan, D. K., & Vijayalakshmi, D. V., "Lightweight Cryptography Algorithms for Internet of Things Enabled Networks: An Overview", *Journal of Physics: Conference Series*, 1717(1), 012072, 2021. doi:10.1088/1742-6596/1717/1/012072

7. Ahmad, N., & Hasan, S. M. R., "A New ASIC Implementation of an Advanced Encryption Standard (AES) Crypto-Hardware Accelerator", *Microelectronics Journal*, 117, 105255, 2021. doi: 10.1016/j.mejo.2021.105255

8. Yang, Z., Bao, Z., Jin, C., Liu, Z., & Zhou, J., "PLCrypto: A Symmetric Cryptographic Library for Programmable Logic Controllers", *IACR Transactions on Symmetric Cryptology*, 2021(3), 170–217. doi:10.46586/tosc.v2021.i3.170-217

9. Bavdekar, R., Chopde, E. J., Bhatia, A., Tiwari, K., Daniel, S. J., & Atul, "Post Quantum Cryptography: Techniques, Challenges, Standardization, and Directions for Future Research", 2022. ArXiv, abs/2202.02826.

10. Goyal, P., Sahoo, A. K., & Sharma, T. K., "Internet of Things: Architecture and Enabling Technologies", *Materials Today: Proceedings*, 34, 719–735. 2021. doi: 10.1016/j.matpr.2020.04.678

11. Kadam, V. R., & Naidu, P. S., "Lightweight Cryptography to Secure Internet of Things (IoT)", *International Research Journal of Engineering and Technology*, 07(05), 6184–6188, 2020.

12. Sharma, R., Jindal, P., & Singh, B., "Study and Analysis of Key Generation Techniques in Internet of Things", *Journal of Discrete Mathematical Sciences and Cryptography*, 23, 373–383, 2020.

13. Wang, Q., & Li, H., "Application of IoT Authentication Key Management Algorithm to Personnel Information Management", *Computational Intelligence and Neuroscience*, 4584072, 2022. doi: 10.1155/2022/4584072

14. Tabassum, T., Hossain, S. K. A., Rahman, M. A., Alhamid, M. F., & Hossain, M. A., "An Efficient Key Management Technique for the Internet of Things", *Sensors* (Basel, Switzerland), 20(7), 2020 doi: 10.3390/s20072049

15. Balogh, S., Gallo, O., Ploszek, R., Špaček, P., & Zajac, P., "IoT Security Challenges: Cloud and Blockchain, Postquantum Cryptography, and Evolutionary Techniques", *Electronics*, 10(21), 2647, 2021 Retrieved from https://www.mdpi.com/2079-9292/10/21/2647

16. Feroz Khan, A. B., & Anandharaj, G., "AHKM: An Improved Class of Hash Based Key Management Mechanism with Combined Solution for Single Hop and Multi Hop Nodes in IoT", *Egyptian Informatics Journal*, 22(2), 119–124, 2021. doi: 10.1016/j.eij.2020.05.004

17. Rana, M., Quazi, A. M., & Islam, R., "A Key Management Scheme for Lightweight Block Cipher in Iot Networks", SSRN, 2022.

18. Priyadharshini, T. C. G., & Mohana, D., "Efficient Key Management System Based Lightweight Devices in IoT", *Intelligent Automation & Soft Computing*, 31(3), 2022.

19. Nafi, M., Bouzefrane, S., & Omar, M., "Matrix-Based Key Management Scheme for IoT Networks", *Ad Hoc Networks*, 97, 102003, 2020. doi: 10.1016/j.adhoc.2019.102003

20. Pamarthi, S., & Narmadha, R., "Adaptive Key Management-Based Cryptographic Algorithm for Privacy Preservation in Wireless Mobile Adhoc Networks for IoT Applications", *Wireless Personal Communications*, 124(1), 349–376. 2022. doi: 10.1007/s11277-021-09360-9

21. Faisal, M., Ali, I., Khan, M. S., Kim, J., & Kim, S. M., "Cyber Security and Key Management Issues for Internet of Things: Techniques, Requirements, and Challenges", *Complexity*, 6619498, 2020. doi: 10.1155/2020/6619498

22. Ramadan, R., "Internet of Things (IoT) Security Vulnerabilities: A Review", *PLOMS AI*, 2(1), 2021. Retrieved from https://plomscience.com/journals/index.php/PLOMSAI/article/view/14

23. Azrour, M., Mabrouki, J., Guezzaz, A., & Kanwal, A., "Internet of Things Security: Challenges and Key Issues", *Security and Communication Networks*, 5533843, 2021. doi: 10.1155/2021/5533843

24. Bharathi, N., & Jayavel, K., "Role of VLSI Design to Build Trusted and Secured IOT Devices- A Methodological Approach", Paper presented at the *2021 5th International Conference on Electronics, Communication and Aerospace Technology (ICECA)*, 2–4 Dec. 2021.

25. Das, I., Jha, K., Debnath, P., & Nath, S., "VLSI and AES Based IoT Security by Modified Random S-Box Generation", Paper presented at the *2022 IEEE International Conference of Electron Devices Society Kolkata Chapter (EDKCON)*, 26–27 Nov. 2022.

26. Kandpal, J., & Singh, A., "Opportunity and Challenges for VLSI in IoT Application", *5G Internet of Things and Changing Standards for Computing and Electronic Systems*, edited by Augustine O. Nwajana, IGI Global, 2022, pp. 245–271, 2022. doi: 10.4018/978-1-6684-3855-8.ch010

27. Li, S., Tryfonas, T., & Li, H, "The Internet of Things: A Security Point of View", *Internet Research*, 26, 337–359, 2016. doi: 10.1108/IntR-07-2014-0173

28. Rajadurai, K., Kavitha, T., & Subashini, V. J., "Application of Modulo Key-Predistribution Protocol", *Research Journal of Applied Sciences, Engineering and Technology* (P- 2040-7459), 11(7), pp. 780–787, Nov 2015.

29. Kavitha, T., & Sridharan, D., "Optimal Resource Key Management Protocol for Clustered Heterogeneous Wireless Sensor Networks", *Malaysian Journal of Computer Science* (0127-9084), Univ Malaya, 26(3), pp. 211–231, 2013.

30. Kavitha, T., & Sridharan, D., "Key Distribution Scheme Using Modulo Operation for WSN", *Information* (P-1343-4500), International Information Inst Publisher, Japan, 16(11), pp. 8213–8228, Nov 2013.

31. Shirur, Y.J.M., Ks, N.I., Sujay, K.S. and Uday, V.N., "Design and Implementation of Synthesizable Two-Level Cryptosystem for High-Security enabled Applications," *2023 International Conference on Intelligent and Innovative Technologies in Computing, Electrical and Electronics (IITCEE)*, Bengaluru, India, pp. 922–926, 2023. doi: 10.1109/IITCEE57236.2023.10091066

32. Bharathi, N. and Jayavel, K.,"Role of VLSI Design To Build Trusted And Secured IOT Devices- A Methodological Approach," *2021 5th International Conference on Electronics, Communication and Aerospace Technology (ICECA)*, Coimbatore, India, pp. 473–479, 2021, doi: 10.1109/ICECA52323.2021.9675878

7 Secure Data Collection, Aggregation, and Sharing in the Internet of Things

Sanjay Mate
Sangam University, Bhilwara, Government Polytechnic, Diu UT, India

Vikas Somani
Sangam University, Bhilwara, India

Prashant Dahiwale
Government Polytechnic, Daman UT, Gujarat Technological University, India

7.1 INTRODUCTION

Data collection is a systematic approach to gathering information, i.e., observations and measurements from devices. Internet of Things (IoT) data collections involve gathering information from devices connected to sensors and the internet. Sensors have a vital role in collecting and transmitting data on the network. Sensors are in connection with gateways or other edge devices, for example, linearity, range, repeatability, reproducibility, resolution, sensitivity, zero drift, and full-scale drift.

IoT data collection includes equipment data, sub-meter data, and environmental data. Equipment data includes the gathering of real-time data to facilitate activities of predictive maintenance. A sub-meter data helps in gathering automated measurement-based information like reading information from meters of data cables, electricity, gas, water, etc. Environmental data gives real-time information about weather parameters like high temperature, atmospheric quality, humidity, and moisture.

For gathering or collecting real-time data there are numerous algorithms and few of them are highly popular like Modified Traveling Path Algorithm (MTPA), A GTAC-DG (i.e., Game Theory and Ant Colony Data Gathering) is one of the popular paths for mobile sink in wireless sensor networks [1]. The MTPA is a heuristic algorithm used to solve optimization problems, particularly those involving finding the shortest path between multiple points. MTPA is based on the classical Traveling Salesman Problem (TSP), which is a well-known NP-hard problem. However, MTPA

DOI: 10.1201/9781003477327-7

modifies the TSP by introducing a new concept called the "traveling path," which allows for extra competent exploration of the search space. In MTPA, the traveling path is a sequence of nodes that are visited in a specific order. The algorithm starts by randomly generating an initial traveling path, which may not essentially be optimal. The algorithm then iteratively improves the traveling path by swapping two nodes and evaluating the resulting path length. The swapping process continues until no further improvements can be made, at which point the algorithm returns the optimal traveling path. One key advantage of MTPA over other optimization algorithms is its ability to handle large-scale problems. Since MTPA only evaluates two nodes at a time, the computational difficulty of the algorithm is much lower compared to other optimization algorithms that evaluate all possible node combinations.

A GTCA-DG algorithm is designed to optimize the performance of wireless sensor networks, particularly in environments where traditional centralized data-gathering approaches are not feasible. The algorithm is based on the principle of self-organization, where every single node in the system acts autonomously based on local information and a set of rules. The nodes are organized into multiple groups, with each group controlled by a leader node. The leader nodes compete with each other in a game-theoretic framework to determine the optimal data-gathering strategy for their respective groups. The algorithm also incorporates ant colony optimization, which is a metaheuristic algorithm inspired by the behavior of ants. In GTAC-DG, each node acts as an artificial ant and moves around the network based on pheromone trails left by other nodes. The pheromone trails indicate the quality of the data gathered at different locations in the network, and the ants use this information to determine where to gather data. The combination of game theory and ant colony optimization allows GTAC-DG to adapt to changing environmental conditions and network dynamics. The algorithm is particularly useful in scenarios where there are limited resources, such as energy or bandwidth, as it minimizes the number of transmissions needed to gather data while maintaining a high level of accuracy.

7.1.1   Secure Data Collection in the IoT

The selection of a suitable approach for secure data collection, aggregation, and sharing in IoT depends on several factors, including the number of devices, the level of security required, and the complexity of data management. A hybrid approach may offer the best solution for many IoT applications, as it combines the benefits of both centralized and decentralized approaches. Additionally, the use of advanced techniques such as homomorphic encryption, secure multi-party computation, and differential privacy may enhance the security and privacy of IoT data. Secure real-time data collection algorithms commonly used in IoT environments:

Role-Based Access Control (RBAC): Implementing RBAC mechanisms in IoT systems allows for fine-grained access control to real-time data collected from IoT devices. It ensures that only authorized entities can access specific data streams, enhancing data privacy and security.

LightweightM2M (LwM2M) with DTLS: LwM2M is a protocol for managing IoT devices and their data. When secured with DTLS, it enables secure real-time data collection, ensuring confidentiality, integrity, and authentication.

Elliptic Curve Cryptography (ECC): ECC is a public-key cryptography algorithm that offers strong security with shorter key lengths. It is commonly used in IoT for secure key exchange, data encryption, and authentication.

Datagram Transport Layer Security (DTLS): DTLS is a variant of TLS specifically designed for UDP-based protocols in IoT. It enables secure real-time data collection over unreliable networks, ensuring encryption, authentication, and integrity.

Constrained Application Protocol (CoAP) with DTLS: CoAP is a lightweight IoT protocol. By incorporating DTLS, it facilitates secure real-time data collection in resource-constrained IoT environments, ensuring confidentiality and integrity.

7.2 DATA AGGREGATION MECHANISM

It is the process of summarizing large-scale data and applying some statistical analysis; it helps to get large dataset information in an expected categorized format. For example, data of universities in the form of number of students enrolled for various courses on the basis of age, sex, income, course tenure, or level. Goods purchased by customers in a shopping mall on the basis of daily essentials, health and fitness, entertainment, luxury items, etc. Tree-based, cluster-based, and centralized are three renowned data aggregation approaches.

7.2.1 TREE-BASED DATA AGGREGATION MECHANISM

Sensor nodes are structured into trees where data aggregation is implemented at intermediate nodes along the tree and aggregated data is carried to the root node. In this procedure, nodes at various hierarchical positions and intermediate nodes can do data aggregation; this mechanism provides an energy-efficient data aggregation tree [2]. Ahaj et al. suggested a design or tree-based aggregation which has significant improvement in energy consumption along with the maintained quality of science. A transmission power allocation process is achieved dynamically resulting in reduced utilization of resources in the data aggregation process [3]. Homa et al. proposed a distributed congestion-aware routing mechanism in tree-based data aggregation, it deals with limitations occurring in data communication, transmission, and aggregation [4]. Mao et al. proposed a probabilistic routing mechanism in tree-based data aggregation, it results in reducing data transmission delay and improving delivery rate [5]. Major advantages of tree-based data aggregation mechanisms are high scalability, high network lifetime, high accuracy, high fault tolerance, high security, low latency, low traffic load, and low energy consumption whereas non-consideration of heterogeneity is disadvantageous for many data aggregation mechanisms. Tree-based data aggregation is a widely used mechanism in the IoT for efficient and effective management of large amounts of data generated by IoT devices. IoT devices are typically connected to a network and generate large amounts of data that need to be processed, stored, and analyzed. Tree-based data aggregation provides an efficient and scalable method for organizing this data. In an IoT context, a tree-based data aggregation mechanism typically involves organizing data generated by IoT devices into a tree structure. The root node of the tree represents the entire data set, while each subsequent level of the tree represents increasingly specific sub-groups of the data. This structure enables efficient navigation, search, and analysis of the data. For example, in an IoT system for monitoring and controlling smart homes, data generated by different sensors and devices (such as temperature sensors, motion detectors, and lights) can be organized into a tree-based structure, with each node representing a specific device or group of devices. This allows for easy aggregation of data from different devices and efficient analysis of the data at different levels of specificity. Tree-based data aggregation is also useful for reducing the quantity of data transmitted over IoT networks, as only relevant data can be selected and transmitted based on the specific needs of the analysis or application. This aids in decreasing the load on the network and conserves resources, improving the overall performance of the IoT system.

Overall, tree-based data aggregation is an important tool for managing and analyzing data in IoT systems, providing efficient and effective methods for organizing, summarizing, and analyzing large amounts of data generated by IoT devices.

7.2.2 CLUSTER-BASED DATA AGGREGATION MECHANISM

The sensor can transmit data to a nearby aggregator or cluster head which aggregates data from all the sensors in its cluster and transmits the short digest to the sink. Each network has several clusters and each cluster consists of a large number of sensors in it. Each and every single cluster has one dedicated header node. A cluster head further reduces bandwidth overhead as the total number of packets to be transmitted is less [6]. The data aggregation mechanism is based on a cluster basis. According to Liu et al., the mechanism cost of communication is decreased among storage and nodes, it also guarantees the tolerance of failure whereas heterogeneity is left behind with less importance. It includes a trust record queue. A trust record queue keeps track of nodes' trust information and detection of malicious activities which captures the nature of trust evaluation [7]. Data aggregation using a Chinese Remainder Theorem (CRT) on the basis of coding algorithms has many good results like reduced traffic load and improved efficiency of data transmission. An aggregator node captures all sensing data; later, it performs the tasks of aggregation and further transmits the results into the decision server [8]. For an IoT applications crossover scheme considered for a mobile ad hoc environment, it ensures network layer and application layer-based failure tolerance, some other advantages like it reduces traffic load, and increases actual residual energy whereas a drawback of high latency [9]. To improve various issues in IoT data aggregation a fuzzy-based cluster routing method is highly helpful. A fuzzy-based method has three stages, the first stage is clustering, the second is cluster head selection and the final stage is parent selection. The overall fuzzy-based cluster routing method extends the packet delivery and lifetime of the network [10]. On the basis of intelligent computing, an optimized cluster-based routing protocol helps to address the issues in communication and interruptions in the communication of IoT [11]. Cluster-based data aggregation is a widely used mechanism in the IoT for organizing and summarizing large amounts of data generated by

IoT devices. The goal of cluster-based data aggregation is to group similar devices together based on their data characteristics, in order to reduce the amount of data that needs to be transmitted, stored, and analyzed. In an IoT context, a cluster-based data aggregation mechanism typically involves dividing IoT devices into clusters based on their data characteristics, like the type of data generated, the frequency of data generation, and the similarity of data patterns. Devices within every single cluster are then treated as a single entity for the purpose of data aggregation and analysis. For example, in an IoT system for monitoring and controlling smart homes, temperature sensors, motion detectors, and lights can be grouped into separate clusters based on their data characteristics. This allows for efficient aggregation of data from similar devices and reduces the amount of data that needs to be transmitted, stored, and analyzed. Cluster-based data aggregation also helps to reduce the load on the network by transmitting only relevant data and enables efficient analysis of the data at different levels of specificity. This improves the overall performance of the IoT system, making it more scalable and responsive to changing needs. Overall, cluster-based data aggregation is an important tool for managing and analyzing data in IoT systems, providing efficient and effective methods for organizing, summarizing, and analyzing large amounts of data generated by IoT devices.

7.2.3 CENTRALIZED DATA AGGREGATION MECHANISM

Communication of sending sensed data begins from the leaf node to the intermediate node further to the header node, data getting aggregated while shifting from the intermediate node to the header node or central node. Header data aggregates data received from all nodes and converts it into one packet. A centralized mechanism technique of modern data aggregation platforms on security assessment has a major advantage of high security and its weakness is low fault tolerance [12]. Another technique of distributed service-oriented architecture has several advantages: high security, considering heterogeneity, and low traffic load whereas the drawback is low fault tolerance [13]. Overall, a centralized node provides high security and it is low fault tolerant. An aggregation of collected hybrid data in IoT is made through various mechanisms like the CRT, Homomorphic Paillier Encryption, and One-way Hash Chain. Centralized data aggregation is a widely used mechanism in the IoT for organizing, summarizing, and analyzing large amounts of data generated by IoT devices. In a centralized data aggregation mechanism, all data generated by IoT devices is transmitted to a central repository, where it is stored and processed. In an IoT context, a centralized data aggregation mechanism typically involves the use of a central server, known as a hub or gateway, which collects data from all IoT devices in the network. The hub is responsible for organizing the data, storing it in a central database, and processing it for analysis. The advantage of a centralized data aggregation mechanism is that it provides a single point of control for the entire IoT system, making it easier to manage and monitor the data. It also enables efficient data analysis, as all data is stored in a single location, and makes it easier to implement security measures, such as encryption and authentication, as all data is transmitted to a single point. For example, in IoT-based smart homes, a centralized data aggregation mechanism can be used to collect data from temperature sensors, motion detectors, and lights, and store it in a central database. This allows for efficient analysis of the data and makes it easy to monitor the state of the smart home and respond to changes. Overall, centralized data aggregation is an important tool for managing and analyzing data in IoT systems, providing efficient and effective methods for organizing, summarizing, and analyzing large amounts of data generated by IoT devices.

7.3   SECURE DATA AGGREGATION

Secure data aggregation is a critical issue in the context of the IoT, as it involves the collection and analysis of sensitive data from multiple sources, including sensors and other devices. It is important to ensure that this data is secure and cannot be intercepted, tampered with, or accessed by unauthorized parties. A few popular methods in secure data aggregation include federal learning, data masking, differential policy, homomorphic encryption, and trusted executed environment. One way to achieve secure data aggregation in IoT is through encryption. Encryption is the process of converting data into a secret code to prevent unauthorized access. End-to-end encryption is a popular encryption technique that encrypts data at the source and decrypts it only at the destination, ensuring that the data is never exposed in transit. Another important aspect of secure data aggregation is authentication. Authentication is the process of verifying the identity of a user or device. In IoT, authentication can be used to ensure that only authorized devices are allowed to access and transmit data. This can be achieved by using secure protocols such as Transport Layer Security (TLS) and Secure Shell (SSH) to establish a secure connection between devices and the data aggregator. Access control is also important in secure data aggregation. Access control is the process of regulating who can access data and resources. It can be used to restrict access to sensitive data and ensure that only authorized users are allowed to view or modify data. Access control can be implemented using RBAC or attribute-based access control (ABAC) to provide granular control over access to data.

Finally, data integrity is crucial in secure data aggregation. Data integrity is the assurance that data has not been tampered with or altered in any way. Data integrity can be achieved by using cryptographic hash functions to create a unique fingerprint of the data that can be used to verify its authenticity.

7.4   DATA SHARING

Data sharing in IoT refers to the process of exchanging information between various IoT devices, applications, and platforms. It is a key aspect of IoT as it enables devices to collaborate and share their data to create new insights, support new use cases, and foster innovation. In IoT, data sharing is facilitated by a variety of technologies such as edge computing, cloud computing, and inter-device communication protocols. Edge computing allows for data processing at the edge of the network, close to the source of data generation, reducing the need for data transmission. Cloud computing enables remote data storage and processing on servers. Inter-device communication protocols, such as MQTT, CoAP, and ZigBee, standardize data exchange between IoT devices. However, as with any data exchange, it is important to ensure the security and privacy of sensitive information when sharing data in IoT. In IoT, edge computing is where data is processed at the edge of the network, close to the source of data generation, reducing the need for data to be transmitted over the network. In IoT, cloud computing is where data is transmitted to a remote server for processing, storage, and analysis. In IoT, Inter-device communication protocols, such as MQTT, CoAP, and ZigBee, allow devices to exchange data with each other in a standardized approach. IoT data sharing happens at various levels. Peer-to-peer, domain-specific sharing, and marketplace data are shared. The major stake of data sharing is in peer-to-peer, followed by domain-specific and marketplace. Peer-to-peer data sharing has a lack of middleman whereas the central system has a concrete structure with a middleman as a common place for data storage and access. In data sharing secure data transmission is highly important, data authenticity, integrity, and confidentiality should be maintained. There are two popular authentication methods: MAC-based scheme and signature-based scheme. Weaken approach in confidentiality, integrity, and authentication may lead to a denial of service attack, absence of mutual authentication and key agreement [14], and replay attack.

7.5   PREVENTION MEASURE TO USE IoT IN AN EFFICIENT WAY

At present in 2023, an IoT can be considered as a mature technology [15, 16]. Preventing attacks on the IoT devices is crucial for ensuring their security and reliability. Given the increasing number of IoT devices in use and the sensitive information they often handle; it is essential to take steps to protect them from potential threats. Here are some precautions that can be taken to secure IoT devices from attack. Secure device authentication, ensuring that only authorized devices can connect to the network and access sensitive data is critical. This can be achieved through the use of strong authentication methods such as password protection and two-factor authentication. Security Information and event management (SIEM) can be preferred to protect from DDOS attacks [17].

In case of encryption of data transmission, all data transmitted between IoT devices and servers should be encrypted to prevent unauthorized access and tampering. In software updates, regular software updates can help fix vulnerabilities and protect against newly discovered threats. In network segmentation, by dividing the network into smaller segments, it becomes more difficult for attackers to access sensitive data and systems. In firewall protection, a firewall can help prevent unauthorized access to the network and protect against malware and other cyber threats. In monitoring and logging, regular monitoring and logging of network activity can help detect and prevent attacks and provide valuable information for post-attack analysis. In educating users, educating users on how to secure their IoT devices and recognizing potential threats is also important. A TOPSIS algorithm helps secure data communication among things against wormhole attacks and it is comparatively superior to HRCA and HBA in view of lost packets, PDR, and throughput [18].

7.6   SUMMARY

To achieve secure data collection, encryption and authentication techniques can be used to protect the data from interception and unauthorized access. Access control mechanisms can be implemented to regulate who can access the data and how it can be used. Data integrity mechanisms can also be used to ensure that the data has not been tampered with or altered in any way. Secure data aggregation in IoT is essential to protect sensitive data from unauthorized access and tampering. Encryption, authentication, access control, and data integrity are all important aspects of secure data aggregation that must be implemented to ensure the security and privacy of IoT data. Secure data sharing is also essential to allow authorized users to access and use the data for various purposes, such as research or monitoring. Techniques such as privacy-preserving data sharing and data masking can be used to ensure that sensitive data is not disclosed to unauthorized parties.

REFERENCES

[1] Raj, P. V. Pravija, Ahmed M. Khedr, and Zaher Al Aghbari. "Data gathering via mobile sink in WSNs using game theory and enhanced ant colony optimization." *Wireless Networks* 26, 2020, pp. 2983–2998.

[2] Dagar, Mousam, and Shilpa Mahajan. "Data aggregation in wireless sensor network: a survey." *International Journal of Information and Computation Technology* 3, no. 3, 2013, pp. 167–174.

[3] Ashaj, Sudad J., and Ergun Erçelebi. "Energy saving data aggregation algorithms in building automation for health and security monitoring and privacy in medical internet of things." *Journal of Medical Imaging and Health Informatics* 10, no. 1, 2020, pp. 204–210.

[4] Homaei, Mohammad Hossein, Ely Salwana, and Shahaboddin Shamshirband. "An enhanced distributed data aggregation method in the Internet of Things." *Sensors* 19, no. 14, 2019, p. 3173.

[5] Mao, Yuxin, Chenqian Zhou, Yun Ling, and Jaime Lloret. "An optimized probabilistic delay tolerant network (DTN) routing protocol based on scheduling mechanism for internet of things (IoT)." *Sensors* 19, no. 2, 2019, p. 243.

[6] Sirsikar, Sumedha, and Samarth Anavatti. "Issues of data aggregation methods in wireless sensor network: a survey." *Procedia Computer Science* 49, 2015, pp. 194–201.

[7] Liu, Yanbing, Xuehong Gong, and Congcong Xing. "A novel trust-based secure data aggregation for internet of things." In *2014 9th International Conference on Computer Science & Education*, pp. 435–439. IEEE, 2014.

[8] Xie, Feng. "CaCa: Chinese remainder theorem based algorithm for data aggregation in internet of things on ships." *Applied Mechanics and Materials* 701, 2015, pp. 1098–1101.

[9] Alkhamisi, Abrar, Mohamed Saleem Haja Nazmudeen, and Seyed M. Buhari. "A cross-layer framework for sensor data aggregation for IoT applications in smart cities." In *2016 IEEE International Smart Cities Conference (ISC2)*, pp. 1–6. IEEE, 2016.

[10] Sankar, S., and P. Srinivasan. "Fuzzy sets based cluster routing protocol for Internet of Things." *International Journal of Fuzzy System Applications (IJFSA)* 8, no. 3, 2019, pp. 70–93.

[11] Sun, Zeyu, Xiaofei Xing, Tian Wang, Zhiguo Lv, and Ben Yan. "An optimized clustering communication protocol based on intelligent computing in information-centric Internet of Things." *IEEE Access* 7, 2019, pp. 28238–28249.

[12] Sándor, Hunor, Béla Genge, and G. Á. L. Zoltán. "Security assessment of modern data aggregation platforms in the Internet of Things." *International Journal of Information Security Science* 4, no. 3, 2015, pp. 92–103.

[13] Zhu, Tao, Sahraoui Dhelim, Zhihao Zhou, Shunkun Yang, and Huansheng Ning. "An architecture for aggregating information from distributed data nodes for industrial internet of things." In *Cyber-Enabled Intelligence*, pp. 17–35. Taylor & Francis, 2019.

[14] Kavitha, T. and D. Sridharan, "Key distribution scheme using modulo operation for WSN", *Information* (P-1343-4500), International Information Inst Publisher, Japan, 16, no. 11, November 2013, pp. 8213–8228.

[15] Calderoni, Luca, Antonio Magnani, and Dario Maio. "IoT Manager: an open-source IoT framework for smart cities." *Journal of Systems Architecture* 98, 2019, pp. 413–423.

[16] Saadeh, Maha, Azzam Sleit, Mohammed Qatawneh, and Wesam Almobaideen. "Authentication techniques for the Internet of Things: a survey." In *2016 Cybersecurity and Cyberforensics Conference (CCC)*, pp. 28–34. IEEE, 2016.

[17] Aziz, Azlan Abd, and Syamsuri Yaacob. "IoT performance and security analysis based on WiFi systems." In *Proceedings of the Multimedia University Engineering Conference (MECON 2022)*, vol. 214, p. 3. Springer Nature, 2023.

[18] Sahraneshin, Tayebeh, Razieh Malekhosseini, Farhad Rad, and S. Hadi Yaghoubyan. "Securing communications between things against wormhole attacks using TOPSIS decision-making and hash-based cryptography techniques in the IoT ecosystem." *Wireless Networks* 29, no. 2, 2023, pp. 969–983.

8 Authentication, Authorization, and Anonymization Techniques in the IoT

V. M. Sivagami, M. Deekshitha, and N. Devi
Sri Venkateswara College of Engineering, Anna University, Chennai, India

S. Swarna Parvathi
Director Experiment Station, (BAEG)-Biological and Agricultural Engineering, University of Arkansas, Fayetteville, AR, USA

8.1 INTRODUCTION TO AUTHENTICATION, AUTHORIZATION, AND ANONYMIZATION TECHNIQUES IN THE IoT

The Internet of Things (IoT) has enabled the unprecedented connection of devices, systems, and people across the world. It has revolutionized the way people interact with each other and with the environment around them. The rise of the IoT has brought with it a range of new security challenges, as the number of IoT devices grows, so do the security risks associated with them. Authentication, authorization, and anonymization are three key techniques used to ensure the security of IoT devices, networks, and data.

8.2 AUTHENTICATION

Authentication is the process of verifying the identity of a user or device [1–3]. It is used to ensure that the user or device attempting to access a system is who they claim to be. Authentication typically involves the use of passwords, tokens, biometrics, or a combination of these methods. Authentication is essential for secure access to sensitive systems and data, and it is used to prevent Malicious attackers from accessing the system or data.

DOI: 10.1201/9781003477327-8

There are several authentication techniques that can be used in IoT systems, including the following:

1. **Passwords**: Passwords are the most common form of authentication. In IoT systems, users can authenticate by entering a unique password to access a device or network resource. Passwords can be stored on the device or on a central server.

2. **Biometrics**: Biometric authentication uses physical or behavioral characteristics, such as fingerprints, facial recognition, or iris scans, to verify the identity of a user. In IoT systems, biometric authentication can be used to grant access to sensitive data or network resources and can be stored on the device or on a central server.

3. **Smart Cards**: Smart cards are physical cards that contain a microprocessor and memory, and are used to store and process authentication information. In IoT systems, smart cards can be used to authenticate devices and users and to grant access to sensitive data or network resources.

4. **Two-Factor Authentication**: Two-factor authentication (2FA) is a multi-step authentication process that requires a user to provide two forms of authentication, such as a password and a smart card, or a password and a biometric scan. In IoT systems, 2FA can provide an additional layer of security, making it more difficult for an attacker to access sensitive data or network resources [4, 5].

8.3 AUTHORIZATION

Authorization is the process of granting users or devices access to certain resources [6, 7]. It is used to ensure that users or devices have permission to access certain functions or data. It is typically based on user roles or privileges, and it is used to control access to sensitive systems and data.

There are several authorization techniques that can be used in IoT systems, including the following.

1. **Role-Based Access Control**: Role-based access control (RBAC) is a form of authorization that grants access to network resources or data based on a user's role or job function. In IoT systems, RBAC can be used to determine what actions a user or device can perform, and what data they are able to access is illustrated in Figure 8.1.

2. **Attribute-Based Access Control**: Attribute-based access control (ABAC) is a form of authorization that grants access to network resources or data based on attributes, such as the location, time of day, or device type, of a user or device [8, 9]. In IoT systems, ABAC can be used to control access to sensitive data or network resources based on specific attributes, such as the location of a device or the time of day represented in Figure 8.2.

3. **Access Control Lists**: Access control lists (ACLs) are a form of authorization that grants or denies access to network resources or data based on the identity of a user, device, or system as illustrated in Figure 8.3. In IoT systems, ACLs can be used to determine which users or devices are able to access specific network resources or data, and what actions they can perform [10–12].

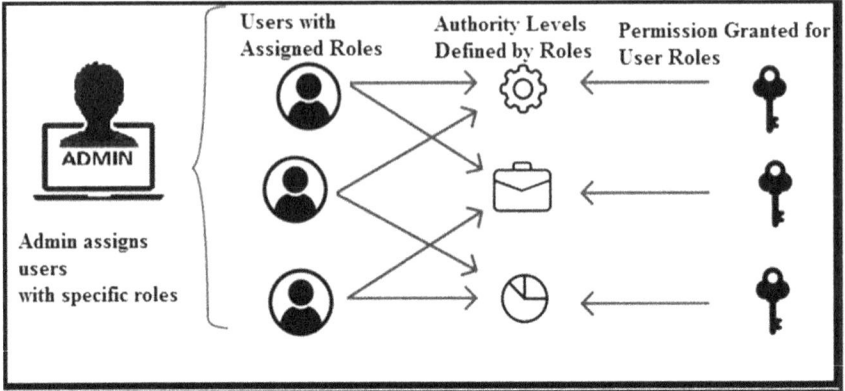

FIGURE 8.1 Role-based access control.

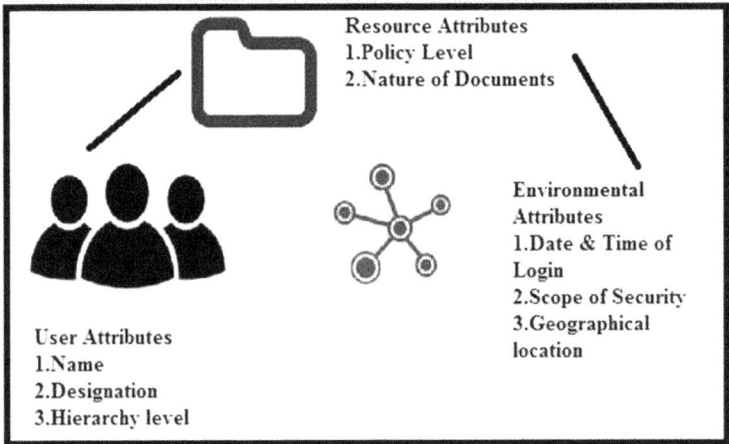

FIGURE 8.2 Attribute-based access control.

8.4 ANONYMIZATION

Anonymization is the process of masking or obscuring data to prevent it from being linked to a specific user or device. It is used to protect the privacy of users and prevent their data from being used for malicious purposes [13–15]. It can be achieved through data encryption, de-identification, and pseudonymization. These three techniques are essential for securing IoT systems and data. Authentication is used to verify user and device identities, authorization is used to control access to resources, and anonymization is used to protect the privacy of users. Together, these techniques help ensure the security of the IoT and protect users and their data from Malicious attackers.

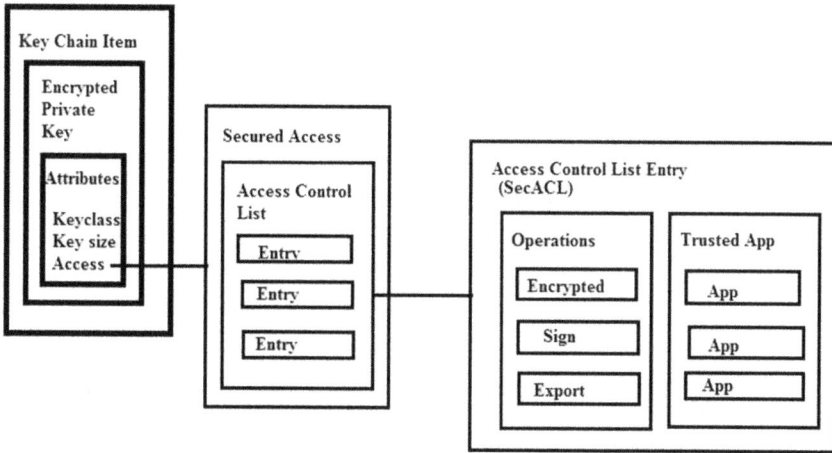

FIGURE 8.3 Access control list.

There are several anonymization techniques that can be used in IoT systems, including the following.

1. **Data Masking**: Data masking is a technique that replaces sensitive information with a substitute value, such as asterisks or random characters, to prevent sensitive information from being accessed or used without the consent of the individual. In IoT systems, data masking can be used to protect sensitive information, such as financial information or personal identification numbers.

2. **Data Encryption**: Data encryption is a technique that uses a mathematical algorithm to convert plain text into a code that can only be decrypted using a key. In IoT systems, data encryption can be used to protect sensitive information, such as personal or financial information, from being accessed or used without the consent of the individual.

3. **Data De-identification**: Data de-identification is a technique that removes or masks personal information from data, such as names, addresses, or social security numbers, to protect the privacy of individuals. In IoT systems, data de-identification can be used to protect sensitive information, such as medical records or financial information.

8.5 VULNERABILITIES IN IoT

The IoT represents a rapidly expanding technology that facilitates the connection of everyday physical objects to the internet. IoT devices have found extensive use across diverse domains, including healthcare and agriculture, gaining millions of users worldwide [11, 16–18]. However, despite the immense potential unlocked by these interconnected devices, they also introduce a substantial risk due to their inherent vulnerabilities. Despite the increasing sophistication of IoT devices, they often

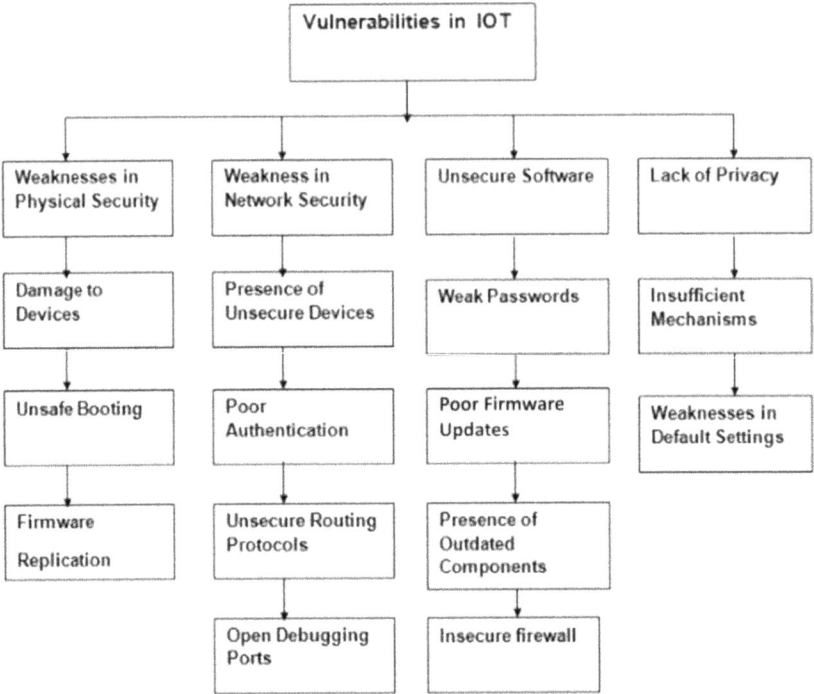

FIGURE 8.4   Vulnerabilities in IoT devices.

lack robust security measures, making them susceptible to Malicious attackers. This security gap can result in severe consequences, spanning from data breaches to physical harm. Figure 8.4 illustrates the different types of vulnerabilities in IoT devices.

Some common vulnerabilities in IoT devices are as follows [19]:

1. **Weak Passwords**: Weak passwords are a common vulnerability in IoT devices, as many users do not take the time to create strong, unique passwords. This can make it easier for attackers to gain unauthorized access to IoT devices, steal sensitive information, or cause harm.
2. **Unsecured Networks**: IoT devices often communicate through unsecured networks, such as Wi-Fi, that are susceptible to hacking and other security threats. If a network is unsecured, attackers can easily intercept data and steal sensitive information.
3. **Software Flaws**: IoT devices are often equipped with software that can contain security vulnerabilities, such as bugs or backdoors that can be exploited by attackers. This can result in data theft, unauthorized access, or system malfunctions.
4. **Hardware Flaws**: IoT devices are also vulnerable to hardware flaws, such as design weaknesses or manufacturing defects, that can be exploited by attackers. This can result in security threats, such as data theft or unauthorized access to sensitive information.

5. **Unsecured APIs**: Many IoT devices use APIs to communicate with other systems and devices, but these APIs can be vulnerable to attack if they are not secured properly. Unsecured APIs can be used by attackers to steal sensitive information, gain unauthorized access to network resources, or cause harm.

6. **Outdated Software**: IoT devices are often not updated regularly, which can result in the use of outdated software that is vulnerable to attack. This can result in security threats, such as data theft or unauthorized access to sensitive information.

8.5.1 LACK OF AUTHENTICATION AND AUTHORIZATION

Among the prevalent vulnerabilities found in IoT devices, a notable one is the absence of proper authentication and authorization mechanisms [1, 2]. Authentication pertains to the procedure of confirming a user's identity, while authorization involves granting access to resources. In the realm of IoT devices, the customary absence of robust authentication and authorization mechanisms exposes them to potential security breaches. This deficiency heightens the possibility of unauthorized individuals gaining entry to the device, potentially leading to the compromise of sensitive data, or even enabling malicious control of the device.

8.5.2 WEAK ENCRYPTION

Another common vulnerability in IoT devices is weak encryption. Encryption is the process of encoding data so that it can only be read by authorized users. Encryption is an important security measure, as it prevents attackers from accessing sensitive data. Unfortunately, many IoT devices use weak encryption algorithms or outdated encryption keys, making them vulnerable to attack. This vulnerability can allow attackers to gain access to data, or even control the device itself.

8.5.3 UNSECURED NETWORK CONNECTIONS

IoT devices often use unsecured network connections, making them vulnerable to attack. An unsecured network connection is open to anyone, allowing attackers to access the device or intercept data. This vulnerability increases the risk of data theft, as attackers can gain access to the device without the user's knowledge. It also increases the risk of physical harm, as attackers can control the device remotely.

8.5.4 INSECURE SOFTWARE

Updates for IoT devices often rely on software updates to fix bugs and add new features. However, these updates can also introduce new vulnerabilities. If the update is not properly secured, attackers can gain access to the device and exploit the vulnerabilities. This vulnerability increases the risk of data theft, as attackers can gain access to sensitive data. It also increases the risk of physical harm, as attackers can control the device remotely.

8.5.5 POORLY DESIGNED MOBILE APPS

Many IoT devices use mobile apps to control and monitor the device. However, these apps can often be poorly designed and contain vulnerabilities. If the app is not properly secured, attackers can gain access to the device and exploit the vulnerabilities. This vulnerability increases the risk of data theft, as attackers can gain access to sensitive data. It also increases the risk of physical harm, as attackers can control the device remotely.

8.6 AUTHENTICATION MANAGEMENT IN AWS CORE

IoT Core is an important aspect of providing secure access to cloud-based services. AWS IoT Core provides an array of authentication mechanisms to ensure secure access to its services and protect the data stored in the cloud.

8.6.1 WHAT IS AWS-IAM?

AWS offers an Identity and Access Management (IAM) directory service [5, 16] that serves as a valuable tool for monitoring cloud users and facilitating the tracking of authentication-related information. IAM plays a pivotal role in overseeing authorization management and handling data associated with 2FA. For instance, business owners can establish multiple "user" accounts for their employees, each requiring either a password or 2FA for authentication. These passwords serve to govern the access permissions granted to users once they have successfully entered a system. The AWS-IAM system provides a means to efficiently manage users' access to systems and their associated capabilities.

8.6.2 AWS-IAM ARCHITECTURE

IAM offers a robust and highly secure access control system [5, 6], providing users with the capability to create and oversee multiple user accounts, each of which can be assigned varying levels of access privileges. This system empowers users to establish groups and allocate specific permissions to them. IAM stands out as a potent authentication mechanism characterized by its adaptable access control capabilities.

Amazon IAM primarily caters to individuals with privileged access to an account responsible for group management and authorization for service modifications, such as system administrators. These administrators utilize the AWS Management Console to initiate and terminate instances. They not only define password regulations for the account, including criteria like length and expiration but also specify permissions that restrict user access to AWS account management resources and govern the actions that users can perform. Figure 8.5 depicts the architecture of AWS-IAM.

Furthermore, administrators have the authority to create roles, groups, and user accounts, and they can assign distinct sets of privileges to each entity. Different groups may receive privileges in distinct manners. For instance, Group B may have the ability to edit and delete all resources, while Group A can modify X, Y, and Z but lacks the capability to delete them.

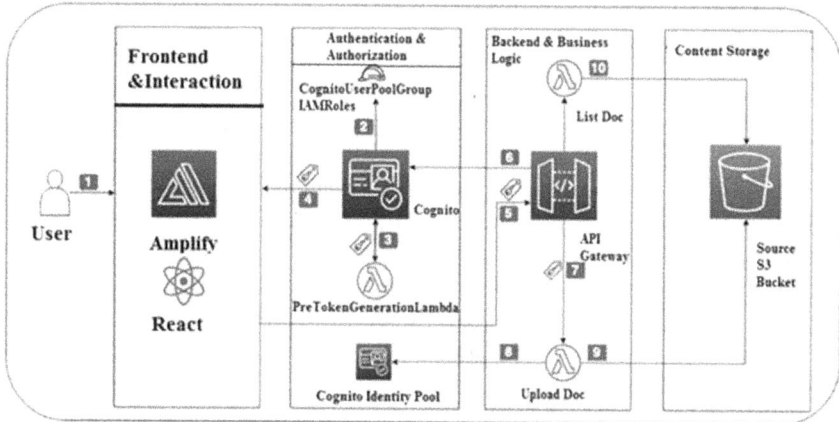

FIGURE 8.5 AWS identity and access management.

Adding users is just the first step in this process. Administrators must regularly verify the status of their IAM tools system to make sure that the correct people have the necessary access and privileges. Furthermore, it is essential to take into account long-term identity management solutions. The ability to delete a user from the system after they leave the company is another skill that system administrators should possess. System administrators ensure that policies are put in place to back up buckets and automatically deny access to specific users in order to ensure overall security. Some of the authentication mechanisms include:

8.6.3 AWS COGNITO

Cognito is an authentication service that can be used to securely authenticate users and manage authentication tokens. It provides a secure authentication process that is easy to integrate with AWS IoT Core. Additionally, it can be used to authenticate users from third-party services such as Facebook and Google.

8.6.4 X.509 CERTIFICATES

X.509 is a public key infrastructure (PKI) standard that enables secure communication between two parties [14, 15]. This authentication mechanism is based on the exchange of digital certificates that are signed by a trusted Certificate Authority (CA). With X.509, the identity of the user is verified and authorization is granted.

8.6.5 AWS SECURITY TOKEN SERVICE (STS)

STS is an authentication service that enables secure communication between two parties. It provides a secure authentication process that is easy to integrate with AWS IoT Core. Additionally, it can be used to authenticate users from third-party services such as Facebook and Google.

8.6.6 AWS Key Management Service (KMS)

KMS, known as Key Management Service, is a cryptographic service that empowers users to have control over their encryption keys. Through KMS, users can securely create, oversee, and store their encryption keys within the cloud environment. Furthermore, KMS seamlessly integrates with AWS IoT Core, facilitating the convenient utilization of these encryption keys for securing data stored in the cloud.

Apart from the authentication mechanisms, AWS IoT Core offers a range of other security features, including encryption, access control, and IAM [12, 20]. These security measures can be employed in conjunction to ensure secure access to cloud-based services and safeguard data stored in the cloud. When addressing authentication management, it is crucial to consider the diverse factors that can impact the security of the system.

- To begin with, the authentication system must possess sufficient robustness to thwart unauthorized access attempts.
- Second, the authentication procedure should prioritize user-friendliness and clarity, ensuring that it is straightforward to employ and comprehend.
- Third, it is imperative to safeguard authentication credentials, preventing easy access by unauthorized individuals.
- Finally, the authentication process should undergo regular assessment to confirm its continued security. When selecting an authentication mechanism, it's crucial to consider the specific security demands of the system.

For instance, IAM, while potent, may not be the ideal choice for applications requiring a high level of security. In such scenarios, alternatives like X.509 certificates or KMS might be more suitable.

8.7 METHODOLOGY FOR AUTHENTICATION, AUTHORIZATION, AND ANONYMIZATION IN IoT SUBSYSTEM

The IoT is an expanding technology at a rapid pace, facilitating the connectivity of physical objects to the internet, as elaborated in [9, 20]. While IoT holds immense promise for enhancing the quality of life and transforming various industries, it presents distinctive security challenges. Among these challenges, authentication, authorization, and anonymization stand out as critical concepts that demand attention when devising secure IoT systems. In this chapter, we will delve into a methodology for implementing these concepts within an IoT subsystem.

To secure all AWS resources and services, the initial step involves considering the solutions provided by IAM. These solutions can be summarized as follows:

- **Multi-Factor Authentication (MFA)**
  Enhancing account and user security can be swiftly and effortlessly achieved by implementing Multi-Factor Authentication (MFA). This involves adding an extra layer of security to your account or user credentials. In this approach, either you or the user provides an access key or password, along with a code that has been securely established by the device.

- **Granular Permission Management**
  Utilizing this finely tuned permission system, you have the flexibility to grant specific permissions to individuals based on their resource needs. For instance, you can define how to provide access to resources like S3 (Amazon Simple Storage Service) and Amazon EC2 (Amazon Elastic Compute Cloud) to various users or groups. Figure 8.6 represents the methodology of implementing AAA in the IoT subsystem.
- **Identity Federation**
  Users who are already familiar with their credentials can access their accounts using IAM's identity federation. For example, you can envision a scenario where you temporarily gain access to your existing AWS account management while connected to the X company network or through another internet service provider.
- **Assurance through Identification Data**
  Enabling the CloudTrail option in your AWS account ensures that you will receive comprehensive log records, including all data generated by the resources within your account. IAM identities are typically associated with these details.
- **Secure AWS Resource Access**
  Through this IAM feature, all potential login passwords for EC2 instances are safeguarded. Additionally, if desired, they can be granted access to your AWS account management application.
- **Shared Access to Your AWS Account**
  The resources can be accessed within your current AWS account and utilize other administrative capabilities without disclosing your password.
- **PCI DSS Compliance**
  AWS IAM fully supports compliance with the Payment Card Industry Data Security Standard (PCI DSS) by ensuring secure data storage, transmission, and handling for both providers and merchants.

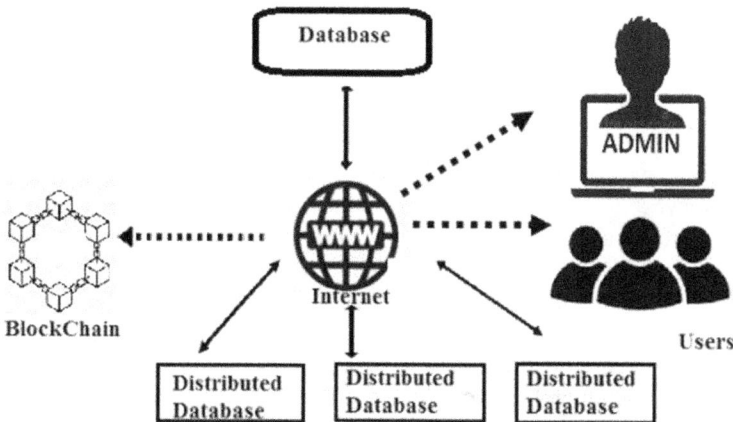

FIGURE 8.6   Methodology of implementing AAA in an IoT subsystem.

- **Password Policy Guidelines**
  The IAM password policy empowers you to remotely reset or modify passwords. It also allows you to define rules, such as password selection criteria or the number of allowed attempts before a password is rejected.
- **Cost-Free Usage**
  AWS IAM is a feature that comes included with your AWS account at no additional cost. Charges only apply when you use an IAM user to access other AWS services. Adding new users, groups, or policies is entirely cost-free.
- **Creation of IAM Policy Conditions**
  AWS users have the flexibility to customize rules by incorporating conditions that impose additional restrictions on resource access. For instance, you can set requirements such as Secure Sockets Layer (SSL) source address ranges, date and time constraints, or SSL encryption. Conditions can, for instance, prevent the termination of an EC2 instance until successful MFA usage, enhancing security for sensitive requests, even though they are typically optional.

8.8 CONCLUSION

Authentication, authorization, and anonymization are important concepts for understanding the security of the IoT. The use of secure authentication, authorization, and anonymization techniques is essential for ensuring the security of the IoT. These techniques help to protect user privacy, secure data, and prevent Malicious attackers from gaining access to systems. Furthermore, they help to ensure that only authorized users can access a system and that unauthorized users are unable to gain access. By implementing these techniques, organizations can ensure the security of their IoT systems and protect user data from Malicious attackers.

8.9 FUTURE RESEARCH DIRECTIONS

Currently, the research into authentication, authorization, and anonymization techniques for the IoT is focusing on improved security, privacy, and scalability.

1. Research will continue to focus on developing new authentication techniques, such as blockchain-based authentication, to increase security and privacy.
2. Research is also carried out into new authorization techniques, such as the use of artificial intelligence and machine learning algorithms, to automate the process of determining which users or devices are allowed to access which resources.
3. Finally, research will focus on developing new anonymization techniques, such as homomorphic encryption and differentially private algorithms, to protect user privacy while maintaining the accuracy of the data.

REFERENCES

1. C. Ni, L.S. Cang, P. Gope, G. Min, "Data anonymization evaluation for big data and IoT environment", *Information Sciences*, 605 (2022), pp. 381–392.
2. M. Aboubakar, M. Kellil, P. Roux, "A review of iot network management: Current status and perspectives", *Journal of King Saud University-Computer and Information Sciences*, 1 (1) (2021), pp. 1–14.
3. M. Mamdouh, A.I. Awad, A.A. Khalaf, H.F. Hamed, "Authentication and identity management of IoHT devices: Achievements, challenges, and future directions", *Computers & Security*, 111 (2021), p. 102491.
4. W. Ren, X. Tong, J. Du, N. Wang, S. Li, G. Min, Z. Zhao, "Privacy enhancing techniques in the Internet of Things using data anonymisation", *Information Systems Frontiers* (2021), pp. 1–12.
5. M. Saqib, B. Jasra, A.H. Moon,, "A lightweight three factor authentication framework for IoT based critical applications", *Journal of King Saud University-Computer and Information Sciences*, 34 (9) (2022), pp. 6925–6937.
6. Z.-K. Zhang, M.C.Yi Cho, C.-W. Wang, C.-W. Hsu, C.-K. Chen, S. Shieh, "IoT security: Ongoing challenges and research opportunities", in *2014 IEEE 7th International Conference on Service-Oriented Computing and Applications*, pp. 230–234. IEEE, 2014.
7. O. Garcia-Morchon, S.S. Kumar, G. Selander, R. Hummen, "Security considerations in the IP-based Internet of Things", *IEEE Communications Surveys & Tutorials*, 18 (1) (2016), pp. 129–163.
8. A. Al-Fuqaha, M. Guizani, M. Mohammadi, M. Aledhari, M. Ayyash, "Internet of Things: A survey on enabling technologies, protocols, and applications", *IEEE Communications Surveys & Tutorials*, 17 (4) (2015), pp. 2347–2376.
9. C. Perera, A. Zaslavsky, P. Christen, D. Georgakopoulos, "Context aware computing for the Internet of Things: A survey", *IEEE Communications Surveys & Tutorials*, 16 (1) (2014), pp. 414–454.
10. L. Caruccio, O. Piazza, G. Polese, G. Tortora, "Secure IoT analytics for fast deterioration detection in emergency rooms", *IEEE Access*, 8 (2020), pp. 215343–215354.
11. K. I. Ahmed, M. Tahir, M.H. Habaebi, S.L. Lau, A. Ahad, "Machine learning for authentication and authorization in iot: Taxonomy, challenges and future research direction", *Sensors*, 21 (15) (2021), p. 5122.
12. V. Gupta, A. Khanna, "A survey on anonymization techniques and challenges in Big Data privacy", in *2019 3rd International Conference on Trends in Electronics and Informatics (ICOEI)*, pp. 1571–1576. IEEE, 2019.
13 H. Al-Refai, A.A. Alawneh, "User authentication and authorization framework in IoT protocols", *Computers* 11 (10) (2022), p. 147.
14. B. Bordel, R. Alcarria, T. Robles, M.S. Iglesias, "Data authentication and anonymization in IoT scenarios and future 5G networks using chaotic digital watermarkin", *IEEE Access*, 9 (2021), pp. 22378–22398.
15. M. Shahzad, M.P. Singh, "Continuous authentication and authorization for the Internet of Things", *IEEE Internet Computing*, 21 (2) (2017), pp. 86–90.
16. X. Yao, F. Farha, R. Li, I. Psychoula, L. Chen, H. Ning, "Security and privacy issues of physical objects in the iot: Challenges and opportunities", *Digital Communications and Networks*, 7 (3) (2021), pp. 373–384.
17. T. Kavitha, G. Senbagavalli, D. Koundal, Y. Guo, D. Jain, eds., *Convergence of Deep Learning and Internet of Things: Computing and Technology*, IGI Global, 2022.

18. U. Ghosh, M. Alazab, A.K. Bashir, A.-S.K. Pathan, eds., *Deep Learning for Internet of Things Infrastructure*, CRC Press, 2021.
19. T. Kavitha and D. Sridharan, "Security Vulnerabilities in Wireless Sensor Networks: A Survey", *International Journal on Information Assurance and Security (1554-1010) (JIAS)*, no. 5, pp. 031–044, 2010.
20. L. Xie, X. Zhang, X. Jia, F. Shang, Z. Xie, "A survey of data anonymization techniques", *ACM Computing Surveys (CSUR)*, 52 (4) (2019), pp. 1–35.

9 Secure Identity, Access, and Mobility Management in the Internet of Things

*S. Angel Deborah, S. Rajalakshmi, and
T. T. Mirnalinee*
Sri Sivasubramaniya Nadar College of Engineering,
Anna University, Chennai, India

9.1 INTRODUCTION

The Internet of Things (IoT) is a network of physical devices, vehicles, home appliances, and other objects that are embedded with sensors, software, and network connectivity to collect and exchange data. IoT has the potential to revolutionize many industries, but it also introduces new security challenges. In today's world, many products and inventions are moving toward the IoT. The number of devices and users using IoT is growing day by day.

One of the biggest challenges in securing IoT [1] is identity management. In traditional IT environments, users are authenticated using strong passwords or two-factor authentication (2FA). However, these methods are not practical for IoT devices, which often have limited computing power and storage capacity. As a result, IoT devices are often vulnerable to brute-force attacks and other forms of unauthorized access. More details on secure identity are discussed in Section 9.2.

Another challenge is access control. In traditional IT environments, access to resources is controlled using role-based access control (RBAC). However, RBAC is not well-suited for IoT environments, where devices are often dynamically added and removed from the network. As a result, IoT devices are often left with open access to sensitive data and resources. Section 9.3 elaborates the access control and management in IoT. Finally, IoT devices need mobility management. The primary goal of mobility management in IoT is to ensure that the devices remain connected to the network, even as they move across different network domains or access points. This is further detailed in Section 9.4.

9.2 SECURE IDENTITY

Secure identity in IoT refers to the process of authenticating and verifying the identity of devices, users, or other entities that are connected to the IoT ecosystem [2].

DOI: 10.1201/9781003477327-9

It involves ensuring that only authorized devices or users can access the network or sensitive data and that the data is protected against unauthorized access or tampering.

Secure identity in IoT typically involves the use of strong authentication mechanisms, such as digital certificates or biometrics, to verify the identity of devices or users [3]. Access control mechanisms, such as RBAC or attribute-based access control (ABAC), are also used to restrict access to sensitive data or systems.

Secure communication protocols, such as Transport Layer Security (TLS) or Secure Socket Layer (SSL), are also used to encrypt and protect data in transit. Additionally, IoT devices should be regularly updated with the latest security patches and firmware updates, and policies for device authentication, authorization, and access control should be regularly reviewed and updated to ensure the ongoing security of the IoT ecosystem.

Overall, secure identity in IoT is essential to protect against cyber threats and ensure the safety and security of sensitive data in the IoT ecosystem [4]. As more and more devices become connected to the internet, the issue of secure identity in the IoT becomes increasingly important. In the IoT, devices need to securely identify themselves and communicate with other devices in a trusted manner. As shown in Figure 9.1, the following includes some key considerations for ensuring secure identity in the IoT.

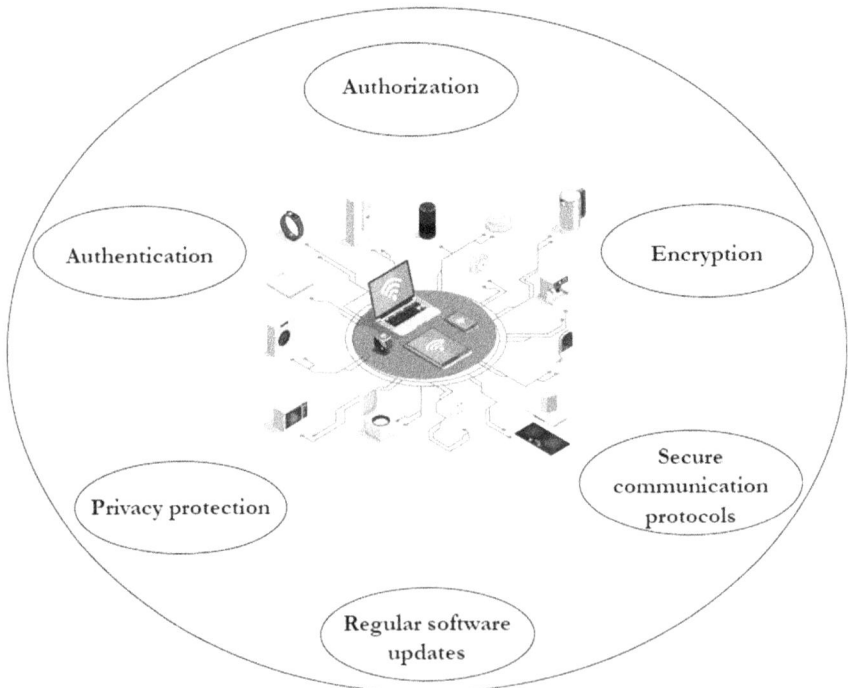

FIGURE 9.1 IoT security challenges.

9.2.1 AUTHENTICATION

Authentication in IoT refers to the process of verifying the identity of a device or user before allowing access to the IoT ecosystem [5]. In IoT, authentication is a critical security measure that ensures that only authorized devices or users can access the network or sensitive data [6]. There are several authentication methods that can be used in IoT, including the following.

Password-based authentication: Devices or users are authenticated using a password or PIN. This method is commonly used in consumer IoT devices, such as smart home devices.

Certificate-based authentication: Devices are authenticated using a digital certificate that is issued by a trusted third-party authority. This method is commonly used in industrial IoT (IIoT) and other high-security environments.

Biometric authentication: Devices or users are authenticated using biometric data, such as fingerprints or facial recognition. This method is becoming increasingly popular in consumer IoT devices, such as smartphones and wearables.

Token-based authentication: Devices or users are authenticated using a token, such as a physical security key or a one-time password (OTP). This method is commonly used in 2FA or multi-factor authentication (MFA) systems.

Overall, authentication is an essential security measure in IoT that ensures the safety and security of the IoT ecosystem. By implementing strong authentication mechanisms, IoT devices and networks can be protected against cyber threats and unauthorized access.

9.2.2 AUTHORIZATION

Authorization in IoT refers to the process of granting or denying access to specific resources or functions within the IoT ecosystem based on the authenticated identity of a device or user. In other words, authorization determines what actions or operations a device or user is allowed to perform within the IoT system [5]. Authorization is a critical security measure in IoT because it ensures that only authorized devices or users can access sensitive data or perform critical functions. There are several authorization methods that can be used in IoT, including the following.

Role-based access control: Access to specific resources or functions is based on the roles and responsibilities of the device or user. For example, an IoT system may have different roles such as administrator, user, or guest, and each role may have different levels of access.

Attribute-based access control: Access to specific resources or functions is based on a set of attributes or properties of the device or user, such as location, time of day, or device type.

Rule-based access control: Access to specific resources or functions is based on a set of predefined rules that are applied to the device or user's identity and the requested resource or function.

Overall, authorization is an essential security measure in IoT that ensures the safety and security of the IoT ecosystem. By implementing strong authorization mechanisms, IoT devices and networks can be protected against cyber threats and unauthorized access, and sensitive data can be kept secure.

9.2.3 ENCRYPTION

Response encryption in IoT refers to the process of encrypting the data that is transmitted from an IoT device back to the cloud or other systems in response to a request. This is an important security measure that helps to protect the privacy and confidentiality of the data being transmitted. Regenerating response encryption involves periodically changing the encryption keys [7] used to encrypt the data being transmitted. This is important because if the encryption keys are compromised, an attacker could potentially decrypt the data and gain unauthorized access to sensitive information. To regenerate response encryption in IoT, the following steps can be taken.

Generate new encryption keys: Generate new encryption keys to replace the old ones that are being used for response encryption.

Distribute new keys: Distribute the new encryption keys securely to all devices and systems that are involved in response encryption.

Update systems: Update all systems and devices to use the new encryption keys for response encryption.

Test encryption: Test the new encryption to ensure that it is working correctly and that there are no compatibility issues with existing systems.

Rotate keys regularly: Repeat the process of regenerating response encryption periodically, such as every six months or once a year, to ensure that the encryption keys are always up to date and secure.

By regularly regenerating response encryption in IoT, organizations can ensure that the data being transmitted is protected against cyber threats and unauthorized access.

9.2.4 SECURE COMMUNICATION PROTOCOLS

Secure communication protocols are essential in IoT because they provide a mechanism for encrypting and protecting data as it is transmitted between devices, gateways, and the cloud [8]. Without secure communication protocols, IoT devices and networks are vulnerable to cyber-attacks, data breaches, and other security threats [9]. There are several secure communication protocols that can be used in IoT, including the following.

Transport Layer Security: TLS is a protocol that provides secure communication over the internet. It is commonly used to secure HTTP and MQTT connections in IoT and is based on the use of public key encryption.

Datagram Transport Layer Security (DTLS): DTLS is a variant of TLS that is designed to work with datagram protocols such as User Datagram Protocol (UDP). It is commonly used to secure CoAP connections in IoT.

Secure Socket Layer (SSL): SSL is an older protocol that is similar to TLS and provides secure communication over the internet. It is still used in some legacy IoT systems.

Message Queuing Telemetry Transport (MQTT): MQTT is a lightweight messaging protocol that is commonly used in IoT. It can be secured using TLS to provide secure communication between devices and the cloud.

Constrained Application Protocol (CoAP): CoAP is a protocol designed for IoT devices that have limited processing power and memory. It can be secured using DTLS to provide secure communication between devices and gateways.

ZigBee: ZigBee is a low-power wireless communication protocol that is commonly used in IoT. It includes security features such as encryption and authentication to provide secure communication between devices.

Overall, secure communication protocols are critical for ensuring the safety and security of IoT devices and networks. By implementing secure communication protocols, organizations can protect against cyber threats, data breaches, and other security risks.

9.2.5 REGULAR SOFTWARE UPDATES

IoT devices should be regularly updated with security patches and firmware updates to fix vulnerabilities and ensure that the latest security features are being used. Regular software updates are essential in IoT for several reasons.

Security: Regular software updates can help to address security vulnerabilities and protect IoT devices and networks from cyber threats. As new security threats emerge, software updates can provide patches or fixes that address these vulnerabilities and help to keep devices secure [9].

Functionality: Software updates can also provide new features and functionality to IoT devices, improving their performance and usability. For example, an update may introduce new sensors, improve energy efficiency, or provide new control options.

Compatibility: As new technologies and protocols emerge, software updates can help to ensure that IoT devices remain compatible with other systems and devices. This can help to prevent interoperability issues and ensure that devices continue to function as expected.

Bug fixes: Regular software updates can fix bugs and improve the stability of IoT devices, reducing the risk of malfunctions, errors, and downtime.

To ensure that IoT devices are kept up to date with the latest software updates, it is important to establish a regular update schedule. This schedule should include regular software maintenance, including updates and patches as needed. In addition to regular software updates, it is also important to ensure that devices are updated securely. This can involve implementing secure update mechanisms that use encryption and authentication to protect against cyber threats and unauthorized access.

Regularly testing updates before deployment can also help to ensure that they do not introduce new security vulnerabilities or compatibility issues.

Overall, regular software updates are critical to the ongoing security, functionality, and compatibility of IoT devices and networks. By implementing a regular update schedule and ensuring that updates are deployed securely, organizations can help protect against cyber threats and ensure that devices continue to function effectively.

9.2.6 PRIVACY PROTECTION

IoT devices should be designed to collect only the data necessary to perform their intended function, and the data should be protected with strong encryption and access controls to prevent unauthorized access or misuse [10]. Privacy protection is an important aspect of IoT because these devices often collect and transmit sensitive personal data such as biometric data, health information, and location data [3]. To ensure privacy protection in IoT, the following measures can be taken.

Data encryption: IoT devices should use encryption to protect data in transit and at rest. This helps to prevent unauthorized access to the data.

Data minimization: IoT devices should only collect and transmit the minimum amount of data necessary for their intended function. This helps to minimize the amount of sensitive data that is at risk of being compromised.

Consent: IoT users should be provided with clear information on what data is being collected and how it will be used. They should also be given the option to provide explicit consent for data collection and usage.

Anonymization: IoT data should be anonymized where possible, to prevent it from being linked to individual users.

Access control: IoT devices should implement access control mechanisms to ensure that only authorized users can access the data. This can include password protection, multi-factor authentication, and RBAC.

Regular updates: IoT devices should be updated regularly to address security vulnerabilities and ensure that they continue to provide adequate privacy protection.

Privacy by design: Privacy considerations should be built into the design of IoT devices from the beginning. This can help to ensure that they are secure and protect user privacy by default.

By implementing these measures, organizations can help to ensure that IoT devices are secure and protect user privacy. It is important to note that privacy protection in IoT is an ongoing process that requires regular review and updates to address new privacy threats and evolving technologies.

Overall, ensuring secure identity in the IoT requires a combination of authentication, authorization, encryption, secure communication protocols, regular updates, and privacy protection. By taking these steps, IoT devices can operate in a trusted and secure manner.

9.3 ACCESS MANAGEMENT IN IoT

Access management in IoT refers to the process of controlling and managing user access to IoT devices, networks, and data [11]. With the proliferation of IoT devices and the increasing amount of data they collect and transmit, access management has become a critical component of IoT security. Access management in IoT involves ensuring that only authorized users have access to IoT devices and data. This involves implementing authentication mechanisms to verify the identity of users, authorization mechanisms to determine what resources and data they are allowed to access, and access control mechanisms to enforce these policies [2].

Effective access management is essential for protecting IoT devices and data from unauthorized access, theft, and tampering. It is also important for protecting user privacy, as IoT devices often collect sensitive personal data. By implementing robust access management policies and mechanisms, organizations can ensure that only authorized users have access to IoT devices and data, helping to prevent security breaches and protect user privacy.

In Figure 9.2, an IoT device communicates with an IoT gateway, which in turn communicates with an IoT platform. The access control module is present in both the IoT device and the IoT gateway, while the authentication and authorization modules are present in the IoT platform [12]. The access control module in the IoT device and IoT gateway manages access to the device and gateway resources, respectively. These modules ensure that only authorized entities can access the device or gateway and that access is granted based on predefined policies. The authentication module in the IoT platform verifies the identity of the entity seeking access, while the authorization module determines the level of access granted based on predefined policies. Together, these modules provide a comprehensive access management system that ensures the security and integrity of IoT devices, gateways, and platforms. Here are some best practices for access management in IoT.

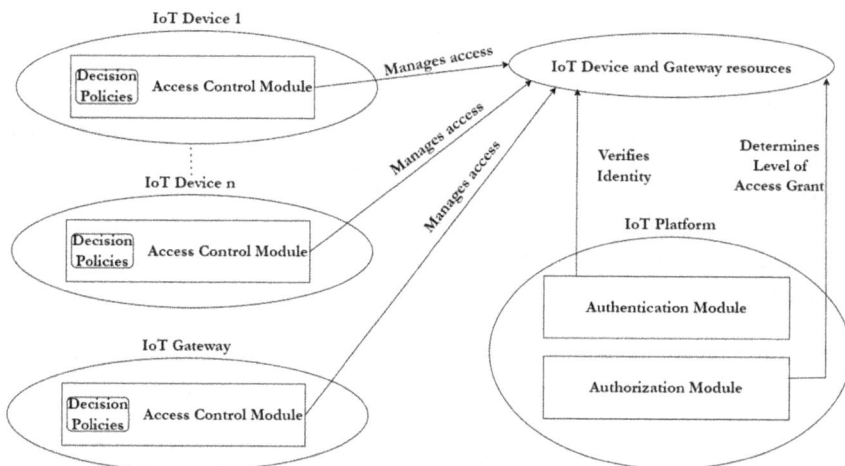

FIGURE 9.2   IoT access management.

Authentication: Users should be required to authenticate themselves before accessing IoT devices and data. This can include the use of passwords, biometric authentication, or other forms of authentication.

Authorization: Once a user is authenticated, they should be granted access only to the resources and data that they are authorized to access. This can be achieved through the use of access control mechanisms such as RBAC.

Identity management: Identity management is the process of managing user identities, including user authentication, authorization, and user account management. IoT systems should have robust identity management systems in place to ensure that only authorized users have access to devices and data.

Secure protocols: IoT devices should use secure protocols such as TLS or DTLS to ensure that user access is secured against unauthorized access.

Regular audits: Regular audits of access logs and access control mechanisms can help to identify and address any potential security issues.

Secure storage and transmission of access credentials: Access credentials such as usernames and passwords should be stored securely and transmitted using secure protocols to prevent unauthorized access [13].

Multi-factor authentication: Multi-factor authentication can provide an additional layer of security by requiring users to provide more than one form of authentication to access devices and data.

Effective access management is critical to the security and privacy of IoT systems. By implementing best practices such as authentication, authorization, identity management, and secure protocols, organizations can help to ensure that only authorized users have access to IoT devices, networks, and data.

9.4 MOBILITY MANAGEMENT IN IoT

Mobility management in IoT refers to the process of managing the movement of IoT devices and ensuring seamless connectivity as they move between different network domains or access points [14]. IoT devices are often designed to be mobile and can move from one location to another, which makes it challenging to maintain their connectivity and ensure seamless communication. Mobility management is, therefore, critical to the functioning of IoT devices and the success of IoT applications. The primary goal of mobility management in IoT is to ensure that the devices remain connected to the network, even as they move across different network domains or access points. This is achieved through a variety of techniques, including the following.

Handover management: This involves ensuring that the device can smoothly transition between different access points or networks without interrupting its communication.

Network discovery and selection: This involves identifying the available networks in the vicinity of the device and selecting the best one based on factors such as signal strength and network capacity.

Resource management: This involves managing the network resources efficiently to ensure that the device can access the resources it needs without causing congestion or other performance issues.

In summary, mobility management in IoT is a critical component of IoT connectivity, enabling seamless communication and movement of IoT devices across different network domains or access points.

9.5 SUMMARY

Secure identity management involves establishing trust and authentication mechanisms between IoT devices and applications to ensure that only authorized entities can access IoT resources. Access management ensures that users and devices are granted access to the appropriate resources based on their roles and privileges. Mobility management enables IoT devices to move seamlessly across different network domains while maintaining connectivity and security.

The key challenges in implementing a secure identity, access, and mobility management framework in IoT include establishing trust between devices and applications, managing access to resources, and enabling seamless mobility across different network domains. To address these challenges, various authentication and authorization mechanisms, access control models, and mobility management techniques have been developed.

Overall, secure identity, access, and mobility management are critical components of a comprehensive security framework for the IoT. By implementing these components, IoT devices and resources can be protected from unauthorized access, and mobility across different network domains can be made secure and seamless.

REFERENCES

1. Kavitha T, Ajantha Devi V, Neelavathy Pari S, Ramanathan S, *Internet of Everything: Smart Sensing Technologies*, Nova Science Publishers, Publication Date: June 17, 2022, doi: 10.52305/PNQM1088
2. Chen Z, Jiang X, Jin Y, Hu F, "A secure authentication and access control scheme for Internet of Things based on blockchain and smart contract", *IEEE Access*, vol. 6, pp. 52607–52617, 2018.
3. Kumar M, Tripathi RK, "A review on security and privacy issues in Internet of Things (IoT)", *Journal of Ambient Intelligence and Humanized Computing*, vol. 12, no. 9, pp. 8631–8648, 2021.
4. Liu L, Chen X, Guizani M, "A survey on IoT service and application security", *Journal of Network and Computer Applications*, vol. 88, pp. 10–28, 2017.
5. Saleem K, Aziz B, Razaque A, "Secure and efficient authentication scheme for Internet of Things using blockchain technology", *IEEE Access*, vol. 7, pp. 131130–131138, 2018.
6. Nie J, Chen J, Zhang N, "A review of Internet of Things for smart home: Challenges and solutions", *Journal of Sensors*, pp. 1–19, 2018.
7. Kavitha T, Kaliyaperumal R, "Energy Efficient Hierarchical Key Management Protocol", in *2019 5th International Conference on Advanced Computing & Communication Systems (ICACCS)*, Sri Eshwar College of Engineering Coimbatore, India, 2019, pp. 53–60. doi: 10.1109/ICACCS.2019.8728343

8. Al-Fuqaha A, Guizani M, Mohammadi M, Aledhari M, Ayyash M, "Internet of Things: A survey on enabling technologies, protocols, and applications", *IEEE Communications Surveys & Tutorials*, vol. 17, no. 4, pp. 2347–2376, 2015.

9. Shafagh H, Burkhalter L, Hithnawi A, Duquennoy S, "Towards a comprehensive security framework for the Internet of Things", *IEEE Internet of Things Journal*, vol. 4, no. 6, pp. 2180–2195, 2017.

10. Seneviratne S, Liyanage M, Zaslavsky A, "Internet of Things (IoT): Review of security and privacy approaches", in *Proceedings of the 2017 IEEE International Conference on Industrial Internet (ICII)*, IEEE Xplore, pp. 58–63, 2017.

11. Wu X, Yang P, Sun Y, "A survey on access control for Internet of Things", *IEEE Access*, vol. 8, pp. 110021–110038, 2020.

12. Alotaibi F, Althuwayeb AA, Almogren A, "A survey of access control in Internet of Things", *Journal of Ambient Intelligence and Humanized Computing*, vol. 13, no. 3, pp. 2773–2803, 2022.

13. Wang H, Jin D, Hu Y, "Internet of Things security: A review", *IEEE Internet of Things Journal*, vol. 7, no. 1, pp. 1–12, 2020.

14. Wang J, Cao J, Zhang Q, "Mobility management in Internet of Things: A comprehensive survey", *IEEE Communications Surveys & Tutorials*, vol. 21, no. 2, pp. 1286–1321, 2019.

10 Privacy-Preserving Scheme for Internet of Things

T. Kavitha
New Horizon College of Engineering, Visvesvaraya Technological University, Bengaluru, India

G. Senbagavalli
AMC Engineering College, Visvesvaraya Technological University, Bengaluru, India

S. Saraswathi
Sri Sivasubramaniya Nadar College of Engineering, Anna University, Chennai, India

10.1 INTRODUCTION

Internet of Things (IoT) capabilities have been revolutionized by the quick growth of the microelectronics sector and 5G/6G communication. IoT applications have brought immense convenience and connectivity to human lives. Due to this, human lives now include a significant amount of social media. Nowadays, common household equipment are connected to the internet in some way, but they have also raised concerns about privacy and data security [1].

The quantity of personal information being gathered and exchanged is increasing quickly as more gadgets are connected to the internet. One of the key issues with IoT, though, is that the technology that is now available cannot simultaneously offer solutions for high endurance, lower power consumption, and strong protection in data processing, all of which are necessary for developing IoT applications.

The adoption and deployment of IoT products, services, mission-critical applications [2, 3], and industries are sometimes constrained by concerns about privacy and security. Security and privacy are the two main cyber concerns present with IoT devices. Confidentiality, Integrity, and Availability make up the first security triangle. The next three factors in the privacy trinity are traceability, likeability, and recognizability [4].

The main objective of adopting security mitigation is to protect confidentiality and privacy while assuring the security of infrastructures, devices, availability of IoT services, IoT users, and data [5]. Users' personal information must be safeguarded, and their right to privacy must be ensured. Depending on the categories of personal information, users may choose suitable privacy protection. Traditional privacy methods fall into two categories: Discretionary Access and Limited Access [6]. For IoT applications, traditional privacy protection systems are insufficient. Modern wireless network technologies and the IoT are emerging, creating new research challenges and economic prospects [7]. It necessitates innovative strategies including distributed cybersecurity controls, models, and judgments that take into consideration attack surfaces, hostile users, and weaknesses in system development platforms. When working with enormous amounts of data in IoT systems, machine learning (ML) approaches can increase the detection of novel assaults [8].

IoT devices in IoT smart applications have resource limitations that raise major privacy issues while doing data processing. These fall under direct and indirect threats. Direct information exposure is possible in every layer of the IoT architecture. The six basic categories of indirect privacy threats are impersonation, membership inference, re-identification, model inversion, property inference, and model theft. Privacy and security breaches from each layer of the IoT are very important before proposing a privacy preservation scheme [9]. According to the learning architecture (distributed, centralized, and federated) and protection strategies (various cryptographic mechanisms), a privacy protection scheme can be proposed [10]. Privacy preservation in an IoT setting has to be done based on the classification of the work under consideration includes four categories: smart domain, data acquisition approach, privacy approach, and based on architectural model [11]. Data-privacy-enhancing technologies can be proposed based on cutting-edge hardware and software solutions that make use of analyzing data from IoT devices while upholding strict data privacy assurances [12].

The next generation of wireless networking is predicted by 2030 [7], when there are projected to be more than 30 billion connections between IoT devices. There is an increasing need for new architectural solutions that can support the predicted bandwidth where these devices will create at rates quicker than they can now handle as the number of devices connecting to the internet rises. So, this chapter discusses a few of the recently proposed privacy preservation scheme that meets the current requirement.

10.2 HYBRID META-HEURISTIC ALGORITHM FOR BUSINESS INTELLIGENCE (BI) APPLICATIONS

Paper [13] recommends a comprehensive architecture which has an intrusion detection engine and two-level privacy for boosting security and privacy in IoT-based BI applications. With the help of a two-level privacy-preservation engine, unauthorized access, inference, and data poisoning attacks are prevented. In the first level of the privacy preservation engine, data authentication is done using blockchain and smart contracts. In the second level, privacy is achieved with the help of the Independent Component Analysis transformation implemented using machine learning to encode the data in different formats. The intrusion detection engine is designed using a gradient tree boosting model that consists of a decision tree and linearly superimposed base classifiers. Figure 10.1 shows the framework of privacy preservation and intrusion detection.

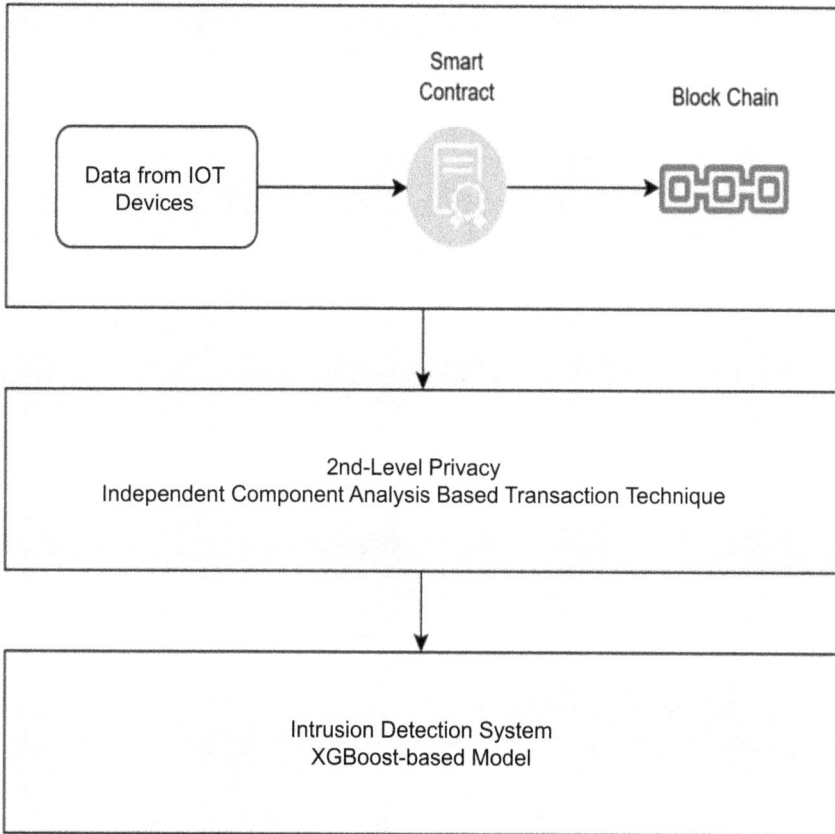

Smart
Contract

Block Chain

Data from IOT
Devices

2nd-Level Privacy
Independent Component Analysis Based Transaction Technique

Intrusion Detection System
XGBoost-based Model

FIGURE 10.1   Framework of privacy preservation and intrusion detection.

10.3 HYBRID META-HEURISTIC ALGORITHM FOR INDUSTRIAL IoT

Paper [14] proposed the privacy preservation model for IIoT using a hybrid optimization algorithm by incorporating the advancements of artificial intelligence techniques. It used three different test cases. This proposed framework, as shown in Figure 10.2, has data restoration and sanitization steps. Data sanitization prevents the leakage of information by covering the sensitive information in IIoT and prevents unauthorized access. The Grasshopper–Black Hole Optimization (G–BHO) algorithm is used to generate the new optimal key. The optimal key generation uses the multi-objective function involving many parameters like correlation coefficient, degree of modification, hiding ratio, and information preservation ratio are utilized for. This optimal key performs a crucial role in performing the data_sanitization and data_restoration. The authors of paper [14] implemented the proposed algorithm using MATLAB and evaluated the results with Jaya Algorithm (JA), Grey Wolf Optimization (GWO), GOA, and BHO techniques.

IPR   DoM   HR   CC

Multi-Objective Function

Optimal Key

IIoT Data

Data Sanitization

Data Restoration

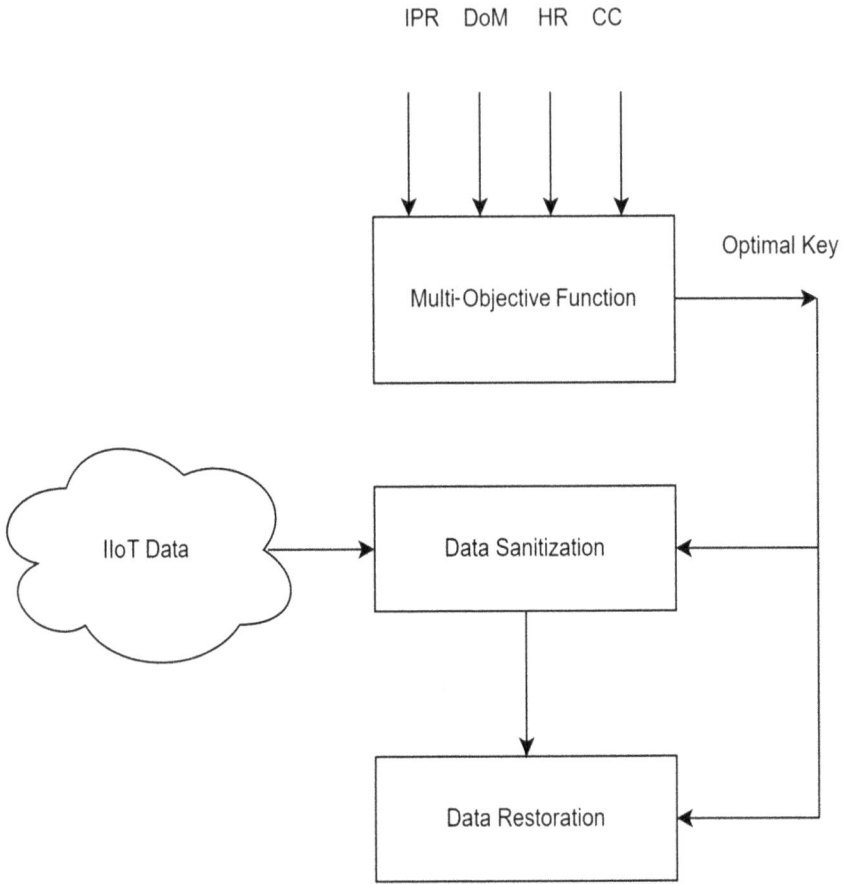

FIGURE 10.2   Framework of a heuristic algorithm.

10.4   HOMOMORPHIC CRYPTOGRAPHIC PRIVACY PRESERVATION SCHEME

A lightweight privacy-preserving scheme is proposed by Abdulrahman et al., which reduces the computation time involved in cryptographic schemes in an IoT. The proposed model uses the general communication IoT architecture where IoT devices are connected with the cloud server. Edge devices involved in this architecture act as fog devices that form the network among IoT devices and cloud servers. IoT devices in the network can avail of any service from the IoT applications in the cloud server through edge nodes. There is a Trusted Authority that manages the registration of all network components. In IoT applications, latency issues result from high bandwidth communication and high cryptographic computation costs. Here, designing IoT applications with high computation costs will be critical to meet their goals. Therefore, in paper [15], the authors proposed a privacy-preserving scheme with low latency and, at the same time, meeting identity privacy, message privacy, mutual authentication, secure unlinkability, and efficiency by using the bilinear pairing cryptography technique. This LPP is based on the computational Diffie–Hellman problem and decisional bilinear Diffie–Hellman problem. It uses the aggregate and batch verification method with the help of a homomorphic cryptographic system to reduce the communication and computation costs involved in cryptographic mechanisms. The four steps of the LPP scheme's implementation are system startup, entity registration, service querying, and service provisioning. Together, they make up the full process.

10.5 FULLY HOMOMORPHIC ENCRYPTION SCHEME

The IoT supported through the Wireless Sensor Networks naturally produces a large set of unfocussed information that has to be processed quickly, depending upon the sensitivity of the information. In such a situation, the packet authentication code and routing protocols are fundamentally mandatory for scalability and interoperability. So, paper [16] proposed a methodology based on packet authentication code to preserve privacy that includes various processes such as grouping of node members, creation of keys, distribution, and development of distinct sensor numbers for authentication. Homomorphic encryption is based on hybrid Ducas and Micciancio (DM) and Gentry, Sahai, and Waters (GSW) is used. In this mechanism, nodes are authenticated well before the data transmission to improve the efficiency of security. The authors did the comparative analysis and proved the efficiency of the proposed hybrid GSW–DM with PAC for privacy preservation to handle the False Data Injection (FDI) attacks.

10.6 PRESERVING DATA AGGREGATION WITH A FAIR ACCESS FRAMEWORK (FAF)

Paper [17] proposed a data aggregation scheme along with FAF that ensures privacy to support data security. The data aggregation process provides a shield for the removal of FDI attacks. The FAF uses a blockchain technology that enables users to provide, terminate, and allocate access to other users. It also pointed up the difficulty with data security and privacy in IoT-related green agriculture. The paper also did detailed compatibility analyses about how the suggested security system can be adapted to the green IoT to ensure security, especially privacy issues.

10.7 MULTI-LEVEL PRIVACY PROTECTION SCHEME

The user's original files he wants to store in the public cloud are transformed. The plain forms of the transformation parameters of the files are stored in the private cloud. Due to internal attacks, the transformation parameters of the file may be compromised. So the file stored in the public cloud is at risk. The transformation parameters [18] are secured through a linear array representation of parameters which is used and stored. To overcome this issue, a multi-level privacy protection scheme is suggested. The user's secret symmetric key is generated by himself/herself. With the help of Advanced Encryption Standard (AES), the linear array of transformation parameters is encrypted using the user's secret key. A 2D matrix of size m × n is created using the transformation parameters represented as a linear array. The majority of deep learning models now in use can only process 2D matrix pictures. The generator part of the Generative Adversarial Networks (GANs) receives the 2D matrix and transforms it into a masked picture. The private cloud houses the masking image. The masked picture is sent to the discriminator part of the GAN to extract the 2D matrix when the data owner wishes to examine the data. The 2D matrix is flattened into a linear array using row-major order, and the transformation parameters are then obtained by decrypting the array using AES and the user's secret key. The transformation parameters are then used to retrieve the original data. The complete process is shown in Figure 10.3.

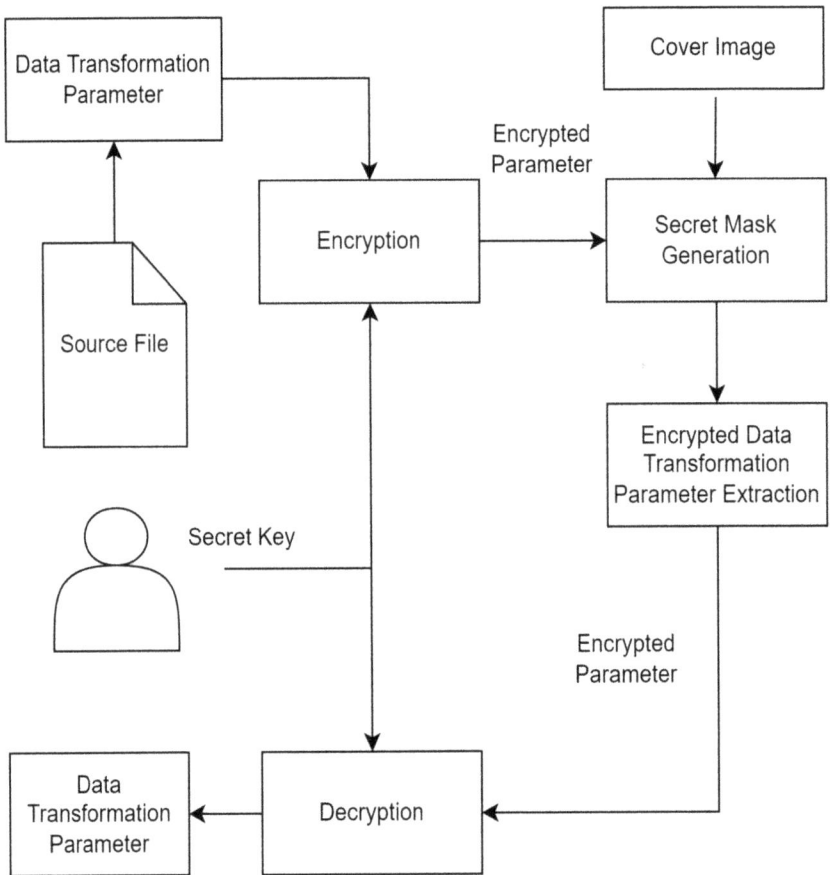

FIGURE 10.3 Multi-level scheme.

10.8 SOFTWARE DEFINED NETWORK BASED SCHEME

As technology advances, challenges in IoT are increasing. The 5G cellular network faces difficulties in providing a stable connection between IoT devices. The authors address these issues through Software Defined Networks (SDNs) which provide 5G connection along with complete secrecy. The proposed architecture consists of a control plane, a data plane, and an application plane. The SDN is present in the control plane [19], the base station, or access points in the data plane. The base station has the local database which has the information about the IoT devices and cluster head (CH). The SDN enforces intrusion detection, authentication, mobility management, and routing. The SDN controller defines the authentication and intrusion detection process. This authentication process validates the certification authority (CA), CH, and IoT devices. The intrusion detection process prevents attacks such as a black hole, packet duplication wormhole, preferential forwarding, resource depletion, and Sybil attacks. Data pre-processing, tensor-based dimensionality reduction, and fuzzy C-means clustering are the three stages that make up this procedure. In the IoT network, a certificate authority-to-cluster-head authentication protocol is used. To reduce computing costs and find potential network breaches, an intrusion detection is developed. The authors conclude that this proposed framework provides modular security for the current IoT Network.

10.9 FUZZY-BASED TRUST PRIVACY-PRESERVING SCHEME

A novel privacy-preserving scheme [20] is constructed on a T-S fuzzy trust theory, and network coding data streams that are routed in optimal clusters formulated more quickly under a kind of camouflage attack by a designed repeated game model to defend against pollution attacks in coding IoT networks.

- Network coding is considered when an IoT device sends a batch of sequenced messages to multiple target nodes.
- With the help of trust evaluation technology, the degree of trustworthiness between IoT devices is evaluated under the premise of ensuring the required stability of IoTs. It is used to defend against malicious nodes in IoTs that create pollution attacks. The trust between IoT devices generated in the previous round of data transmission is considered to set the trusted routing device in the next round of data transmission.
- The network controller uses the game model to create the best cluster for sending data to the trusted privacy-preserving method. The game model is able to strike a compromise between resource usage and network performance, ensuring maximum network performance while using fewer resources. Here, network performance metrics cover things like energy usage and defense assault capabilities.

The discrete logarithm's difficulty determines how secure this system is. Under the condition of data confidentiality, this created repeated game model features a subgame-perfect Nash equilibrium and can increase resource utilization effectiveness. This plan is more time- and energy-efficient.

10.10   AUTHENTICATION SCHEMES OVER A MULTI-CRYPTOSYSTEM

Industries are primarily concerned with the integrity and authenticity of the data created and sent by IIoT devices since it is a network of highly connected heterogeneous devices. The IIoT network allows for the deployment of devices by various factories to gather a variety of data. As a result, these devices could employ different system settings. The devices' public and private keys, together with all of the system settings, were pre-stored in the devices, according to the authors. Devices will gather data after the deployment and send it to cloud data centers. Multiple cryptographic methods are constantly used for data authenticity due to the heterogeneity of IIoT. The data is sent through a legitimate device that has not been tampered with in any way, ensuring the message's validity and integrity.

Paper [21] constructed a novel authentication scheme over a multi-cryptosystem called ASMC (RSA-based, DL-based, ECC-based, and lattice-based cryptosystems) for the heterogeneous IIoT environment. The underlying ring signature mechanism in ASMC that satisfies the condition of privacy guarantees the security aim of unforgeability and anonymity.

10.11   DATA PRIVACY IN MAC

By adding a security field to the MAC header of IEEE 802.15.4 MAC standards, new Medium Access Control (MAC) and routing algorithms have been developed to guarantee end-to-end network efficiency. The ElGamal public key cryptosystem is advanced [22] for protecting IoT data by creating the best private key as well. This work proposes to develop a novel hybrid optimization method called the Cuckoo Mated-Lion method (CM-LA) that combines the concepts of the Lion Algorithm (LA) and Cuckoo Search Algorithm (CS) for this optimum key selection.

Through a security-enabled MAC frame structure, as shown in Figure 10.4, the user here transmits and receives messages over an IoT channel in a secure way. For this reason, security (private key xi) is enabled in the MAC layer payload, which includes the message to be sent or received. Additionally, in the MAC layer, the private key xi and the cipher text produced using the ElGamal cryptosystem's public key yi are sent as message (fmi1; mi2g) to the MAC payload. Here, the cipher text (ci1; ci2) is carried by the MAC payload, which also conducts encoding on the transmitter side and transmits encoded cipher data. The cipher text (ci1; ci2) transferred from the transmitter side is decoded and provided as decoded cipher data by the MAC frame on the receiver side. The decryption procedure then takes place using the optimal private key xi to attain the original data (mi1; mi2).

10.12   TRUSTED EXECUTION ENVIRONMENTS AND MACHINE LEARNING

The research work [23] enhances security and privacy in IoT-based systems by isolating hardware peripheral drivers. It makes use of Open Portable Trusted Execution Environment (OP-TEE) technology to protect hardware peripheral device driver

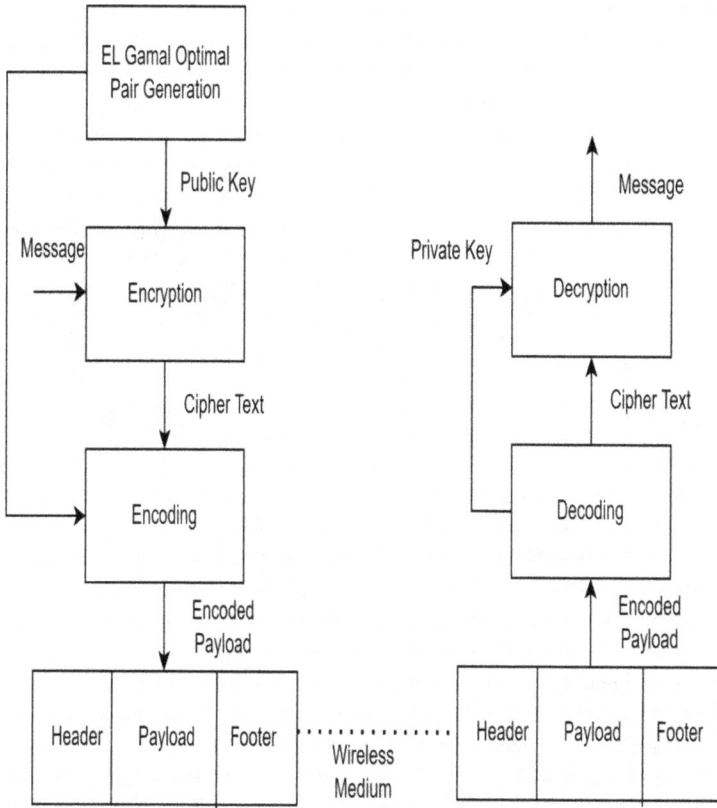

FIGURE 10.4   Process at MAC layer to ensure data privacy.

software using Arm TrustZone and ML techniques to stop unintentional disclosure of critical peripheral data to an unreliable cloud provider or OS.

10.13   CANELLA: PRIVACY-AWARE END-TO-END INTEGRATED IoT DEVELOPMENT

Canella [24] is an entire IoT development environment enhanced with cutting-edge privacy-preserving components for creating IoT apps that are conscious of privacy. It aids in supporting developers in understanding the behavior of their code, overcoming privacy difficulties and complying with privacy and data protection legislation, reducing the amount of time required to implement privacy into an IoT application, and reducing the cognitive burden associated with doing so.

10.14   FUTURE RESEARCH DIRECTIONS

The security of personal data acquired by personal IoT devices on current networks depends on developers and individuals.

- Many personal IoT devices lack cryptographic functionality due to power restrictions and the desire to preserve energy. It is encouraged to research whether encryption at the physical layer is viable given advancements in batteries and microprocessors [7].
- Mitigating the security risk through contemporary technology in low-powered personal IoT devices is limited. Research must be done on the aspect of how contemporary technology can be incorporated [7].
- Low-power IoT devices struggle to maintain the speed of transactions of 5G/6G which restricts the adoption of blockchain technology [7].
- At the moment, individual confidentiality of data coming from personal IoT devices is not protected by the existing standards. The security and privacy of personal IoT devices depend on standardized IoT standards that must be developed [7].
- At present, consumer IoT is widely used and it generates tonnes of fine-grained, comprehensive data about preferences, personalities, and consumers' daily behaviors. Consumers are unable to manage the gathering and processing of their sensitive and personal data due to a lack of openness and support. In order to improve the privacy issues that arise from IoT, the following need to be solved [25]:
  - the technical expertise needed to comprehend privacy notifications;
  - the lack of transparency and control over personal data; and
  - the absence of specialized privacy guidance for everyone.
- Organizations and academics want a method of comprehending how privacy issues affect IoT and customer desire to utilize IoT. Paper [26] framed a scientific framework for analyzing privacy issues in the IoT space. As a result of this effort, the use of mobile users' information privacy concerns (MUPIC) in the IoT field opens the door for more research.
- The research opportunities based on the criteria notice and discovery mechanisms; risk inference; user engagement and incentives; privacy preferences and privacy policies management; consent management; enforcement points and compliance; and real scenario deployment and assessment are identified in [27].
- The article in [28] pointed out the problems and difficulties of IoT privacy solutions based on machine learning (ML) and deep learning (DL). Reducing the latency and increasing the throughput of NN training on encrypted data is a big challenge.
- Current systems use DL without comprising user data, which makes the solution computationally efficient. The efficiency can be improved further with the help of quantum computing techniques and parallel learning to make the solution competitive with effective cost optimization [28].
- Several issues [28] are still open, such as network pruning and the inter-play between different malicious activities. The balance between real IoT applications' quality of service and privacy protection is still a challenge. The evaluation and assessment of privacy solutions in practical contexts remains an ongoing topic.

- While most research has focused on 6G and the development of data communication, the full extent of privacy and security implications still need to be explored [7].

10.15   SUMMARY

In this era of connectivity, IoT is an essential aspect of our daily life. IoT is key to increasing efficiency and convenience. While the number and sophistication of these devices increase, security and reliability in the context of IoT devices become more and more critical. Privacy is one of the core issues with IoT. With these devices gathering incredible amounts of personal information, it is vital to implement privacy-aware designs that maintain individual anonymity while not hindering operation.

Strong encryption methods and anonymization protocols allow people to have better authority over their privacy, even maintaining hassle-free device interaction. Strong security tools are key to keeping out unauthorized users and protecting the system against attacks within an IoT network. Adding robust authentication methods like multifactor authentication and biometric authentication can boost the general security posture. Also, regular monitoring and timely patch-up of vulnerabilities are extremely important in terms of risk reduction. Trust is another significant element that needs to be dealt with in an IoT domain. It is essential to use an advanced level of crypto-algorithm or technology such as digital signature or blockchain for validating the data passed between devices and ensuring integrity and credibility. These solutions both protect against alteration of the data and can support auditing transparently in order to meet regulatory requirements. In addition to this, implementing specific guidelines and rules for the manufacturing of IoT devices will create an overall safer atmosphere. Implementing security by design promotes adherence to industry standards at the design stage, which can prevent vulnerabilities before they occur.

REFERENCES

[1] W. Detres, M. M. Chowdhury and N. Rifat, "IoT Security and Privacy," *2022 IEEE International Conference on Electro Information Technology (eIT)*, Mankato, MN, USA, 2022, pp. 498–503, doi: 10.1109/eIT53891.2022.9813933

[2] T. Kavitha and G. Senbagavalli, "The Future of Travel in public Bus Service: How a Mobile Bus Ticketing System Is Revolutionizing the Public Travel," *2023 International Conference on Applied Intelligence and Sustainable Computing (ICAISC)*, Dharwad, India, 2023, pp. 1–7, doi: 10.1109/ICAISC58445.2023.10200016

[3] T. Kavitha, S. Kumari, S. A. Kamble, D. N. Rachana and C. K. DhruvaKuma, "Design of IoT Based Smart Coin Classifier Using OpenCV and Arduino," *2022 IEEE 2nd International Conference on Mobile Networks and Wireless Communications (ICMNWC)*, Tumkur, Karnataka, India, 2022, pp. 1–5, doi: 10.1109/ICMNWC56175.2022.10031997. E-ISBN: 978-1-6654-9111-2.

[4] R. Veluvarthi, A. Rameswarapu, K. V. Sai Kalyan, J. Piri and B. Acharya, "Security and Privacy Threats of IoT Devices: A & Short Review," *2023 4th International Conference on Signal Processing and Communication (ICSPC)*, Coimbatore, India, 2023, pp. 32–37, doi: 10.1109/ICSPC57692.2023.10125863

[5] N. Chaurasia and P. Kumar, "A Comprehensive Study on Issues and Challenges Related to Privacy and Security in IoT," *e-Prime - Advances in Electrical Engineering, Electronics and Energy*, vol. 4, 100158, 2023, ISSN: 2772-6711, doi: 10.1016/j.prime.2023.100158

[6] S. K. A. Yaklaf, A. S. Elmezughi, S. M. H. Naas and N. B. Ekreem, "Privacy, Security, Trust and Applications in Internet of Things," *2023 IEEE International Conference on Advanced Systems and Emergent Technologies (IC_ASET)*, Hammamet, Tunisia, 2023, pp. 01–06, doi: 10.1109/IC_ASET58101.2023.10150619

[7] J. Cook, S. U. Rehman and M. A. Khan, "Security and Privacy for Low Power IoT Devices on 5G and Beyond Networks: Challenges and Future Directions," *IEEE Access*, vol. 11, pp. 39295–39317, 2023, doi: 10.1109/ACCESS.2023.3268064

[8] J. Sen and S. Dasgupta, "Data Privacy Preservation on the Internet of Things," *Information Security and Privacy in the Digital World*, ISBN: 978-1-83768-196-9, June 2023.

[9] S. B. Sadkhan and Z. Salam, "Security and Privacy in Internet of Things - Status, Challenges," *2021 4th International Iraqi Conference on Engineering Technology and Their Applications (IICETA)*, Najaf, Iraq, 2021, pp. 308–312, doi: 10.1109/IICETA51758.2021.9717785

[10] E. Rodriguez, B. Otero and R. Canal, "A Survey of Machine and Deep Learning Methods for Privacy Protection in the Internet of Things," *Sensors*, 23, pp. 1–24, 2023.

[11] J. E. Rivadeneira and J. Silva, R. Colomo-Palacios, A. Rodrigues and F. Boavida, "User-Centric Privacy Preserving Models for a New Era of the Internet of Things", *Journal of Network and Computer Applications*, vol. 217, 103695, 2023, ISSN: 1084-8045, doi: 10.1016/j.jnca.2023.103695

[12] Y.-T. Tsou, "The Next Big Thing: IoT Applications with Data Privacy-Enhancing Technologies," *2023 International VLSI Symposium on Technology, Systems and Applications (VLSI-TSA/VLSI-DAT)*, HsinChu, Taiwan, 2023, pp. 1–1, doi: 10.1109/VLSI-TSA/VLSI-DAT57221.2023.10134305

[13] R. Kumar, P. Kumar, A. Jolfaei and A. K. M. N. Islam, "An Integrated Framework for Enhancing Security and Privacy in IoT-Based Business Intelligence Applications," *2023 IEEE International Conference on Consumer Electronics (ICCE)*, Las Vegas, NV, USA, 2023, pp. 01–06, doi: 10.1109/ICCE56470.2023.10043450

[14] M. Kumar, P. Mukherjee, S. Verma, et al., "A Smart Privacy Preserving Framework for Industrial IoT Using Hybrid Meta-Heuristic Algorithm," *Scientific Reports*, vol. 13, 5372, 2023, doi: 10.1038/s41598-023-32098-2

[15] A. M. A. Alamer, S. A. M. Basudan and P. C. K. Hung, "A Privacy-Preserving Scheme to Support the Detection of Multiple Similar Request-Real-Time Services in IoT Application Systems," *Expert Systems with Applications*, vol. 214, 2023, 119005, ISSN 0957-4174, doi: 10.1016/j.eswa.2022.119005

[16] A. Chaudhari and R. Bansode, "A Privacy Preservation Strategy Using Hybrid Fully Homomorphic Encryption Scheme in IoT," *International Journal of Cooperative Information Systems*, vol. 32, no. 03, 2350007, 2023, doi: 10.1142/S0218843023500077

[17] M. M. Jaber, et al., "PPDA-FAF: Maintaining Data Security and Privacy in Green IoT-Based Agriculture", *International Journal of Cooperative Information Systems*, 2250007, 2022, doi: 10.1142/S0218843022500071

[18] A. K. Budati, S. R. Vulapula, S. B. H. Shah, A. Al-Tirawi and A. Carie, "Secure Multi-Level Privacy-Protection Scheme for Securing Private Data over 5G-Enabled Hybrid Cloud IoT Networks," *Electronics*, vol. 12, 1638, 2023, doi: 10.3390/electronics12071638

[19] I. Appiah, X. Jiang, E. Boahen and E. Owusu, "A 5G Perspective of an SDN-Based Privacy-Preserving Scheme for IoT Networks," *International Journal of Communications, Network and System Sciences*, vol. 16, pp. 169–190, 2023, doi: 10.4236/ijcns.2023.168012

[20] L. Cao and M. Zhu, "Fuzzy-Based Privacy-Preserving Scheme of Low Consumption and High Effectiveness for IoTs: A Repeated Game Model," *Sensors*, vol. 22, 5674, 2022, doi: doi.org/10.3390/s22155674

[21] Z. Tan, J. Jiao and M. Yu, "A Privacy Preserving Authentication Scheme for Heterogeneous Industrial Internet of Things", *Security and Communication Networks*, vol. 2022, Article ID 9919089, 15 pages, 2022, doi: https://doi.org/10.1155/2022/9919089

[22] G. Kalyani and S. Chaudhari, "Data Privacy Preservation in MAC Aware Internet of Things with Optimized Key Generation," *Journal of King Saud University - Computer and Information Sciences*, vol. 34, no. 5, pp. 2062–2071, 2022, doi: 10.1016/j.jksuci.2019.12.008

[23] P. Yuhala, "Enhancing IoT Security and Privacy with Trusted Execution Environments and Machine Learning," *2023 53rd Annual IEEE/IFIP International Conference on Dependable Systems and Networks - Supplemental Volume (DSN-S)*, Porto, Portugal, 2023, pp. 176–178, doi: 10.1109/DSN-S58398.2023.00047

[24] A. Aljeraisy, O. Rana and C. Perera, "Canella: Privacy-Aware End-to-End Integrated IoT Development Ecosystem," *2023 IEEE International Conference on Pervasive Computing and Communications Workshops and other Affiliated Events (PerCom Workshops)*, Atlanta, GA, USA, 2023, pp. 279–281, doi: 10.1109/PerComWorkshops56833.2023.10150254

[25] S. Tokas, G. Erdogan, K. Stølen, *"Privacy-Aware IoT: State-of-the-Art and Challenges,"* Proceedings of the 9th International Conference on Information Systems Security and Privacy - Volume 1, ICISSP, SciTePress, INSTICC, 2023, pp. 450–461, doi: 10.5220/0011656400003405

[26] P. M. Chanal and M. S. Kakkasageri, "Security and Privacy in IoT: A Survey," *Wireless Personal Communications*, vol. 115, pp. 1667–1693, 2020.

[27] J. E. Rivadeneira, J. Silva, R. Colomo-Palacios, A. Rodrigues and F. Boavida, "User-Centric Privacy Preserving Models for a New Era of the Internet of Things," *Journal of Network and Computer Applications*, vol. 217, 103695, 2023, ISSN: 1084-8045, doi: 10.1016/j.jnca.2023.103695

[28] E. Rodriguez, B. Otero and R. Canal, "A Survey of Machine and Deep Learning Methods for Privacy Protection in the Internet of Things," *Sensors*, 23, 2023.

11 Trust Establishment Models in IoT

V. Rajeswari and K. Sakthipriya
Karpagam College of Engineering, Anna University,
Chennai, India

11.1 INTRODUCTION

The "Internet of Things", IoT for short, is a technology that is revolutionising the world by leaps and bounds. The IoT is a combined space for technologies that encompass objects that are connected through the internet to perform real-time tasks controlled through appropriate hardware/software combination. The recent developments in the fields of embedded systems, communication technologies, robust software systems, mini- and micromechanical systems, and manufacturing technologies have made IoT employed across various use cases.

Many approaches have been developed and are being considered for streamlining the trust establishment process among IoT devices. The techniques can be grouped in multiple ways based on the communication point of interest in the network, viz. Human to Thing, Thing to Thing, or Thing to Things, has been suggested by various researchers. The approach to building a viable and secure trust establishment model should consider the network resources and the reliability, consistency, and speed of the transaction to arrive at a universally acceptable model. In evaluating the trust establishment models, various factors like the research already carried out, the developments that have taken place in communication technologies, and the explosive growth in computing power need to be accounted for. Another important aspect to consider in a model for adaptation will be the trade-off between the complexity of the model and the security requirement for the application. This also becomes a dominating factor since IoT devices are becoming a part of even very low-cost household appliances. The dominance of IoT, AI, and ML in the integration of engineering, industrial, and business processes brings out more and more challenges in establishing trust among IoT devices and will be the most researched area in the near future. These issues need to be understood in detail to propose or adopt a trust establishment model for an IoT application.

In the IoT domain, various objects like sensors, electro-mechanical actuators, card readers, PoS systems, and RFID tags are interconnected. These may be part of a utility system or stand-alone objects. For example, a refrigerator may contain sensors as well as RFID tags to measure parameters as well as to connect to an IoT network. Further, the network may be enabled through different technologies like Bluetooth, 3G, Wi-Fi, or any proprietary network technology. The path for communicating with these objects to hosts or controllers is through the internet and hence the complexity in ensuring data security and guaranteeing quality of service.

DOI: 10.1201/9781003477327-11

The IoT technology provides an enormous possibility of services that can be offered to the public in various segments like health care, automated customer care, autonomous home security, transport services, monitoring agricultural and environmental issues, and automated toll-gates. One can understand the difficulty involved in ensuring security considering the heterogeneous nature and high speed of execution required in such applications. Unless a trustful ecosystem is ensured, these applications cannot deliver their intended outcomes. Trust establishment at various levels is to be seamlessly integrated and executed at a fast pace since these tasks are handled in real-time.

11.2 SECURITY REQUIREMENTS FOR IoT

Security requirements for IoT are unique in the sense that they are based on the heterogeneity of devices connected in the IoT domain. Apart from the variety of devices, the requirement of connectivity everywhere or ubiquity is another concern that impacts the security parameters. In addition, the limitation of resources in the IoT devices, the autonomy of such compact devices and their inherent nature to be moving around also necessitate certain security issues to be primarily addressed. After considering all such factors, it is also important to see that the system is scalable since the domain keeps continuously evolving.

It has been suggested by various researchers and IoT system developers to split the end-to-end security system in the IoT domain into various components of the security-ensuring mechanism. The common approach to IoT security is formed by the following components as suggested by (Al-Fuqaha et al., 2015; Airehrour, Gutierrez, & Ray, 2016) and others.

Authentication: Authentication is the process by which the elements or devices connected in an IoT ecosystem are identified. This identification is the first step that happens in any secure system. It is kind of like getting the answer to a "who are you" type question from the device that wants to be connected to the IoT system. It is possible that a breach of security can happen here through an identity theft and a mechanism is to be provided to detect such impersonation.

Access Control: Access control mechanisms ensure that only authorised entities have access to the resources of the respective IoT ecosystem. In other words, unauthorized entities are barred from entering the network and illegally using the resources. The resource may be highly sensitive data from an IoT device or a precious network connection and routing and when accessed by an unauthorised entity, can cause irreparable damage to the ecosystem. An access control or authorisation mechanism can be a single-level or a multi-level process depending on the security requirement of the system.

Availability: An IoT network ecosystem is expected to be available at all times since it may be catering to entities and devices that can form a critical part as in a medical emergency system or a fire safety system. It is possible that a pointed attack on such a network can incapacitate the same and render it inaccessible to devices that need to communicate in an emergency.

An obvious solution will be to provide redundancy which is usually the method adopted. But considering resource constraints inherent in the IoT domain, other alternatives can be devised as well and implemented where required.

Confidentiality: In the IoT domain, when most of the actions take place in real-time, it is essential that utmost secrecy is maintained in the end-to-end communication of data between the authorised nodes/entities of the system concerned. Though encryption can be a solution for the same, it is also essential that unauthorised elements do not hijack the data or other network resources. In addition, if confidentiality is not ensured, the entire ecosystem is exposed to various attacks and threats that exist in cyberspace. In the case of a military or defence-level IoT system, such loss of confidentiality may lead to even a war-like situation. In the commercial domain, the customer base will quickly erode if confidentiality is lost. Data theft, unauthorised data usage, impersonation, and other cyber-frauds arise because of poor design of confidentiality in any online system.

Integrity: In any networked system, the integrity of data communication among nodes is a critical factor. It is given utmost importance since an intruder can manipulate the data and cause disastrous consequences through the inadvertent usage of manipulated data. The system should be capable of detecting if the data has been subjected to tampering or routed through malicious channels. The very purpose of maintaining the integrity intact is to ensure that the data sent reaches the destination as is, through the specified route only. It is also essential to detect attempted capture and manipulation of data en route to further improve upon the integrity check protocols employed.

Privacy: Privacy ensures hiding the communication between two entities of the IoT system from the other unrelated devices/entities connected. It is done to keep the data exchange between the sender and receiver, a strictly private process. This prevents data from being exposed to malicious attacks by mischievous, hostile entities. It also prevents sensitive data from being read by entities that do not form part of the IoT ecosystem concerned. The implementation can be done through simple encryption techniques or other complex algorithms depending on the IoT system under consideration.

Trust: Any application working in an IoT ecosystem will be useful only if the Trust element is high in that. Trust is something the user fundamentally expects in an IoT application environment. The total existence of Trust emanates from the combined effect of all the factors mentioned above in the security requirements. The IoT applications operate in a highly dynamic, ever-changing environment or ecosystem such as the nature of the devices, enabling resources like connecting networks and power constraints. Because of this, the possibilities of hostile and malicious attacks on such applications are very high. In order to mitigate this, it is essential that the entire ecosystem is built with a highly trustworthy set of the above-mentioned security requirements. For instance, privacy can be ensured by adopting techniques that ensure anonymity combined with methods that help localised anonymous tracking of data packets. Existing practices like message integrity codes can be used to achieve the required integrity. There are such well-established methods in use to take care of each of the above security requirements. Any trust management system will be overseeing the successful implementation of such modular checking methods to ensure the trustworthiness of the IoT application concerned. The trust management system should be able to catch any deviation from anywhere in the security requirements and isolate such instances before they could adversely affect the IoT ecosystem. It may also be noted that the disturbance in any part of the ecosystem should immediately be localised and isolated so that it does not lead to a crashing down of the entire IoT ecosystem. These important features that ensure trust in IoT applications can be integrated into an IoT trust framework.

11.3   IoT TRUST FRAMEWORK

In ensuring trust in the IoT domain, factors like what constitutes trust, is trust quantifiable, if so, what parameters are required to measure trust, and how to evaluate the trust when implemented are very important. In other words, defining trust, measuring trust and evaluating trust are to be clearly spelt out to design networks and devices that conform to these. The IoT Trust Framework was developed specifically for this.

The IoT Trust Framework forms the basis for establishing trust in the domain of IoT applications. The framework has been established by the Open Trust Alliance (OTA) which is an initiative of the Internet Society that formed the IoT Trust Working Group in January 2015. The IoT Trust Framework initially focused on the need to review the security, privacy, and sustainability of smart home and wearable electronic IoT devices and technologies. Later with the arrival of the concept of Industry 4.0, the framework expanded to cover industrial IoT-related security and trust issues. The OTA-IoT Trust Framework is designed to provide a set of strategic principles for securing the data that is generated, transmitted, and utilised in an IoT environment. The framework is the culmination of a collaborative effort of various state agencies and industrial partners in addition to end-user forums.

The IoT Trust Framework includes a set of strategic principles necessary to help secure IoT devices and their data when shipped and throughout their entire life cycle. Through a consensus-driven multi-stakeholder process, criteria have been identified for connected home, office, and wearable technologies including toys, activity trackers, and fitness devices.

The Framework addresses four key areas (https://www.internetsociety.org/iot/trust-framework/).

> **Security Principles**: Applicable to any device or sensor and all applications and back-end cloud services. These range from the application of a rigorous software development security process to adhering to data security principles for data stored and transmitted by the device, to supply chain management, penetration testing and vulnerability reporting programs. Further principles outline the requirement for life-cycle security patching.

User Access and Credentials: Requirement of encryption of all passwords and usernames, shipment of devices with unique passwords, implementation of generally accepted password reset processes, and integration of mechanisms to help prevent "brute force" login attempts.

Privacy, Disclosures, and Transparency: Requirements consistent with generally accepted privacy principles, including prominent disclosures on packaging, point of sale and/or posted online, capability for users to have the ability to reset devices to factory settings, and compliance with applicable regulatory requirements including the EU GDPR and children's privacy regulations. Also addresses disclosures on the impact on product features or functionality if connectivity is disabled.

Notifications and Related Best Practices: Key to maintaining device security is having mechanisms and processes to promptly notify a user of threats and action(s) required. Principles include requiring email authentication for security notifications and that messages must be communicated clearly for users of all reading levels. In addition, tamper-proof packaging and accessibility requirements are highlighted.

Since the framework is formed with the stakeholders as partners in the initiative, it takes into account the concerns of different IoT application providers like home automation, wearable electronics, and autonomous transportation (Yan, & Holtmanns, 2008). Various criteria have been developed to address the issues that can arise due to the heterogeneity of devices, data form and communication, and the software platforms used in each IoT application. The detailed information about the IoT Trust Framework is available on the OTA website. Anyone desirous of knowing about the basic structure of trust establishment and understanding the process of implementing it can access the URL https://otalliance.org/IoT.

11.4 CONCEPT AND DEFINITIONS OF TRUST

Trust can be defined as an entity that exists between a trustor and a trustee. For example, in an online transaction, a person buying goods from Amazon is a trustor and Amazon Inc. is the trustee. When this definition is extended to the IoT ecosystem a number of concerns and situations arise and lead to the requirement of a complex trust establishment system.

The IoT space is unique in the sense that it involves a heterogeneous mix of devices, software systems, hardware systems, network protocols, geographic boundaries, etc., apart from the variety of people using or interacting in the ecosystem (Djedjig et al., 2018). Ensuring security in such a situation where security can be compromised in any layer or link in the process becomes an enormous challenge. Trust being the key component in the overall execution, there is a need for properly formulating policies to mitigate the uncertainty and risk that prevail. When proper trust exists in the transaction, reliability, confidentiality, and privacy are assured to the maximum possible extent (Djedjig, Tandjaoui, & Medjek, 2015). This will enhance the user experience and IoT will find application in many areas.

The regular ways in which security and confidentiality are seen to happen are mostly by means of access control mechanisms. This happens through authentication and authorisation techniques. Cryptography enables secrecy to the transaction and adds additional dimensions like confidentiality to the process. These techniques will prove to be insufficient in the IoT domain where the sheer variety of devices and the vast nature of data types through sensing elements makes the task difficult. Most of the IoT systems are embedded with purpose-driven low-end microcontrollers with limited computing power. The role of a trust establishment in the system from end to end will provide a viable solution. The formation of a trust between the trustor and the trustee through the various elements of identification, access control, and authorisation in addition to the security protocol in the network will be the basis for the system (Djedjig, Tandjaoui, Medjek, & Romdhani, 2017).

Researchers have worked extensively on the concept of trust formation and have suggested defining trust according to different perspectives. The different definitions that have been proposed are mentioned here to understand the concept. Most of these definitions revolve around things like faith, risk-taking, past experience, and expectations of the trustor in reference to the trustee. Corritore, Kracher, and Wiedenbeck (2003, p. 740) defined trust as "an attitude of confident expectation in an online situation of risk that one's vulnerabilities will not be exploited". Chang, Dillon, and Hussain (2005) defined trust as the belief that the trustee has in the trustee's capability to deliver a quality of service in a given context and in a given time slot. As per Buttyan and Hubaux (2007), trust is about the ability to predict the behaviour of another Party. Aljazzaf, Perry, and Capretz (2010, p. 168) defined trust as "Trust is the willingness of the trustor to rely on a trustee to do what is promised in a given context, irrespective of the ability to monitor or control the trustee, and even though negative consequences may occur". Daubert, Wiesmaier, and Kikiras (2015) defined trust in the context of IoT for different components of the IoT ecosystem. For the device, it is the trusted computing and computational trust, for the entity it refers to the expected behaviour of participants such as persons or services, and for data trust, it may be derived from untrusted sources by aggregation or may be created from IoT services.

According to researchers, the different definitions pave the way for developing trust evaluation models that are used for ensuring the trustworthiness of an IoT application and lifting up the user's confidence (Djedjig, Romdhani, Tandjaoui, & Medjek, 2017). Also, the definitions provide scope for incorporating and innovating mechanisms through which the trust can be improved and expanded to take care of future applications. It is also possible that new definitions will evolve in future based on user experience and developments in technology.

11.5   IoT TRUST EVALUATION MODELS

In ensuring trust in the IoT domain, there is a need to decide if a particular transaction is trustable or not based on certain considerations (Yan, Zhang, & Vasilakos, 2014). The context in which the transaction takes place, the layer at which it happens, and the security already in place for the transaction are all playing a part in the process. If we consider different elements or devices that participate in an IoT transaction, then according to Airehrour, Gutierrez, and Ray, a model that can estimate the reliability level between the devices can be used as a guideline to carry out the intended transaction. The model can locate issues of concern and address the areas where a low value of trust can occur. This low value of trust being the weak link in the chain can degrade the operational efficiency and usability. This model helps to make an informed decision through trust management, which according to Wang et al., is a service mechanism that self-organizes a set of items based on their trust status to take an informed decision.

Due to the presence of different contexts, different properties, different purposes and different elements in the IoT domain, trust models are also formed accordingly. This results in different trust models being proposed by researchers. The models basically involve processes of extracting, evaluating, and transmitting trust information and a mechanism used to make the final decision. Some of the trust models are discussed in this context, elucidating their structure and schemes for trust determination.

The trust model Type A, proposed by Airehrour et al. as given in Figure 11.1 is based on analytical techniques used to evaluate a trust score. Analytical techniques like Game theory, fuzzy logic methods, probability, neural network methods, swarm intelligence, and graph theory are used to analyse the trustworthiness of the elements of the IoT application. This may include the device, hardware, network protocol, communication medium, and control software. The model is expected to provide a secure, trusted routing for the IoT application by applying the analytical processes that are given in the trust model. The authors used the corresponding analytical algorithm for a particular model and grouped them as fuzzy logic trust model, neural network trust model, swarm intelligence trust model, etc. This mode of model classification is solely based on which kind of analytical algorithm is used to verify the trustworthiness of the system.

The trust model Type B, as proposed by Nunoo-Mensah, Boateng, and Gadze, suggests classifying the methods into three categories, viz. bio-inspired, socio-inspired, and analytical. This idea of classifying the methods as bio-inspired and socio-inspired in addition to Type A classification is based on the fact that new computational algorithms have been developed to resemble biological and social behaviours. Some such algorithms are evolutionary algorithms, ant colony–based algorithms, artificial intelligence, machine learning, and algorithms used in social networks. Figure 11.2 illustrates how this trust model is organised.

In the Type C model, Figure 11.3, as proposed by Moyano et al., the trust evaluation is considered as two process classes. The first class takes into consideration the steps involved in decision making and the second class is about evaluating the same. These classes have been named the trust decision model and the trust evaluation model.

The trust decision model has policy models and negotiation models as their subdivisions. The policy model as the name implies, is about framing policies that give access to resources for the IoT application at different levels when the requests adhere to said policies. This gives a measure of how safe the transaction can be. The policies are formed as per the context of the application and the system resources available.

The negotiation model ensures hand-shaking between any two entities in the IoT application chain and the exchange of relevant data as per the policy. This exchange of data in the process checks the credentials of the elements that are negotiating and

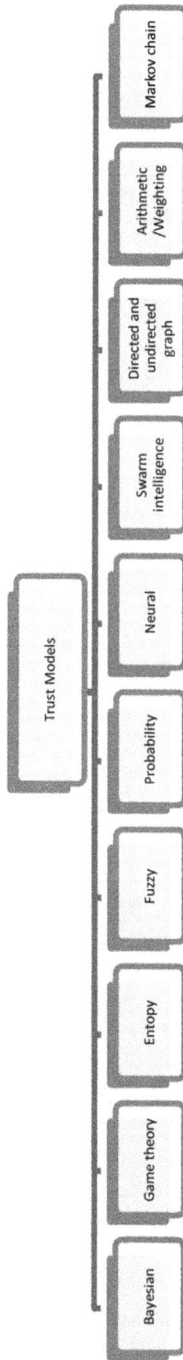

FIGURE 11.1 Trust model Type A.

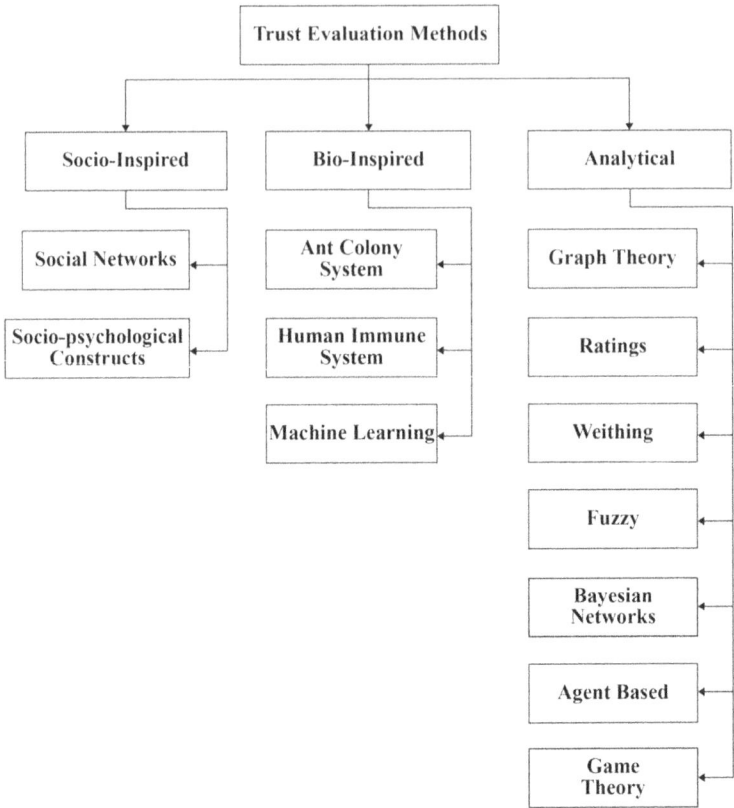

FIGURE 11.2   Trust model Type B.

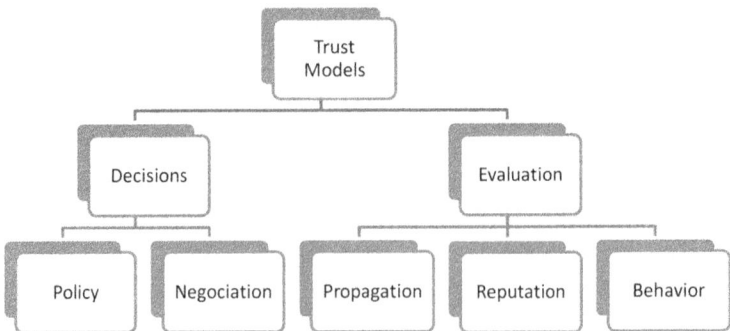

FIGURE 11.3   Trust model Type C.

verifies the adherence to the policy. At the end of the negotiation phase, the decision to make a trusted connection or not is made and when the decision to proceed with the trusted connection, the next phase of evaluating the trustworthiness is invoked.

The trust evaluation models are meant to quantify the trust so that it can be either accepted above a threshold level or rejected if below. For calculating and quantifying the trust value, attributes such as reliability, honesty, and integrity of the entities that connect are considered.

The evaluation model uses three sub-models, viz. behaviour model, propagation model, and reputation model, to accomplish the determination of trust value. The behaviour model is based on trust values assigned between a trustor and trustee as per their relationship based on metrics according to the context. In the propagation model, an element or entity in the connection disseminates the trust information to the other elements or entities so that all the elements are provided the trust information uniformly. In the reputation model, each element in the connection exchanges the trust information it has with the other element and then collaboratively works to evaluate the trustworthiness.

In addition to the three models A, B, and C there are other models proposed by other researchers as well. Guo, Chen, and Tsai (2017) have proposed models that are more elaborately organised or sub-models or derivatives of the previously listed models. The models of interest are the composition model with sub-classifications: the Quality of Service model and the Social model, the Aggregation model which is an extended version of the reputation model, the Update model which is designed to be updated from time to time, and the Formation model which can be either a single-trust model or a multi-trust model. The type D model shown in Figure 11.4 represents the classification of trust models as proposed by Chen, R., & Guo, J. (2014).

11.6 SUMMARY

In this chapter, various aspects of trust establishment related to an ecosystem that is specifically suitable for the IoT environment are presented. The IoT systems are radically different from other conventional IT systems. The difference is caused due to the fact that it consists of a heterogeneous mix of devices, networks, communication protocols, and control software. Issues like security, privacy, availability, and confidentiality, which are of paramount importance for IoT applications, have been highlighted. The concept of Trust in IoT applications and how it can be modelled, evaluated, and ensured is presented through the related works of eminent researchers in this field. Different evaluation models proposed by researchers are provided with details of the building blocks in them. It can be seen that there is a variety of possibilities to develop models based on how TRUST is perceived for a given IoT application. It is also to be seen how optimally a specific model can be implemented in an IoT ecosystem by considering the resource constraints. It is obvious that trust establishment and management are extremely important in the future as more and more tasks will be performed through autonomous IoT systems. The rapid developments in computing capabilities, sensor technologies, and ultra-high-speed communication networks will lead to a world full of IoT ecosystems that is trustworthy and highly reliable.

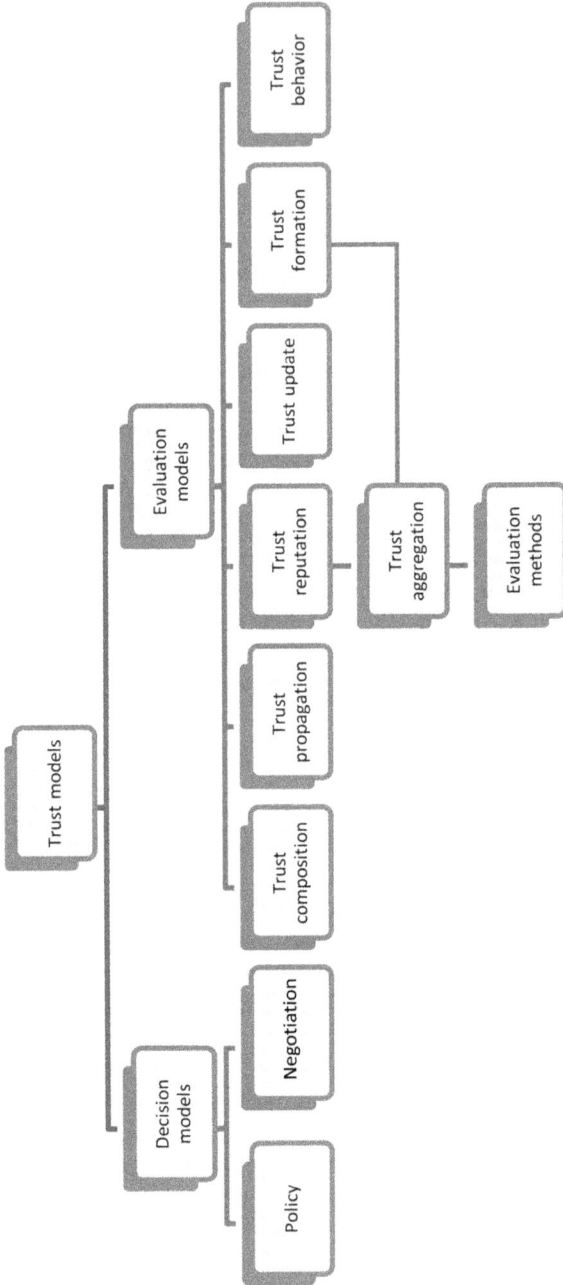

FIGURE 11.4 Trust model Type D.

REFERENCES

Airehrour, D., Gutierrez, J., & Ray, S. K. (2016). Secure routing for internet of things: A survey. *Journal of Network and Computer Applications, 66*, 198–213.

Al-Fuqaha, A., Guizani, M., Mohammadi, M., Aledhari, M., & Ayyash, M. (2015). Internet of Things: A survey on enabling technologies, protocols, and applications. *IEEE Communications Surveys & Tutorials, 17*(4), 2347–2376.

Aljazzaf, Z. M., Perry, M., & Capretz, M. A. (2010, September). Online trust: Definition and principles. In *Computing in the Global Information Technology (ICCGI), 2010 Fifth International Multi-Conference on* (pp. 163–168). IEEE.

Buttyan, L., & Hubaux, J. P. (2007). *Security and Cooperation in Wireless Networks: Thwarting Malicious and Selfish Behavior in the Age of Ubiquitous Computing.* Cambridge University Press.

Chang, E., Dillon, T. S., & Hussain, F. K. (2005, July). Trust and reputation relationships in service-oriented environments. In *Information Technology and Applications, 2005. ICITA 2005. Third International Conference on* (Vol. 1, pp. 4–14). IEEE.

Chen, R., & Guo, J. (2014, May). Dynamic hierarchical trust management of mobile groups and its application to misbehaving node detection. In *Advanced Information Networking and Applications (AINA), 2014 IEEE 28th International Conference on* (pp. 49–56). IEEE.

Corritore, C. L., Kracher, B., & Wiedenbeck, S. (2003). On-line trust: concepts, evolving themes, a model. *International Journal of Human-Computer Studies, 58*(6), 737–758.

Daubert, J., Wiesmaier, A., & Kikiras, P. (2015, June). A view on privacy & trust in IoT. In *2015 IEEE International Conference on Communication Workshop (ICCW)* (pp. 2665–2670). IEEE.

Djedjig, N., Romdhani, I., Tandjaoui, D., & Medjek, F. (2017). Trust-based defence model against MAC unfairness attacks for IoT. *ICWMC, 2017*, 127.

Djedjig, N., Tandjaoui, D., & Medjek, F. (2015, July). Trust-based RPL for the Internet of Things. In *Computers and Communication (ISCC), 2015 IEEE Symposium on* (pp. 962–967). IEEE.

Djedjig, N., Tandjaoui, D., Medjek, F., & Romdhani, I. (2017, April). New trust metric for the RPL routing protocol. In *Information and Communication Systems (ICICS), 2017 8th International Conference on* (pp. 328–335). IEEE.

Djedjig, N., Tandjaoui, D., Romdhani, I., & Medjek, F. (2018). Trust management in Internet of Things, doi: 10.4018/978-1-5225-5736-4.ch007

Guo, J., Chen, R., & Tsai, J. J. (2017). A survey of trust computation models for service management in internet of things systems. *Computer Communications, 97*, 1–14.

Website. (n.d.). URL: https://www.internetsociety.org/iot/trust-framework/

Yan, Z., & Holtmanns, S. (2008). Trust modeling and management: From social trust to digital trust. *IGI Global*, 290–323.

Yan, Z., Zhang, P., & Vasilakos, A. V. (2014). A survey on trust management for the Internet of Things. *Journal of network and computer applications, 42*, 120–134.

12 Secure Communication Protocols for the IoT

Prathibha Kiran and S. Shilpa
AMC Engineering College, Visvesvaraya Technological
University, Bengaluru, India

12.1 INTRODUCTION TO IoT COMMUNICATION PROTOCOLS

In the last 30 years, the internet has undergone significant development and adoption for effective communication. Its use as a means of connecting various devices has resulted in the rise of the "Internet of Things (IoT)" technology, which facilitates machine-to-machine (M2M) connectivity. These connections incorporate sensors, RFID, Wi-Fi, data networks, activators, LTE, WLAN, and other devices that process and exchange information seamlessly and without human intervention, thereby improving efficiency and accuracy in computer networking [1].

Given the diversity of devices and protocols in the IoT ecosystem, selecting the right protocol is a challenging task. It gives a clear understanding of the necessity of the IoT system messages and choosing a standard and efficient data format that meets these requirements. It is essential to select a protocol that ensures data security and integrity, especially for financial transactions, file transfers, email, and other sensitive communications.

Ensuring the utilization of protected communication protocols is essential in order to safeguard data from the risks of theft and fraud, which can be detrimental to businesses. Hence, it is important to have a basic understanding of secure internet protocols (IPs) to protect small business data. These protocols should be selected and implemented with care, ensuring that they match the specific needs of the organization.

- To manage devices and allow the interchange of data over the network.
- They are at the network's link, network, transport, and Application layers.
- The data exchange frames, data encoding and addressing schemes for devices, and routing of packets from the source to the destination are defined by these protocols.
- Other control operations such as sequence control, flow control, and retransmission of lost packets to be managed.
- Initiate network connectivity and coupling to applications of various networking nodes.

DOI: 10.1201/9781003477327-12

12.2   EXPLORING MESSAGING PROTOCOLS FOR THE IoT

The collection of data and exchange of messages are essential in building an IoT system, which necessitates the utilization of IoT devices. There are various factors to consider in selecting a messaging protocol or communication for device interconnectivity. To begin with, it's important to evaluate the hardware features of IoT devices as well as the protocols used in the data link layer. Additionally, IoT devices encompass a wide range of bandwidth capabilities. Significant time delays may occur if application layer protocols are utilized to capture data that exceeds the physical data rates. Therefore, effectively handling the physical data rates at the Data Link layer is crucial when considering messaging protocols [2].

Lightweight Communication Protocol would benefit smart devices that generate data at high velocity. The efficiency of IoT devices and their applications are significantly affected by the Messaging Protocol employed. By utilizing an appropriate messaging protocol, one can effectively reduce latency and network traffic, thus enhancing the consistency of an IoT application. For heterogeneous IoT environments, there is no universal protocol. Choosing the right messaging protocol for heterogeneous IoT environments includes multiple factors they are the specific needs of the IoT application, software limitations, device capabilities, and the typical size of data transmissions [3].

When creating an IoT application and deploying smart devices, it is important to take into account the key characteristics and functionalities of available protocols. This analysis focuses on shedding light on the widely used middleware protocols or messaging protocols for building IoT systems. Due to the extensive range of protocols that changes in their objectives, structure, capabilities, and limitations, it becomes challenging to create a link between these protocols and those that interface at the application level. Hence, comprehending only the data link layer protocols is inadequate when it comes to creating IoT applications. Comprehending communication Protocols at the data link layer is also important when considering protocols at the application level.

It is essential to have knowledge about the data link protocols that interact indirectly with those at the application layer through the OSI model. To make an optimal choice of data link layer protocols for IoT devices, it is important to consider system requirements like Quality of Service (QoS), bandwidth, interoperability, compatibility, and security. The selection should be derived from the specific application's needs and objectives, ensuring that the chosen protocol aligns closely with the requirements of the application.

12.3   A REALISTIC COMPARISON MESSAGING PROTOCOLS FOR INTERNET OF THINGS SYSTEMS

There exists a wide range of smart systems available, all equipped with tools, suitable components, and software libraries that adhere to communication standards. These resources efficiently support diverse applications and effectively operate in different environments. In an increasingly competitive market and with constant advancements in software and hardware technologies, designers encounter the challenge of navigating through a multitude of choices to select the optimal components. This chapter will explore the available Messaging Protocols for designers of IoT systems, which aim to streamline communication between application processes and separate them from transport and network protocols. A few messaging protocols offer specific functionalities, such as middleware services or QoS information tagging, which enhance capabilities for storing and directing useful information.

Numerous Messaging Protocols are available as open-source programming libraries and are incorporated into IoT platforms that provide pre-built solutions encompassing a large portion of an end-to-end IoT system. Choosing the suitable messaging protocol for a project can be a complex task due to the plethora of options available. They include XMPP, CoAP, MQTT, STOMP, DDS, AMQP, OPC UA, WAMP, LwM2M, HomeKit, and Weave. An IoT system designer can select one of these pre-existing communication programming libraries to incorporate non-standard messaging protocols into their design. Simultaneously, it is possible to build an IoT system without relying on specific messaging protocols by utilizing general-purpose transport protocols such as QUIC, TCP, or UDP to execute application programs. Therefore, it is essential to note that this method is more appropriate for relatively simple systems within an enterprise network and for domestic applications.

Using e-mail protocols or social media platforms such as Twitter as a transport layer for message exchange is an option for IoT system designers. However, because of the high occurrence of spam, these alternatives might not be the most suitable choice. Typically designed for human communication, e-mail protocols may not always be an adequate method for smart devices and services to communicate. Therefore, we have decided to exclude these protocols in our comparison and instead will concentrate on analyzing internet protocols at each layer.

12.4 INTERNET PROTOCOLS AT THE APPLICATION LAYER

The main purpose of internet protocols is to outline the process by which the application interacts with lower-layer protocols [4] for transmitting data across the network [5]. Meanwhile, port numbers are employed to facilitate application addressing [6]. In the following discussion, we will explore some of the Application layer protocols.

> **HTTP (Hypertext transfer protocol)**: Within the TCP/IP suite, there exist several application layer protocols such as HTTPS, HTTP, FTP, and Telnet. To ensure efficient message transmission, the specific port designated for each protocol must be utilized. Failure to use the proper receiver port would result in the message not being received as intended.
>
> When utilizing HTTP, the messages are sent through port number 80, and specifically web servers will only acknowledge messages transmitted via this port. The HTTP port employs the HTTP protocol to transmit application data stacks to the lower layer, with URLs serving as a means to designate the destination address.

The Secure Socket Layer (SSL) or Transport Layer Security (TLS) variant of HTTP, known as HTTPS, operates on port number 443. An HTTPS URL is utilized to indicate the destination address, with the top-level domain (TLD) assigned begin.org, the domain name, for instance, being wikipedia. org, and the subdomain name sometimes begin.in. The data stack is then transmitted to the input receiver port [3].

Each application layer port is interconnected with a specific protocol and is assigned a number based on that protocol for transmission and reception purposes. Accurately identifying the correct port and protocol is crucial for ensuring effective communication within the TCP/IP suite of the OSI model.

Message Queue Telemetry Transport (MQTT): MQTT is a widely used secure protocol used in IoT security. It was invented by Dr. Andy Stanford-Clark and Arlen Nipper in 1999. Message Queuing Telemetry Transport (MQTT) is a client-server communicating messaging transport protocol. The MQTT runs over TCP/IP or over other conventions of a model that provide requested, lossless, two-way associations [7, 8].

Features of MQTT

- This protocol is uncomplicated and lightweight, facilitating quick and effortless data transfer between the nodes.
- MQTT is specifically meant for devices and networks that are limited in terms of resources like low bandwidth, high latency, or unreliable connections
- The data packet usage is minimized to reduce network usage.
- Optimum power consumption conserves connected device batteries, ideal for low-battery consumption devices such as wearables and mobile phones.
- MQTT depends on messaging techniques, thus, it is extremely fast and reliable.
- It is well-suited mostly for any smart applications.

Constrained Application Protocol (CoAP): It is a web transfer protocol that has been advanced exclusively for utilization with constrained devices, such as microcontrollers, and networks that are limited in the form of power or ability to transmit data effectively. It is one of the commonly used protocols for securing IoT and its associated applications, it has earned a reputation for being highly effective and reliable.

Features of CoAP

- Just like HTTP, CoAP is built on the principle of the REST architectural design. It indicates that clients can utilize a standard method like PUT, GET, POST, and DELETE to interact with server resources that are identified by unique URLs.
- Due to its suitability for microcontrollers, CoAP is highly compatible with the IoT, making it an ideal option for scenarios where there is a need for millions of cost-effective nodes.
- CoAP utilizes minimal resources both on the device and network by relying on UDP and IP in exchange for a complicated transport stack.

- CoAP is evaluated to be among the most secure protocols available, primarily because its DTLS parameters are set to 3,072-bit RSA keys, which provides enhanced security.

LWM2M (Lightweight Machine-to-Machine Communication Protocol): The Lightweight Machine-to-Machine Communication (LWM2M) protocol, specified by the Open Mobile Alliance (OMA). It is an application layer protocol developed for the transfer of service data and messages. It is primarily utilized in M2M applications and enables device management functionalities within cellular or sensor networks [9].

The term "lightweight" is the name of the communication protocol explained in the references paper [10] signifies that it operates without relying on system resource calls during execution. This means that it does not rely on invoking system software functions or APIs, such as displaying menus or accessing network functions.

The LWM2M protocol facilitates communication between an LWM2M client, present in an IoT device, and an LWM2M server located at the M2M application and service capability layer. This protocol is characterized by its compact header and efficient data model. It is generally used in conjunction with the CoAP for improved efficiency.

The LWM2M protocol allows M2M devices to establish LAN connectivity and demonstrates how constrained devices can connect with M2M applications and services using the standardized specifications provided by LWM2M. Assuming the presence of M2M devices, the protocol enables efficient communication between these devices.

- The M2M area network functions as a Personal Area Network (PAN) for connectivity between devices and the M2M gateway. Devices in the M2M domain use various network technologies such as Wi-Fi, cellular, and ZigBee IP to connect through the PAN.
- 10s of bytes communicate between a device and the PAN.
- Communication between LWM2M objects (right-hand side). LWM2M client refers to object instances as per the OMA standard LWM2M protocol. A client object sends a request or receives a response from the LWM2M server over the access and CoRE networks.
- CoRE network, for example, 3GPP or other networks, for IP connectivity
- Communication from an object instance using interface functions. The functions are bootstrapping; registration, deregistering or updating a client and its objects; reporting the notifications with new resource values; and service and management access through the server.
- Utilization of the CoAP, DTLS, and UDP protocols by the object or resource [11].
- 100s of bytes communicate between objects at the client or server for plain text or JSON or binary TLV format data transfer.

12.5   INTERNET PROTOCOLS AT THE TRANSPORT LAYER

TCP (Transmission Control Protocol): TCP is a transport layer protocol utilized by web browsers (HTTP application layer protocols), File Transfer Protocols (FTP), and email programs (SMTP application layer protocol) [10].

UDP (User Datagram Protocol):
1. UDP is a connectionless protocol.
2. It is helpful for applications that are time-sensitive, have small data units to exchange, and want to avoid the overhead of connection setup.

Transport layer protocols provide end-to-end message transfer capability independent of the underlying network. TLS, earlier known as Secure Socket Layer is the Design Principles for Web Connectivity protocol used for securing the TCP-based internet data interchanges. DTLS is the TLS for datagram [12].

Features of DTLS are the following:
- DTLS provisions for three types of security services—integrity, authentication, and confidentiality.
- DTLS protocol derives from TLS protocol and binds UDP for secured datagram transport.
- Applications that require security, for example, tunneling applications (VPN), those that tend to run out of file descriptors or socket buffers, or those that are sensitive to delays (thus use UDP), are all suitable for the implementation of DTLS.

12.6 INTERNET PROTOCOLS AT THE NETWORK/INTERNET LAYER

The primary function of these communication protocols is to enable the transfer of IP datagrams from the sender network to the receiver network. The datagrams contain source and destination addresses, which guide them from the sender to the receiver across multiple networks.

Internet Protocol version 4 (IPv4): IPv4 is a widely deployed internet protocol [13], responsible for identifying devices on a network using hierarchical addressing schemes.

Internet Protocol version 6 (IPv6): It is an advanced version of the IP and serves as the successor to IPv4. It introduces advancements and enhancements over its predecessor [14].

At the Link Layer, internet protocols govern the physical transmission of data across the network's underlying physical layer or medium, such as copper wire, coaxial cable, or radio waves. These protocols dictate how packets are encoded, signaled, and sent by the hardware devices linked to the medium, such as coaxial cables.

Internet Protocol Version 6 (IPv6):
Following are the features of IPv6 protocol.
- Provisions a larger addressing space
- Permits hierarchical address allocation, and thus route aggregation across the internet, and limits the expansion of routing tables
- Provisions additional optimization for the delivery of services using routers, subnets, and interfaces
- Manages device mobility, security, and configuration aspects
- Expanded and simple use of multicast addressing
- Provisions jumbo grams (big size datagram)
- Extensibility of options

IPv6 over Low-power Wireless Personal Area Networks (6LoWPAN):
The adaptation layer of the 6LoWPAN protocol facilitates the reception and transmission of data between the internet layer and the adaptation layer. Before transmitting to the IPv6 internet layer, the data stack utilizes the 6LoWPAN protocol at the adaptation layer. For connectivity, an IEEE 802.15.4 WPAN device incorporates a 6LoWPAN interface serial port [13].

In the referenced paper [5], 6LoWPAN is described as an adaptation-layer protocol specifically designed for IEEE 802.15.4 network devices. These devices, acting as nodes in a mesh network, possess low-speed and low-power capabilities. So as to work within the limitations of low-power devices, it is essential to impose restrictions on the data size for each instance. Therefore, it can be accomplished by employing methods such as data compression and fragmentation techniques [15].

The key features of 6LoWPAN include header compression, fragmentation, and reassembly. When data is fragmented for transmission, the first fragment's header consists of 27 bits, encompassing the datagram size (11 bits) and a datagram tag (16 bits). Subsequent fragments have an 8-bit header that includes the datagram size, datagram tag, and offset. A 60-second duration can be set as the time limit for reassembling fragments.

Features of 6LoWPAN

- Specifies the IETF recommended methods for reassembly of fragments, and IPv6 and UDP (or ICMP) headers compression (6LoWPAN-hc adaptation layer), neighbor discovery (6LoWPAN-nd adaptation layer), and
- Supports mesh routing ICMP (Internet Control Message Protocol). Routers or other devices on the network send error messages or relay the query messages. 6LoWPAN can be implemented using Berkley IP implementation with the operating system TinyOS or 3BSD or other implementations for IoT nodes from Sensinode, Hitachi, or others. The IPv6 network layer has two options, viz. RH4 routing header and hop-to-hop header RPL option.
- 6LoWPAN is utilized to carry data packets in the form of IPv6 over various networks.
- Provides end-to-end IPv6 and hence provides direct connectivity to a wide variety of networks including direct connectivity to the internet.
- 6LoWPAN is utilized for protecting the communications from the end-users to the sensor network.
- 6LoWPAN security for IoT uses AES-128 link layer security which is defined in IEEE 802.15.4 for its security. Link authentication and encryption are utilized to provide security and additional security is provided to TLS mechanisms, which runs over TCP.

12.7 INTERNET PROTOCOLS AT THE PHYSICAL LAYER

802.3 Ethernet: 802.3 Ethernet is a set of standards that defines wired Ethernet networks and provides a range of data rates from 10 Mbps to more than 40 Mbps.

ZigBee: It is a protocol that operates on the IEEE 802.15.4 specification and is intended for creating PANs using small, low-power digital radios. It is generally utilized in applications such as medical device data collection, other low-power, low-bandwidth requirements, and home automation. ZigBee provides a wireless connection suitable for small-scale projects that require close-range communication. Compared to other Wireless Personal Area Network (WPAN) like Bluetooth or Wi-Fi, ZigBee is simpler, more cost-effective, and ideal for applications such as home energy monitors, traffic management systems, wireless light switches, and various consumer and industrial equipment that necessitate short-range, low-rate wireless data transfer.

ZigBee, an advanced protocol for securing IoT devices and applications, is considered state-of-the-art. It enables efficient M2M communication, covering distances of 10–100 meters, specifically designed for low-power embedded devices such as radio systems. Furthermore, is a cost-effective wireless technology that operates on an open-source platform [16].

Features/Advantages of IoT with ZigBee

- ZigBee provides standardization at all layers, which enables compatibility between products manufactured by different companies.
- Due to its mesh architecture, devices tend to connect with every device in the vicinity. This helps in expanding the network and making it more flexible.
- ZigBee uses "Green Power" which facilitates lower energy consumption and cost.
- ZigBee helps in the scalability of networks as it supports a high number (about 6,550) of devices [4].

12.8 CRYPTOGRAPHIC PROTOCOLS

The IoT is a network of interconnected smart devices that generates vast amounts of valuable data. However, it also presents significant security challenges, primarily due to its lack of human interaction. As technology advances, ensuring security in the IoT field has become a major concern [17]. Cryptographic solutions [18, 19] have been recently employed to address these security issues, but with the emergence of quantum computers, therefore traditional cryptographic methods based on mathematical problems are no longer sufficient.

Quantum computers have the capability to decrypt any personal data, emphasizing the importance of utilizing cryptographic solutions that are resistant to quantum attacks in securing IoT networks. Real-time and highly efficient communication is crucial for IoT-enabled devices. However, cloud computing, which was once relied upon, is no longer capable of meeting the demands for low latency and high computing efficiency. New computing models namely edge, dew, and fog computing, have been introduced to enhance cloud computing capabilities.

A mutual authentication protocol proposed in [20] for dew-assisted IoT is discussed. However, the scheme is found to have drawbacks, including a lack of forward security and user anonymity. A new authenticated key agreement (AKA) protocol named e-SMDAS, to overcome these limitations and which has been proven to be secure under the eck security model.

12.9   CHALLENGES

The IoT plays a critical role in our daily lives, handling valuable information that requires stringent security measures [21]. Effective authorization and authentication protocols are necessary to ensure IoT security. It is vital to prioritize access control, ensuring that only authorized individuals are granted access to sensitive IoT information in alignment with policy enforcement. Efficient access control mechanisms that are advanced and user-friendly are necessary to ensure the protection of data in IoT. This chapter focuses on addressing the restriction of traditional access control techniques by exploring protocol-based and hybrid access control methods in IoT. It highlights the limitations and advantages of integrating protocol-based and hybrid access control and emphasizes the significance of access control in large-scale IoT systems. These access control approaches should be optimized to operate efficiently within extensive IoT environments.

12.10   FUTURE OPPORTUNITIES

The utilization of secure communication protocols undoubtedly contributes to data protection and streamlined integration, shielding networks against data tampering and malfunctions, and enhancing network compatibility. However, these protocols present certain difficulties. The incorporation of IoT devices into the network poses a challenge due to their inherent limitations in the form of memory, processing capacity, and energy resources. A single universal protocol that can cater to the diverse needs of an IoT network cannot be identified as it is not feasible [22]. Therefore, the susceptibility to security breaches is heightened, and the allocation of power and resources becomes inefficient and uneven across the various protocol sets. Moreover, the different protocols necessitate distinct network architectures that may not be suitable for all devices or adaptable to modifications.

12.11   SUMMARY

The IoT requires a secure, flexible, and reliable infrastructure. Recent developments have brought deep learning to the forefront of technological advancements, sparking interest worldwide. An overview of protecting the IoT environment against cyberattacks and malware was presented in this chapter. Our objective is to detect threats in IoTs by utilizing a combination of hybrid deep learning algorithms and technologies such as SDN and blockchain. In conclusion, securing IoT environments requires utilizing essential hybrid models of deep learning.

REFERENCES

[1] S. P. Jaikar and K. R. Iyer, "A Survey of Messaging Protocols for IoT Systems," *International Journal of Advanced in Management, Technology and Engineering Sciences*, vol. 8, no. II, pp. 510–514, 2018.

[2] E. Al-Masri et al., "Investigating Messaging Protocols for the Internet of Things (IoT)," *IEEE Access*, vol. 8, pp. 94880–94911, 2020.

[3] K. Mahmood, S. Shamshad, S. Kumari, M. K. Khan and M. S. Obaidat, "Comment on 'Lightweight Secure Message Broadcasting Protocol for Vehicle-to-Vehicle Communication'," *IEEE Systems Journal*, vol. 15, no. 1, pp. 1366–1368, March 2021.

[4] K. Li, B. Lang, H. Liu and S. Chen, "SSL/TLS Encrypted Traffic Application Layer Protocol and Service Classification", *CS & IT Conference Proceedings*, Vol. 12, pp. 6, 2022.

[5] S. Roy, S. Chatterjee, A. K. Das, S. Chattopadhyay, N. Kumar and A. V. Vasilakos, "On the Design of Provably Secure Lightweight Remote User Authentication Scheme for Mobile Cloud Computing Services," *IEEE Access*, vol. 5, pp. 25808–25825, 2017.

[6] L. Thungon, S. Sahana and Md. Iftekhar Hussain, "A Lightweight Certificate-Based Authentication Scheme for 6LoWPAN-Based Internet of Things", *The Journal of Supercomputing*, vol. 79, no. 11, pp. 12523–12548, 2022.

[7] B. Mishra and A. Kertesz,, "The Use of MQTT in M2M and IoT Systems: A Survey," *IEEE Access*, vol. 8, pp. 201071–201086, 2020.

[8] A. R. Alkhafajee, A. M. A. Al-Muqarm, A. H. Alwan and Z. R. Mohammed, "Security and Performance Analysis of MQTT Protocol with TLS in IoT Networks," *4th International Iraqi Conference on Engineering Technology and Their Applications (IICETA)*, Najaf, pp. 206–211, 2021.

[9] N. Oweidat, et al., "Kefaya's Origins," *The Kefaya Movement: A Case Study of a Grassroots Reform Initiative*, 1st ed., RAND Corporation, JSTOR, pp. 3–16, 2008.

[10] A. Aijaz and A. H. Aghvami, "Cognitive Machine-to-Machine Communications for Internet-of-Things: A Protocol Stack Perspective," *IEEE Internet of Things Journal*, vol. 2, no. 2, pp. 103–112, April 2015.

[11] A. Thantharate, C. Beard and P. Kankariya, "CoAP and MQTT Based Models to Deliver Software and Security Updates to IoT Devices over the Air," *International Conference on Internet of Things (iThings) and IEEE Green Computing and Communications (GreenCom) and IEEE Cyber, Physical and Social Computing (CPSCom) and IEEE Smart Data (SmartData)*, pp. 1065–1070, 2019.

[12] M. Asadzadeh, A. Payandeh and M. B. Ghaznavi-Ghoushchi, "TSSL: Improving SSL/TLS Protocol by Trust Model," *Security and Communication Networks*, vol. 8.9, pp. 1659–1671, 2015.

[13] P. Wu, Y. Cui, J. Wu, J. Liu and C. Metz, "Transition from IPv4 to IPv6: A State-of-the-Art Survey," *IEEE Communications Surveys & Tutorials*, vol. 15, no. 3, pp. 1407–1424, 2013.

[14] R. A. Nathi and D. S. Sutar, "Embedded Payload Security Scheme using CoAP for IoT Device", *International Conference on Vision Towards Emerging Trends in Communication and Networking (ViTECoN)*, pp. 1–6, 2019.

[15] L. Alqaydi, C. Y. Yeun and E. Damiani, "Security Enhancements to TLS for Improved National Control," *2017 12th International Conference for Internet Technology and Secured Transactions (ICITST)*, 7, pp. 274–279, 2017.

[16] W. -C. Tsai, T. -H. Tsai, G. -H. Xiao, T. -J. Wang, Y. -R. Lian and S. -H. Huang, "An Automatic Key-Update Mechanism for M2M Communication and IoT Security Enhancement," *2020 IEEE International Conference on Smart Internet of Things (SmartIoT)*, pp. 354–355, 2020.

[17] F. Xiaorong, L. Jun and J. Shizhun, "Security Analysis for IPv6 Neighbor Discovery Protocol," *2013 2nd International Symposium on Instrumentation and Measurement, Sensor Network and Automation (IMSNA)*, pp. 303–307, 2013.

[18] T. Kavitha, S. J. S. Priya and D. Sridharan, "Design of Deterministic Key Pre Distribution Using Number Theory," *Proceedings of 3rd International Conference on Electronics Computer Technology (ICECT)*, IEEE, vol. 5, pp. 134–137, 8–10 April 2011.

[19] T. Kavitha and R. Kaliyaperumal, "Energy Efficient Hierarchical Key Management Protocol," *2019 5th International Conference on Advanced Computing & Communication Systems (ICACCS)*, Sri Eshwar College of Engineering Coimbatore, India, 2019, pp. 53–60. doi: 10.1109/ICACCS.2019.8728343

[20] A. Lohachab, A. Lohachab and A. Jangra, "A Comprehensive Survey of Prominent Cryptographic Aspects for Securing Communication in Post-Quantum IoT Networks," *Internet of Things*, vol. 9, pp.100174, 2020.

[21] A. Altaher, S. Ramadass and A. Ali, "A Dual Stack IPv4/IPv6 Testbed for Malware Detection in IPv6 Networks," *2011 IEEE International Conference on Control System, Computing and Engineering*, pp. 168–170, 2011.

[22] Cruz, Mauro & Rodrigues, Joel & Lorenz, Pascal & Šolić, Petar & Al-Muhtadi, Jalal & Albuquerque, Victor. "A proposal for bridging application layer protocols to HTTP on IoT solutions", *Future Generation Computer Systems* vol. 97, 145–152, 2019.

13 Adaptive Security in the Internet of Things

M. Esther Hannah and N. Vanitha
Women's Christian College, University of Madras,
Chennai, India

13.1 INTRODUCTION

13.1.1 THE EVOLUTION OF IoT SECURITY

The evolution of IoT security has been driven by the growing adoption of IoT devices and the recognition of significant security risks. Initially, security was often overlooked in favor of functionality and connectivity, resulting in vulnerable devices. However, as IoT deployments expanded, high-profile incidents like the Mirai botnet attack in 2016 highlighted the urgent need for improved security measures in the IoT ecosystem.

The evolution of IoT security has unfolded through various strategies and initiatives.

Standardization and Best Practices: Industry organizations and standards bodies have established guidelines to provide a common security framework for IoT devices.

Encryption and Authentication: Encryption and authentication mechanisms are crucial for ensuring data confidentiality and integrity and for verifying device and user identities.

Secure Firmware and Software Updates: Security updates are essential for addressing vulnerabilities, and manufacturers focus on secure delivery methods for updates.

Access Control and Network Segmentation: Network segmentation with strict access controls minimizes attack surfaces and limits the impact of compromised devices.

Threat Intelligence and Analytics: Proactive threat identification and mitigation are achieved through threat intelligence and advanced analytics, enabling rapid response to security incidents.

Privacy Protection: IoT systems must adhere to privacy regulations like GDPR, leading to the adoption of privacy-by-design principles and secure handling of personal data.

Collaboration and Information Sharing: Stakeholders collaborate to share information and best practices, creating a stronger security ecosystem to combat emerging threats.

A general overview of the evolution of IoT security based on the timeline of advancements in the field is given in Table 13.1.

DOI: 10.1201/9781003477327-13

TABLE 13.1
Evolution of IoT Security

Timeline	Advancements in IoT Security
Pre-2010	• IoT concept and early deployments with limited security considerations • Initial recognition of security challenges but lacking comprehensive solutions
2010–2013	• Increasing adoption of IoT devices and growing awareness of security risks • Research focuses on identifying vulnerabilities and potential threats in IoT systems • Limited standardization efforts and best practices development
2014–2016	• High-profile security incidents raise concerns about IoT vulnerabilities • Industry organizations and standards bodies start developing guidelines for IoT security • Encryption and authentication techniques gain prominence in securing IoT communications
2017–2018	• Heightened focus on secure firmware and software updates • More research efforts towards secure delivery mechanisms for patches and updates • Access control and network segmentation approaches gain attention
2019–2020	• Increasing emphasis on threat intelligence and analytics for proactive security measures • Advancements in anomaly detection and machine learning techniques for IoT security • Emerging research on privacy protection in IoT systems
2021–Present	• Continued standardization and best practices development in IoT security • Advancements in secure device provisioning and authentication protocols • Collaborative approaches, information sharing, and cooperative defense mechanisms

Roman et al. [1] showed clearly how the research community has focused on prioritizing specific areas of IoT security and elucidating their significance in the last ten decades.

The evolution of IoT security is an ongoing process, driven by continuous advancements in technology and the ever-evolving threat landscape. As IoT deployments continue to expand into critical infrastructure, healthcare, transportation, and other sectors, it is imperative to remain vigilant and proactive in developing and implementing robust security measures to safeguard the integrity, privacy, and availability of IoT systems.

13.1.2 INADEQUACY OF TRADITIONAL APPROACHES

Many IT security teams keep focused on how to prevent a certain cyberattack and proceed to implement a response system for a particular incident rather than a response system which is continuous. Thus, such systems get easily compromised and miss out on a secure environment. Hence in new scenarios, these "prevent and detect" mechanisms fail to bring out the needed security in various frameworks. This pushes us to look for remedial solutions from adaptive security models.

Adaptive security architecture is a useful framework that helps organizations classify existing and potential security investments to ensure a balanced approach. Security organizations should evaluate their existing investments and competencies to determine whether they are sufficient for their future challenges, recommends Gartner [2].

13.1.3 THE NEED FOR ADAPTIVE SECURITY

The expanding network of IoT devices has revolutionized various industries, enabling advanced automation, real-time data analysis, and enhanced user experiences. However, this connectivity also creates an intricate web of potential entry points for Malicious attackers, making IoT security a paramount concern. Traditional security measures designed for conventional computing systems often fall short in addressing the unique characteristics of IoT deployments. Conventional security mechanisms, rooted in static and predefined architectures, are often inadequate in addressing the dynamic and heterogeneous nature of IoT environments.

Adaptive security approaches in the IoT domain offer a compelling solution to address the shortcomings of traditional security measures. Adaptive security recognizes the ever-evolving threat landscape and the need for real-time risk assessment and response. Context-aware security policies dynamically adapt to the changing conditions and requirements of IoT environments. Proactive risk assessment techniques, driven by machine learning and data analytics, enable early detection of anomalies and emerging threats. Continuous monitoring and timely security updates ensure that IoT systems remain resilient and protected against emerging vulnerabilities.

13.2 SECURITY ISSUES IN IoT

The Internet of Things (IoT) extends beyond traditional computing devices and smartphones, encompassing a wide array of objects that can be connected to the internet through an on/off switch. This interconnectedness implies that the IoT encompasses a vast range of devices, resulting in a significant accumulation of user data. However, the sheer volume and diversity of these connected "things" also increase the risk of data theft or unauthorized access by cybercriminals. As the proliferation of connected devices continues to expand, the corresponding security challenges also escalate in magnitude. Some of the security issues in IoT are as follows.

Standardization and Authentication: The lack of standardization in IoT presents a major security challenge due to the diversity of models, protocols, systems, devices, and platforms, making compatibility and interoperability difficult and creating vulnerabilities exploited by attackers. Additionally, inadequate authentication measures in many IoT devices, along with the absence of robust security features, expose them to potential security risks. This is exacerbated by the failure to implement strong authentication methods, utilize secure gateways, and establish an efficient public key infrastructure (PKI).

Network and Physical Security: IoT devices, often operating on resource-constrained embedded systems, face heightened vulnerability due to their specialized hardware and software configurations. This vulnerability creates openings that attackers can exploit, especially in networks with inadequate security measures. Additionally, the compact size of IoT devices hinders robust physical security, making them susceptible to physical attacks like tampering, theft, or destruction. Such attacks can result in unauthorized access to sensitive data, system disruptions, and data loss, highlighting the pressing need for enhanced security measures in the IoT ecosystem.

Inadequate Data Protection: Inadequate data protection represents a significant security challenge for IoT devices, given their generation and collection of vast amounts of data, which subsequently become vulnerable to attacks. This includes personal data, financial data, and other sensitive data. Failure to implement access control and data authentication can result in unauthorized access by Malicious attackers, leading to potential misuse and harmful consequences.

Visibility and Control: The inherent nature of IoT devices is to operate seamlessly in the background, often without the active involvement or awareness of the user. Consequently, comprehending their behavior and exerting control over their actions can become challenging. A prime example is a smart camera within the IoT ecosystem, where it may transmit data to a cloud service unbeknownst to the user. The lack of visibility into the device's operations poses difficulties in identifying and mitigating potential malicious activities.

Vulnerabilities and Privacy: IoT devices often harbor vulnerabilities that remain unaddressed due to various reasons, including the unavailability of patches and challenges in patch installation. The sheer volume of data generated by these devices is staggering, as illustrated by a Federal Trade Commission report titled "Internet of Things: Privacy & Security in a Connected World," which revealed that even a small number of households can produce 150 million discrete data points daily. This abundance of data creates numerous entry points for potential hackers, leaving sensitive information exposed to potential breaches and damage.

13.3   ADAPTIVE SECURITY FUNDAMENTALS

Adaptive security principles are a set of guiding principles that enable the development and implementation of dynamic and responsive security measures in various domains, including the IoT. Figure 13.1 shows the fundamentals of IoT adaptive security.

These principles aim to address the challenges posed by rapidly evolving threats and the dynamic nature of modern technology ecosystems. Adaptive security is a proactive and dynamic approach that involves continuous monitoring and real-time analysis to detect and respond to emerging threats. This comprehensive strategy encompasses the monitoring of network traffic, device behavior, and system logs to swiftly identify anomalies and potential security breaches. Context is a key

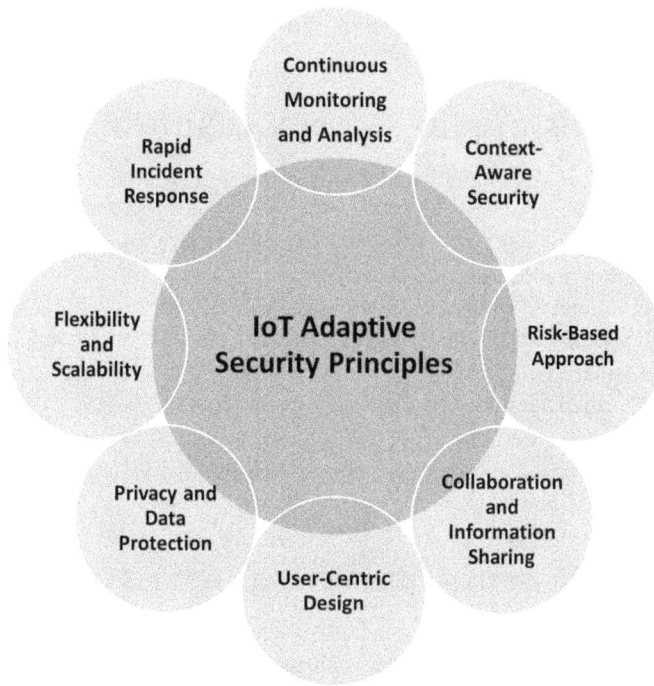

FIGURE 13.1 Adaptive security fundamentals.

consideration, taking into account factors like user behavior, network conditions, and device characteristics to dynamically adjust security measures, ensuring the appropriate level of protection and minimizing false alarms.

Moreover, adaptive security adopts a risk-based approach, conducting thorough risk assessments to pinpoint potential threats and vulnerabilities and then allocating resources based on the assessed level of risk and potential impact. Collaboration and information sharing among various stakeholders, including security professionals, researchers, and industry experts, play a pivotal role in this approach, harnessing collective expertise to effectively address emerging threats. It also places a strong emphasis on user experience and usability, with security measures designed to be transparent, non-intrusive, and user-friendly, ensuring that security enhancements do not disrupt system functionality or usability. Additionally, adaptive security principles encompass privacy and data protection considerations, implementing privacy-enhancing technologies, encryption, secure data handling practices, and compliance with privacy regulations to safeguard user privacy and protect sensitive information. These adaptable security measures are designed to flexibly scale and evolve to accommodate emerging technologies, shifting threat landscapes, and increasing system complexity. Swift and effective incident response mechanisms, including automation and well-defined response plans, are integral components to minimize the impact of security incidents. By adhering to these adaptive security principles, organizations can develop resilient and effective security strategies tailored to the ever-evolving threat landscape, providing robust protection for IoT and other technology ecosystems.

13.4   STAGES OF ADAPTIVE SECURITY ARCHITECTURE

Gartner [3] lists four stages that make the adaptive security architecture:

Predict: Determine risk, anticipate assaults and malware, and implement baseline systems and posture.
Prevent: Harden and isolate structures to save you from safety breaches.
Respond: Check out incidents, layout coverage changes, and conduct retrospective analysis.
Defect: Prioritize dangers, and illness and include incidents.

Figure 13.2 shows the stages that are incorporated into the various models of adaptive security for IoT and its applications.

13.5   ADAPTIVE SECURITY FRAMEWORK FOR IoT

According to Wang et al. [4], the adaptive security framework for IoT mainly includes context-aware computing, decision-making process, and dynamic enforcement, which are discussed in the following subsections.

Context-Aware Computing: Adaptive security solutions in IoT take into account the context in which devices operate. This includes factors such as device location, user behavior, network conditions, and application requirements. By considering contextual information, security measures can be dynamically adjusted and tailored to specific situations, optimizing protection while minimizing false positives and negatives.
Decision-Making Process: The IoT decision-making process involves several key steps for effective system design and management. It starts by defining project objectives and ensuring alignment with desired outcomes. Next, requirements are identified, covering functionalities, data handling, connectivity, security, scalability, and integration with existing systems. Assessing technological options is crucial, considering factors like connectivity

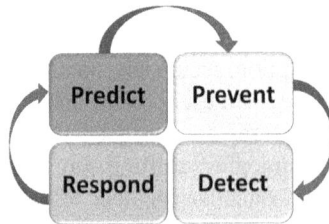

FIGURE 13.2   Stages of adaptive security in IoT.

methods, sensor capabilities, cloud services, edge computing, and data analytics tools. Security and privacy considerations, including authentication, encryption, and compliance with data protection regulations, are paramount.

A cost–benefit analysis helps evaluate financial implications. Standards and interoperability are essential for seamless data exchange within the IoT ecosystem. Engaging stakeholders ensures alignment with organizational goals and addresses specific concerns. Regular evaluation and adaptation support continuous improvement to meet evolving business needs and technological advancements.

Dynamic Enforcement: Dynamic enforcement in the context of IoT refers to the ability to dynamically apply and adjust security measures and policies based on the changing conditions and requirements of the IoT environment. It encompasses the dynamic adaptation of security controls, access rights, and enforcement mechanisms to address emerging threats, vulnerabilities, and operational requirements. This dynamic enforcement capability ensures that security measures remain effective and responsive amid dynamic IoT landscapes.

By continuously monitoring and analyzing the IoT ecosystem, organizations can swiftly identify and respond to security incidents, enforce access controls, and implement policies that align with the ever-changing nature of IoT deployments. This proactive and adaptable approach significantly enhances the overall security posture of IoT systems and effectively mitigates risks. An adaptive security framework for IoT is shown in Figure 13.3, which combines context-aware computing, dynamic enforcement, and the decision-making process.

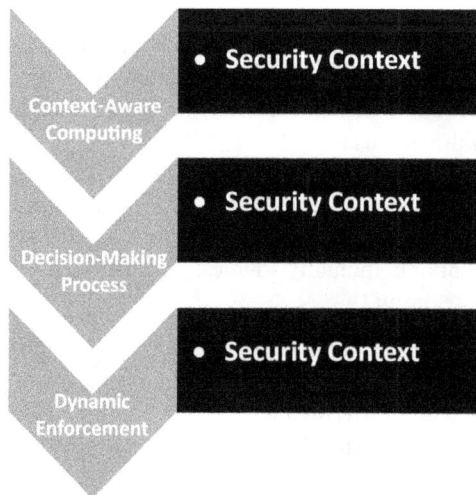

FIGURE 13.3   IoT adaptive security framework.

13.6 ADAPTIVE SECURITY METHODOLOGIES

Adaptive security methodologies for IoT proactively respond to evolving threats by continuously monitoring and analyzing IoT environments. They employ real-time threat intelligence, behavior analysis, machine learning, and automated responses to mitigate risks and safeguard sensitive data. These methodologies also ensure dynamic security controls, timely updates, robust authentication, and privacy protection, collectively enhancing IoT security against the ever-changing threat landscape.

- **AI and Machine Learning for Threat Detection**
  - **Malware-Prediction Models**
    Machine learning plays a crucial role in cybersecurity by enabling the creation of malware-prediction models. These models are trained to detect and predict malware by learning various parameters from harmful files, allowing for proactive blocking of potential threats.
  - **Threat-Hunting Triggered Inconsistencies**
    Machine learning aids in recognizing inconsistent patterns in transmitted data, which can trigger threat-hunting processes. This approach enhances anomaly detection and reduces false positives, ultimately improving the effectiveness of cybersecurity AI models.
- **Behavior Analysis and Anomaly Detection**
  Behavior analysis in IoT involves monitoring and analyzing normal patterns to establish expected behavior and detect any anomalies as potential security risks, including unauthorized access attempts and unusual data traffic. Anomaly detection complements this by identifying deviations from established norms, using statistical analysis, machine learning, or rule-based systems to detect patterns indicative of security incidents. Combining these techniques proactively identifies and responds to IoT security threats, enhancing overall security by providing early detection and reducing the risk of successful attacks. Pacheco et al. [5] proposed a methodology for enhancing intrusion detection through anomaly behavior analysis, effectively authenticating sensors, and detecting sensor attacks with high accuracy and low false alarms.
- **Software-Defined Networking (SDN) for Security**
  SDN is a network architecture method that decouples the network control aircraft from the data plane, making an allowance for centralized control and management of community sources. SDN offers several benefits when applied to IoT security:

  - Segmentation and Isolation
  - Policy-Based Access Control
  - Dynamic Security Provisioning
  - Network Monitoring and Threat Detection
  - Rapid Incident Response

- Kalkan et al. [6] proposed a role-based security controller architecture for IoT devices in the SDN environment.

- **Blockchain for Trust and Integrity**
  Blockchain-based systems offer tamper-proof information, ensuring the integrity and reliability of trust records. Smart contracts, executed on the blockchain, automate and enforce trust agreements among IoT devices, enabling secure and autonomous interactions without intermediaries. Lahbib et al. [7] developed a trust architecture using blockchain, which assesses trust evidence to assign a trust score to each device. This architecture securely stores and shares trust data by embedding it in all blockchain transactions across the network, enhancing IoT device collaboration and security.

13.7 IoT ADAPTIVE SECURITY ESTIMATION CRITERIA

When evaluating the effectiveness of adaptive security measures in IoT, it is important to consider metrics such as detection rate, response time, false positive and negative rate, adaptability and scalability, resource utilization, compliance with standards and regulations, and cost-effectiveness.

These evaluation metrics are essential for assessing the performance of adaptive security systems in IoT environments. The detection rate measures the system's ability to identify security threats, while response time evaluates its speed in initiating countermeasures. False positive and negative rates assess the accuracy of alerts. Adaptability and scalability gauge the system's flexibility in responding to evolving threats and IoT expansion. Resource utilization ensures efficient resource allocation and compliance with standards and regulations checks alignment with security guidelines. Cost-effectiveness considers the financial implications. Monitoring these metrics enables organizations to optimize security, ensuring a resilient framework for IoT security.

13.8 ROLE OF MACHINE LEARNING AND AI

One of the main applications of artificial intelligence and machine learning in IoT security is to identify risks and anomalies in behavior, network connections, or data structure. This can help detect and prevent attacks such as denial of service, malware, or data breaches. Artificial intelligence and machine learning can analyze large amounts of data from different sources and use algorithms to find patterns, relationships, or differences that indicate a crime or crime. For example, AI and machine learning can monitor device health, firmware updates, or authentication tests and flag any suspicious activity that indicates a security breach, hacking, or unauthorized access.

"Advanced machine learning and artificial intelligence (AI) concepts will drive the network analysis with adaptive security," Cearley [8] said. User and entity behavior analytics (UEBA) is an example that is now gaining attention for profiling the work of users. Thus, the unusual patterns of users and organizations can be discovered as unusual behavior can cause harm. For example, organizations can know if their users are visiting websites which they don't generally download or visit. "For next-generation security to be most effective, intelligent security measures must be integrated into an organizational structure," says Cearley.

13.9 CHALLENGES AND OPPORTUNITIES

The design of Adaptive Security for IoT [9, 10] has been constantly encountering challenges that require a comprehensive approach that considers the unique characteristics of IoT deployments. It involves developing robust security frameworks, leveraging advanced technologies, promoting industry standards, and fostering collaboration among stakeholders to overcome the complexities associated with IoT adaptive security. Some of the common challenges and the open opportunities with a few future directions are discussed in the following subsections.

13.9.1 CHALLENGES OF ADAPTIVE SECURITY

Kumar et al. [11] classified the security attacks based on the layers that make up IoT and summarized the security methods on how they address the security issues in the IoT.

There are several challenges associated with implementing IoT adaptive security measures. These challenges include the following:

- Heterogeneity of IoT Devices
- Scalability and Complexity
- Resource Constraints
- Privacy Concerns
- Interoperability and Standardization
- Human Factors

Implementing adaptive security measures in IoT faces several formidable challenges. These include coping with the diverse array of IoT devices, each with distinct hardware, operating systems, and communication protocols, necessitating compatibility and integration efforts. Scalability and complexity issues arise due to the large number of interconnected devices generating vast data streams, demanding robust infrastructure and efficient algorithms. Resource-constrained IoT devices require security measures that don't overwhelm limited processing power, memory, and energy resources. Ensuring privacy in the collection and transmission of sensitive data is crucial, despite the security and privacy balance being a challenging aspect. Interoperability and standardization difficulties hinder seamless integration across IoT components, while human factors like user awareness and education pose a challenge in promoting secure practices and engagement [12]. Overcoming these challenges is essential for effective adaptive security in IoT environments.

13.9.2 FUTURE DIRECTIONS FOR ADAPTIVE SECURITY FOR IoT

Xue et al. [13] analyzed the challenges that are encountered by the combination of blockchain and edge computing systems.

Liu et al. [14] presented many foundational technologies that are capable of enabling edge computing to be used in varying environments. Some of the future application areas with adaptive security for IoT can be found with cloud computing, SDN, information-centric networks, virtual machine (VM) and containers, smart devices, network slicing, and computation offloading paradigms.

The future of adaptive security for IoT holds several exciting directions and advancements. Here are some potential future directions.

- **Machine Learning (ML) and Artificial Intelligence (AI)**: By integrating ML and AI techniques into adaptive security systems one can enhance their ability to detect and respond to emerging threats in real-time.
- **Behavior-Based Security**: By establishing baselines and detecting anomalies in device behavior, adaptive security can become more effective in identifying potential security breaches and taking appropriate actions.
- **Blockchain-Based Security**: Blockchain technology holds promise for enhancing the security and integrity of IoT systems. By leveraging the decentralized and immutable nature of blockchain, future adaptive security solutions can provide transparent and tamper-proof transaction records, secure identity management, and trusted data exchanges between IoT devices.
- **Automated Security Orchestration**: Future adaptive security systems can leverage automation and orchestration capabilities to respond to security incidents in real-time. Automated incident response, threat containment, and remediation processes can help mitigate risks swiftly and minimize the impact of security breaches.
- **Integration of Edge Computing**: As edge computing becomes more prevalent in IoT deployments, adaptive security solutions will need to adapt and integrate with edge devices and edge computing platforms. Distributing security intelligence and decision-making to the edge can enhance real-time threat detection and response capabilities.

These future directions represent potential advancements in adaptive security for IoT, aiming to address the evolving challenges and ensure the security and operation of IoT systems.

13.10 SUMMARY

This chapter delves into the essential components of adaptive security for the IoT. It outlines key principles such as continuous monitoring, context awareness, risk prioritization, collaboration, user-centric design, privacy protection, flexibility, and rapid incident response. The Gartner-defined stages of adaptive security, including Predict, Prevent, Respond, and Defect, are highlighted as critical for developing effective security strategies. The adaptive security framework for IoT encompasses context-aware computing, decision-making processes, and dynamic enforcement. Methodologies discussed include the role of AI and machine learning in threat detection, behavior analysis, SDN, and blockchain for enhancing IoT security. Evaluation criteria, including detection rate, response time, and adaptability, are emphasized. Finally, the chapter addresses the challenges of IoT security, such as device diversity and privacy, while presenting future opportunities such as AI integration and edge computing for advancing adaptive security in the IoT landscape.

REFERENCES

1. R. Roman, J. Lopez, and S. Gritzalis, "Evolution and trends in the security of the Internet of Things," *IEEE Computer*, vol. 51, pp. 16–25, 2018, doi: 10.1109/MC.2018.3011051
2. Neil MacDonald and Peter Firstbrook, "Designing an adaptive security architecture for protection from advanced attacks," *Gartner Research*, 2014.
3. R. van der Meulen, "Build adaptive security architecture into your organisation," *Gartner Research*, 2017.
4. E. K. Wang, et al., "Mdpas: Markov decision process based adaptive security for sensors in internet of things," *Genetic and Evolutionary Computing: Proceeding of the Eighth International Conference on Genetic and Evolutionary Computing*, October 18–20, 2014, Nanchang, China. Springer International Publishing, 2015.
5. J. Pacheco and S. Hariri, "Anomaly behavior analysis for IoT sensors," *Transactions on Emerging Telecommunications Technologies*, vol. 29, no. 4, e3188, 2018.
6. K. Kalkan and S. Zeadally, "Securing Internet of Things with software defined networking," *IEEE Communications Magazine*, vol. 56, no. 9, pp. 186–192, Sept. 2018, doi: 10.1109/MCOM.2017.1700714
7. A. Lahbib, et al., "Blockchain based trust management mechanism for IoT," *2019 IEEE Wireless Communications and Networking Conference (WCNC)*. IEEE, 2019.
8. D. W. Cearley, B. Burke, and M. J. Walker, "Top 10 strategic technology trends for 2016," *Gartner Research*, 2016.
9. T. Kavitha, V. Ajantha Devi, S. Neelavathy Pari, and S. Ramanathan, *Internet of Everything: Smart Sensing Technologies*, Nova Science Publishers, Publication Date: June 17, 2022, doi: 10.52305/PNQM1088
10. T. Kavitha, G. Senbagavalli, D. Koundal, Y. Guo, and D. Jain (eds.), *Convergence of Deep Learning and Internet of Things: Computing and Technology*. IGI Global, 2023, doi: 10.4018/978-1-6684-6275-1. ISBN: 9781668462751.
11. S. A. Kumar, T. Vealey, and H. Srivastava, "Security in Internet of Things: Challenges, solutions and future directions," *2016 49th Hawaii International Conference on System Sciences (HICSS)*. IEEE, 2016.
12. E. B. Nightingale and P. Orwick, "Nineteen cybersecurity best practices used to implement the seven properties of highly secured devices in Azure Sphere," *Microsoft*, July 2020.
13. H. Xue, et al., "Integration of blockchain and edge computing in Internet of Things: A survey," *Future Generation Computer Systems*, vol. 144, pp. 307–326, 2023.
14. Y. Liu, et al., "Toward edge intelligence: Multiaccess edge computing for 5G and Internet of Things," *IEEE Internet of Things Journal*, vol. 7, no. 8, pp. 6722–6747, 2020.

14 Covert Communication Security Methods

A. Babiyola
Meenakshi Sundararajan Engineering College,
Anna University, Chennai, India

C. Chitra
PSNA College of Engineering and Technology,
Anna University, Chennai, India

V. Hari Santhosh
IoT Cloud Architect at trinamiX (BASF)

14.1 INTRODUCTION

The Internet of Things (IoT) is one of the exciting new technologies that has, in recent times, attracted a lot of attention not just from businesses but also from academic institutions. The IoT refers to a network of networked computer devices that are able to share information with one another, have conversations in both directions, and carry out certain activities without the need for human interaction. Every day, there are more and more things that are linked to the Internet as a part of the IoT. In order to create a new, intelligent era of the Internet, it seeks to seamlessly integrate the physical and digital worlds. These items include technology, software, networks, and sensors that aid in communication. This technology has tremendous economic value for organizations and enables diverse applications in energy, healthcare, and other sectors [1]. Although their broad applications are diverse, it is evident from a security standpoint that fraudsters are increasingly focused on them and employing them for shady purposes.

Sensitive information is increasingly being transmitted across wireless channels in the IoT [2]. However, not all vulnerabilities in IoT systems can be fixed with the methods of traditional cryptography. Therefore, covert wireless communication can provide increased security for IoT devices by preventing an adversary from detecting a user's message. With this expansion, there has been an increase in attacks on information-centric business operations over the past decade [3]. This requires protecting the security and confidentiality of sensitive and vital information at multiple levels. The higher the success percentage of any endeavor, the greater the overall success in this aspect. In order to address these challenges, many security measures, including cryptography, steganography, watermarking, and hidden channels, have been proposed [4]. Cryptography hides the contents of a message from an attacker, but not the fact that it occurs. Steganography and watermarking [5] are two important ways to hide private information in audio and video files so that it can't be found or taken away [6]. They could hide the fact that there is a message in the data they send. But the secret route is not like them.

The exchange of data or information through a covert channel (CC) constitutes covert communication [7]. A covert channel is a type of computer attack or threat that enables communication between objects and processes that were previously unable to communicate owing to system or network restrictions. A covert channel transmits information to an outside recipient over a network. A covert route offers a significant risk to the transferred data. The primary threats consist of data theft, malicious attacks, and data loss. A covert channel is a mechanism through which a sender and a receiver establish a compromised logical link between the systems executing at their respective ends in order to exchange sensitive data without a third party detecting it [8]. Covert channels are meant to be hidden in the flow of information on a legal logical channel, like TCP or UDP. Due to the mechanism it employs for illegal data transfers, it is difficult to identify a covert channel. Nevertheless, it is detectable by monitoring system performance.

In 1973, Lampson was the first to use secret channels. The communication channel between the two processes was allowed to exist, but not in the way that it does in practice. A number of covert receivers receive information from one or more covert senders in this scenario. There must be an agreed-upon pattern of communication and a shared resource that can be used by both the sender and the receiver.

The United States Department of Defence published its definition of a covert channel in 1985, and it reads as follows: "communication path that can be exploited by a process to transmit information in violation of system security requirements." [9]. According to Murdoch [10], a covert channel is a means of stealing information from a computer system via a route that has not been authorized for that purpose, in contravention of the system's access control policy. For the purpose of understanding the dynamics of covert communications, in 1983, G.J. Simmons presented the "prisoners' problem" as a common and perhaps ideal case study for exploring the inner workings of covert communications [11].

The two primary categories of covert communication systems are covert information transfer and covert wireless communication, as shown in Figure 14.1. The transmission behaviors are not concealed by the clandestine transmission of information. Encryption makes this a realistic option. Covert wireless communication places a higher value on privacy than traditional encryption. The goal is to ensure

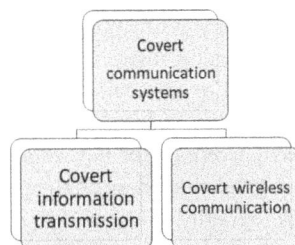

```
        ┌─────────────────┐
        │     Covert      │
        │  communication  │
        │     systems     │
        └─────────────────┘
         ┌──────┴──────┐
┌────────────────┐ ┌──────────────────┐
│    Covert      │ │  Covert wireless │
│  information   │ │  communication   │
│  transmission  │ │                  │
└────────────────┘ └──────────────────┘
```

FIGURE 14.1 Categories of covert communication system.

anonymity or stealth in communication so as to reduce the probability of being discovered. There have been derivations for discrete memory-less channels, spread spectrum, AWGN channels, and covert broadcasting [12].

Covert wireless communications can be used for the security of IoT systems in a variety of contexts, including military, intelligence operations, online social networks, and other security applications. To avoid leaving a record of their actions, soldiers in an ad hoc military network prefer to use covert channels of communication. Medical implants are not designed with the idea that outsiders will be able to listen in on private patient conversations. The location privacy of users can be preserved in another circumstance through the usage of covert communication.

Covert Channel:

- Hide information in common resources to facilitate communication between two computer processes that are not permitted to speak to one another. Confidential data is usually transmitted over covert channels, which have a high success rate when the data burst is small. It's not easy to find.
- Covert channels can run continuously for a long time and leak a significant amount of secret information.
- It may compromise a system that is normally secure, even one that has been formally verified.

Overt Channel:

- A suitable path for data flow within a computer system or network for data transfer.
- It can be used to build a covert channel by exploiting idle parts of the overt channels.

In fact, "covert" is derived from "overt" which means "open." Covert is the opposite of overt, therefore it refers to something that is concealed or undercover. An overt and covert communication network is shown in Figure 14.2. Therefore, covert channels have a dual role in network communication. At a certain time, it may pose a threat to a particular entity. It can also be used unconventionally to achieve secrecy and anonymity for another person. It can be used to protect confidentiality and strengthen the security of vital communications.

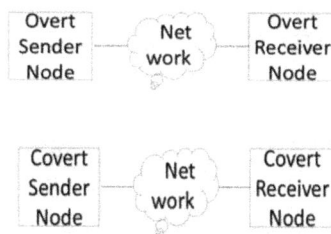

FIGURE 14.2 Covert communication network.

Use a network's covert channel to send secret messages to the intended recipient in a way that will surprise them but not draw attention to themselves. Covert channels in networks are becoming more common thanks to the Internet. A covert communication channel, as its name suggests, [13] conveys confidential communications in defiance of the norms governing networked communication. In addition to securing the content of communications, the use of network covert channels is designed to protect the identity of communicating parties [14].

14.1.1 PROPERTIES OF COVERT CHANNEL

This section includes an indicator of whether the covert channel generated is direct or indirect. Specifically, there are four aspects of the information-hiding approach that must be taken into consideration. This will make it possible to describe the characteristics of the achieved covert channel.

As seen in Figure 14.3, two main communication scenarios can be analyzed in the context of network steganography [15]. The primary scenario, sometimes referred to as the end-to-end situation, is the most prevalent. In this scenario, the individuals involved in the communication process openly engage with each other while simultaneously transmitting concealed information. In this particular instance, the explicit and implicit channels of communication exhibit complete congruence. In the second scenario, known as the Man-in-the-Middle (MitM) scenario, the hidden communication is facilitated by intermediate covert nodes, resulting in the utilization of just a segment of the overall end-to-end overt communication line.

Thus, in theory, the overt sender and overt recipient are oblivious to the steganographic transfer of data. Hybrid scenarios can indeed be observed, wherein the overt sender/receiver assumes the role of the secret sender/receiver, while the second covert participant is situated at an intermediary node.

Furthermore, it is imperative to clarify whether the covert channel is categorized as direct or indirect. This distinction pertains to whether the overt communication transpires directly from the covert sender to the covert recipient, or if it involves one or more intermediaries in the process [16]. In the scenario of a direct pathway, the traffic containing concealed information is transmitted directly from the sender to the receiver, both of whom may serve as intermediaries. In cases where the transmission of covert data from the secret transmitter to the covert receiver does not occur directly, an indirect covert channel is established, with the covert receiver being located downstream from the intended destination. In this scenario, the individual transmitting the

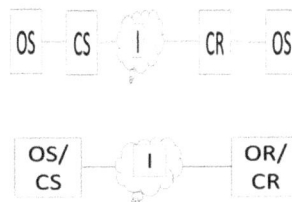

FIGURE 14.3   Communication scenarios.

concealed information opts to send it to an intermediary host. Subsequently, the intermediary host inadvertently transfers the concealed data to the intended recipient, primarily as a result of the features inherent in the overt traffic protocol. This observation suggests the presence of two distinct channels of visible network traffic that are utilized for the transmission of concealed information.

The first channel connects the transmitter of covert data to an unknowing intermediary, while the second connects the intermediary to the receiver. Indirect espionage involves discreetly acquiring secret or sensitive information via intermediaries. Due to the lack of a transmission line between the covert transmitter and receiver, channels make communication more discreet. In contrast, indirect paths are sometimes more complicated to establish and have lower capacity. In an indirect distribution route, intermediary needs and standards must be defined.

The availability of covert channels may be improved by providing a more precise description of the adversary. Four characteristics may be used to classify various types of adversaries: Active and Passive are the two types of adversaries, which depend on the kind of attack being made: An active opponent may create, alter, remove, or delay communication traffic whereas a passive adversary can just monitor and evaluate it.

Internal attackers may physically participate in covert information sharing or control a component of it. Depending on their powers, enemies might be global or partial. Global adversaries can watch the whole channel, unlike partial adversaries. Adversaries may be invariant or adaptable depending on how their resources vary. The capabilities of an invariable enemy are fixed once an assault begins. Adaptive opponents may change resource allocation mid-attack.

Variations in security goals and channel features give each covert channel a distinct supposed opponent. Security analyses of covert channels always presume an opponent can't actively breach one. Access to the originating end's file system and programming skills are needed to create a hidden channel. Assessing current different security aims and channel features gives each covert channel a different opponent. In covert channel safety analysis, the adversary is assumed to not actively breach the channel. Technical expertise and source-end file system access are needed to create a hidden channel. Analysis of existing secret pathways is needed to find new ones. A process's "covert channel." is any mechanism to convey data without the system's detection. Overall, hidden channels communicate data in unusual ways. Performance drops may indicate hidden channels, but today's powerful computers can analyze so much data that the effect is insignificant and the detection becomes harder. In order to detect their deployment, scientists have identified data stream characteristics that suggest a hidden channel.

The two primary modes of covert communication are Covert Storage Channels (CSCs) and Covert Timing Channels (CTCs) [17]. A shared resource is modified by one process (directly or indirectly) while it is read by another. Covert channels that store information in the protocol header and/or the PDU are used in network steganography. In contrast, timing channels conceal information by retransmitting the same PDU many times or switching up the packet sequence. There is a serious cyber security issue since all computer systems are vulnerable to this attack. Figure 14.4 depicts the many covert communication pathways.

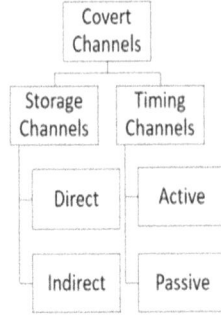

FIGURE 14.4   Covert communication channels.

Covert Storage Channels: Covert data transmission occurs between a sender and a recipient at CSCs. Most of the time, a CSC will change the values of unused packet header fields (such as "1" for "on" and "0" for "off"). Adding information to a message in a simple additional method. CSCs allow for the transport of massive volumes of data, but they are usually easy to find out.

Covert Timing Channels: CTCs control system resources like inter-packet delays and retransmissions to facilitate the transfer of information from the sender to the receiver. Additionally, CTCs may be categorized as either active or passive. The information is encoded by passive CTCs using the timing of active traffic. CTCs that are actively in use create their own traffic that looks like real traffic. CTCs are often harder to detect than CSCs but have better throughput.

14.2   IoT SECURITY OVERVIEW

Steganography hides communications in cover text items like software binary code or pictures. Internet-based covert network channels are another kind. Instead of steganography, covert network channels use network protocols. The well-known physical layer covert communication spread spectrum protects wireless communications from interference and eavesdropping [18]. Covert channels hide within authorized logical channels, providing them a simple yet effective means to transport data between end applications without alerting firewalls or intrusion detectors. Covert data transfer may assist sensitive applications. This secures protocol stack layers. Small bursts of sensitive data operate best on covert channels. Covert channels hide inside the communication flow of an authorized logical channel like TCP or UDP [19]. Channel packets contain confidential data that only recipient applications can decipher. Nobody else monitoring network traffic can tell whether this information is in legitimate channel packets. Covert IoT channels threaten consumers' security and privacy, as shown recently. Though this topic of research has yet to be fully explored, the security community is paying attention to this problem. The IoT security solutions developed and recommended in the literature are not all successful or meet all security standards. The storage and processing efficiency of conventional cryptography systems adopted for IoT applications is generally great. Their diversity and scalability are restricted. In contrast, blockchain technology's distributed design helps address scalability and heterogeneity issues:

Security and Privacy issues

i. **Network Mobility**: It is crucial to link IoT gadgets to a user network that communicates with other systems. Some IoT gadgets may have security flaws that put the user's network at risk. This is due to the fact that it offers hackers a way into the system.

ii. **Privacy of Sensitive Data**: The IoT is made up of a huge variety of different devices, each of which has its own software as well as hardware. Various attacks, including replay, MitM, cheating, and password guessing, can be used against some of them. Sensitive information may be disclosed due to unauthorized access and manipulation. Some devices transmit personal information such as mobile number, username, address, date of birth, credit card details, and patient records. IoT communications must always be protected from potential attacks.

iii. **Lack of Sufficient Safety Measures**: Devices that are part of the IoT, such as a personal computer or a smartphone, are linked to the system. In this kind of situation, the absence of security raises the possibility that private information may be disclosed. The information that is sent and received by IoT devices is open to public view.

14.2.1 Requirement for Security

It explains the numerous security standards that must be met by the IoT ecosystem in addition to the general security requirement.

- **Techniques of Authentication**: This technique verifies the identities of communication devices (i.e., IoT devices). Mutual identity verification is required prior to commencing secure communication and must be performed in advance. Smart IoT devices, various categories of servers (servers, fog, edge cloud), different classes of users, service gateway nodes, and providers may require authentication [20].
- **Possession of Integrity**: It guarantees the authenticity and reliability of the information. No unauthorized changes or inputs should be possible in a received message. Information must be treated as confidential at all times.
- **Protection of Private Information**: With this feature, sensitive information is protected from attackers. In another sense, it is privacy since it prevents attacks on transmitted data [21].
- **Nonrepudiation Property**: This ensures that no entity will deny the message's validity. Because this service confirms the message's originality and the data's integration into the text, it makes it almost hard for malicious groups to challenge the message's validity and authenticity.
- **Source Non-repudiation**: It acts as confirmation of the sender's legitimacy. This proves that a trustworthy source delivered the message.
- **Guaranteed Arrival at Final Destination**: This ensures the legitimacy of the recipient. This verifies that the intended recipient received the message.

Blockchain technology's major drawbacks in real-time and energy-constrained IoT applications are the proof-of-work technique's energy consumption and latency. However, software-defined networking (SDN) transfers greedy operations like cryptographic tasks from constrained IoT devices to high-performance servers (called SDN controllers), optimizing network resources, computation costs, and energy consumption. Centralized SDN cannot solve IoT-scale issues.

14.3   SECURITY PROTOCOLS CLASSIFICATION

A covert channel is any information transmission method a process may employ to avoid system security. In summary, covert channels communicate data against system design.

Covert routes are always an option for encryption because of privacy and legal considerations, even when unavailable. The first scenario emphasizes privacy. People don't want their discussions aired. The second issue is hidden messaging. Despite being encrypted, an email is not confidential. If the two entities want to keep their connection private, this information may be significant.

14.3.1   BENEFITS OF IoT WIRELESS PROTOCOLS

IoT wireless protocol serves higher data throughput with lesser complexities. These protocols become the first choice solution for personal and industrial usage.

- Affordability
- Faster data transfer and smoother operations
- Easy setup and deployment in the IoT infrastructure
- Robust security and data privacy

The protocols and standards used in the IoT may be broken down into two main groups. This includes the following:

1. Data protocols for IoT (Presentation/Application layers)
2. IoT Network protocols (Datalink/Physical layers)

Table 14.1 gives the classification of IoT protocols and Tables 14.2 and 14.3 provides its features based on Datalink and Application Layers.

Attackers have traditionally studied network-based sensitive data transmission. Secret communications reveal information, compromising system security. Covert channels have been used for decades to transmit data via communication networks. Covert channels are constantly improved to be harder to detect. These channels are neither steganographic nor encrypted. Since covert attacks might be malevolent, they are a security concern. Covert channels provide hidden communication in most communication systems. Hackers, attackers, and trustworthy parties seeking covert communication use them.

A covert channel's main purpose is to conceal information or improve the safety of essential communication. However, similar to other forms of security, covert

TABLE 14.1

Classification of IoT Protocols

Application Protocols	Network Layer Protocols
Extensible Messaging and Presence Protocol (XMPP)	Bluetooth
Message Queuing Telemetry Transport (MQTT)	Ethernet
Constrained Application Protocol (CoAP)	Wi-Fi
Simple Object Access Protocol (SOAP)	ZigBee
Hypertext Transfer Protocol (HTTP)	Thread
	Cellular networks (4G or 5G)

TABLE 14.2

Classification of IoT Protocols: Application Protocols and Its Features

Application Protocols	Features
Extensible Messaging and Presence Protocol (XMPP)	Decentralized, open source, and secure protocol developed using XML language. Best for consumer-oriented IoT deployments
Message Queuing Telemetry Transport (MQTT)	A transport protocol that uses TCP/IP or another protocol to facilitate client–server message exchange
Constrained Application Protocol (CoAP)	A web transfer protocol. It is possible to apply this method with confined nodes and constrained networks
Simple Object Access Protocol (SOAP)	Decentralized, open source, and secure protocol developed using XML language
Hypertext Transfer Protocol (HTTP)	Hypertext refers to a kind of organized text that creates connections (hyperlinks) between sections of textual nodes. Transferring hypertext uses the Hypertext Transfer Protocol (HTTP)

TABLE 14.3

Classification of IoT Protocols: Network Layer and Its Features

Network Layer Protocols	Features
Bluetooth	Short-range wireless communication preferably for 2.4GHz networks
Ethernet	First standardized in 1983 as IEEE 802.3. Widely used in homes and industry, and interworks well with wireless Wi-Fi technologies
Wi-Fi	A conventional wireless network protocol used for home and commercial usage
ZigBee	A mesh wireless communication protocol. Short range, highly interoperable, low power consumption
Thread	A wireless protocol specially built for IoT devices. It uses less power
Cellular networks (4G or 5G)	In 4G, speed and low latency are prioritized in response to consumer needs
	5G builds on the progress made by LTE Cat M and NB-IoT by delivering even faster download rates (up to 10 Gbps), wider coverage, and much longer battery life

channels have their drawbacks. As a means of establishing connections that are, in theory, forbidden by the security policy, they provide a covert and secure communication path. Once this happens, the compromised machine and its master are vulnerable to information leaks and asynchronous command channels [22].

A CC in a network protocol is created by exploiting one of many design flaws, such as the feature's optionality, the feature's ability to produce a random value, or the feature's dual nature (i.e., the same function may be accessed in numerous ways). Consequently, it is possible to use some of these features to create new covert channels within the protocol.

14.4   RECENT SECURITY PROTOCOLS IN IoT

IoT is designed, and it finds application in many areas like health monitoring, smart cities, self-driving cars, hospitality, and tourism. But there is a risk because of their security issues. Hackers will take control of devices that are not secured properly and attack the network as a whole [23]. Therefore, security issues must be taken care of and attention paid to reduce the attacks and prevent any breach in security [24].

Representational State Transfer (REST), Message Queuing Telemetry Transport (MQTT), Constrained Application Protocol (CoAP), and Modular Object-Oriented Bus (MODBUS) are the security protocols [25].

MQTT: MQTT is the most widely used encryption standard in the IoT. It operates across TCP/IP and other networks through a client–server message exchange. This protocol makes data transfer quick and easy. Power and battery life are both enhanced. Common MQTT security mechanisms include transport encryption through SSL/TLS, virtual private networking via VPN, and authentication via username/password. A client identification at the application layer is included in data packets.

CoAP: Constrained Application Protocol (CoAP) is a specialized web transfer protocol for use with limited nodes and networks. CoAP minimizes device and network resources. It supports HTTP, multicast, minimal overhead, and restricted networks. CoAP minimizes device and network resources. It supports HTTP, multicast, and minimal overhead. DTLS can secure CoAP, which is insecure [26].

MODBUS: MODBUS uses a master–slave system where the master provides functional instructions to the slaves and the slaves return the output.

14.5   CHALLENGES AND OPPORTUNITIES

As with any system of choice, covert communication techniques also come with their fair share of challenges, which can be seen as opportunities for further research. Covert communication protocols, which are used to hide or conceal the presence of communication, can have several limitations. These include the following.

   Security vs. Quality: The quality of the user experience may suffer if steps are implemented to prevent or restrict covert communication. In order to prevent an attack, the organization responsible for cleaning up network traffic may, for example, arbitrarily delay or delete packets.

Performance vs. Accuracy: The processing capacity of the security enforcement systems, such as middleboxes, can affect the behavior of the traffic and require an excessive amount of resources. This could affect the limiting and elimination process' accuracy and raise the cost of overhead.

Cost vs. Complexity: The sophistication of a covert channel can require complex mitigation methods, such as developing new software, acquiring new hardware, or altering network infrastructure. This can lead to costly expenditures and maintenance costs.

Blockage vs. Functionalities: Blocking or limiting a network protocol can impede its use and reduce its functionality. This trade-off should be carefully evaluated, especially for networks that operate on a larger scale.

Risk vs. Security: Eliminating hidden channels may not be practicable or essential. It may be safer to limit the channel's capabilities.

The same choices apply to detection, but channel limitation and network traffic sanitization demand more resources to avoid performance bottlenecks or network functionality [27]. Covert communication systems have drawbacks.

For example, steganography can be vulnerable to steganalysis, which is the process of detecting hidden data within other data. Covert channels can also be vulnerable to detection and disruption by network administrators.

Tunneling can be vulnerable to attacks that target the encapsulated protocol [28]. Cryptography can be vulnerable to attacks that target the encryption algorithm or key used. Therefore, it is important to carefully evaluate the situation and the scenario before implementing these covert communication protocols in IoT systems.

Some of the difficulties in covert communication may be overcome with the use of machine learning algorithms, which may provide new means of disguising the existence of communication. However, it's important to note that these algorithms also have limitations and trade-offs, and it is important to carefully evaluate the situation and the scenario before implementing them.

Examples include Generative Adversarial Networks (GANs) and Auto encoders, which generate or compress apparently harmless data that is identical to the data being used to convey the message in order to hide its existence. Because of this, it may be difficult for attackers to recognize that there is communication taking place. By embedding messages into a time series of data, Recurrent Neural Networks (RNNs) may be used to disguise the existence of communication, making it more difficult for attackers to detect.

On the other hand, a Generative Pre-Trained Transformer (GPT) can be fine-tuned to generate human-like text, which can be used to conceal the presence of communication by generating seemingly innocent text that is similar to the text being used to transmit the message. Adversarial Machine Learning can also be used to fool machine learning models by adding small perturbations to the data, which can be used to conceal the presence of communication by adding small perturbations to the data being used to transmit the message, making it difficult for attackers to detect [27].

However, it's important to note that these algorithms have their own limitations and trade-offs, such as the complexity of the covert channel and the resources available, and the attackers will also develop new techniques to detect and defeat these machine learning-based concealment methods [29]. Therefore, it is important to carefully evaluate the situation and the scenario before implementing them.

14.6 SUMMARY

The future of covert communication is likely to involve the development of new technologies and techniques to conceal the presence of communication. With the increasing use of IoT and other connected devices, the need for secure and covert communication will continue to grow. As a result, research and development in this area are likely to continue to advance.

Some possible future developments in covert communication include the following:

- The use of artificial intelligence and machine learning to conceal the presence of communication, for example, machine learning algorithms could be used to analyze network traffic and automatically identify and conceal the presence of communication.
- The use of quantum encryption [30] and communication techniques to secure covert communication. Quantum encryption uses the properties of quantum mechanics to secure communication, and quantum communication can be used to transmit information in ways that are difficult to detect.
- The use of blockchain technology to secure and conceal the presence of communication. By encrypting data and spreading it throughout a network, blockchain technology may keep conversations safe. It prevents attackers from easily intercepting and reading data.
- It's important to note that as the technology evolves, the attackers will also have more sophisticated tools to break the security, so there will be an ongoing "arms race" between security researchers and attackers.

In summary, the future of covert communication is likely to involve the continued development of new technologies and techniques to conceal the presence of communication and make it more secure. However, it is important to keep in mind the trade-offs and limitations of these technologies and to carefully evaluate the situation and scenario before implementing them.

REFERENCES

1. Z. Liu, J. Liu, Y. Zeng, and J. Ma, Security and privacy in the wireless internet of things: emerging trends and challenges. *IEEE Wireless Communications*, 2018, doi:10.1109/MWC.2017.1800070
2. J. Premalatha, I. T. J. Swamidason, P. Harshavardhanan, S. P. Anandaraj, and V. Jeyakrishnan, Analytical review on secure communication protocols for 5G and IoT networks. *International Journal of Adhoc and Ubiquitous Computing*, vol. 40, no. 1–3, pp. 50–66, Published Online: June 14, 2022, doi: 10.1504/IJAHUC.2022.123527
3. K. T. Nguyen, M. Laurent, and N. Oualha, Survey on secure communication protocols for the Internet of Things. *Ad Hoc Networks*, vol. 32, pp. 17–31, 2015.

4. W. Mazurczyk, S. Wendzel, S. Zander, et al., *Information Hiding in Communication Networks: Fundamentals, Mechanisms, Applications, & Countermeasures.* Wiley-IEEE, Hoboken, 2016.

5. K. Cabaj, L. Caviglione, W. Mazurczyk, S. Wendzel, A. Woodward, and S. Zander, The new threats of information hiding: The road ahead. *IT Professional*, vol. 20, no. 3, pp. 31–39, 2018.

6. S. Wendzel, L. Caviglione, W. Mazurczyk, A. Mileva, et al., A generic taxonomy for steganography methods. *Institute of Electrical and Electronics Engineers (IEEE)*, 2022, doi: 10.36227/techrxiv.20215373.v2

7. P. Rajba and W. Mazurczyk, Exploiting minification for data hiding purposes. In *Proceedings of the 15th International Conference on Availability, Reliability and Security*, 2020, doi: 10.1109/ACCESS.2021.3077197

8. W. Mazurczyk and S. Wendzel, Information hiding: challenges for forensic experts. *Commun ACM*, vol. 61, pp. 86–94, 2017. doi: 10.1145/3158416

9. Computer Security Center (US), *Computer Security Requirements: Guidance for Applying the Department of Defense Trusted Computer System Evaluation Criteria in Specific Environments.* Dod Computer Security Center, 1985.

10. S. J. Murdoch, Covert channel vulnerabilities in anonymity systems (No. UCAM-CL-TR-706). University of Cambridge, Computer Laboratory, 2007.

11. G. J. Simmons (1984). The prisoners' problem and the subliminal channel. In Advances in Cryptology (pp. 51–67). Springer, Boston, MA.

12. K. Cabaj, P. Żórawski, P. Nowakowski, M. Purski, and W. Mazurczyk, Efficient distributed network covert channels for Internet of Things environments, *Journal of Cybersecurity*, vol. 6, no. 1, 2020, tyaa018, doi: 10.1093/cybsec/tyaa018

13. K. T. Nguyen, M. Laurent, and Nouha Oualha, Survey on secure communication protocols for the Internet of Things, *Ad Hoc Networks*, vol. 32, pp.17–31, 2015, ISSN 1570-8705, doi: 10.1016/j.adhoc.2015.01.006

14. L. Caviglione, Trends and challenges in network covert channels countermeasures. *Applied Sciences*, vol. 11, pp. 1641, 2021, doi: 10.3390/app11041641

15. A. Mileva, A. Velinov, and D. Stojanov, New covert channels in Internet of Things. *SECURWARE 2018: The Twelfth International Conference on Emerging Security Information, Systems and Technologies*, 30–36, 2018, ISBN: 978-1-61208-661-3.

16. R. Gurunath, M. Agarwal, A. Nandi, and D. Samanta, An overview: Security issue in IoT network. In *2018 2nd International Conference on I-SMAC (IoT in Social, Mobile, Analytics and Cloud) (I-SMAC) I-SMAC (IoT in Social, Mobile, Analytics and Cloud) (I-SMAC), 2018 2nd International Conference on*, pp. 104–107. IEEE, 2018.

17. J. Tian, G. Xiong, Z. Li, and G. Gou, A survey of key technologies for constructing network covert channel. *Security and Communication Networks*, 2020, doi: 10.1155/2020/8892896

18. Z. Liu, J. Liu, Y. Zeng, and J. Ma, Covert wireless communications in IoT systems: Hiding information in interference. *IEEE Wireless Communications*, vol. 25, no. 6, pp. 46–52, December 2018, doi: 10.1109/MWC.2017.1800070

19. Harris, Kyle S., "Exploiting the IoT through network-based covert channels", 2022. Theses and Dissertations. 5322. https://scholar.afit.edu/etd/5322

20. K. Nguyen, L. Maryline, and N. Oualha, Survey on secure communication protocols for the Internet of Things. *Ad Hoc Networks*, vol. 32, no. 10, 2015, doi: 10.1016/j.adhoc.2015.01.006

21. D. E. Kouicem, A. M. Bouabdallah, and H. Lakhlef, Internet of Things security: A top-down survey. *Computer Networks*, vol. 141, pp. 199–221, Elsevier, April 2018. doi: 10.1016/j.comnet.2018.03.012

22. D. E. Kouicem, A. Bouabdallah, and H. Lakhlef, Internet of things security: A top-down survey. *Computer Networks*, vol. 141, pp. 199–221, 2018.

23. A. K. Das, S. Zeadally, and D. He, Taxonomy and analysis of security protocols for Internet of Things. *Future Generation Computer Systems*, vol. 89, pp. 110–125, 2018.

24. A. K. Das, S. Zeadally, and D. He, Taxonomy and analysis of security protocols for Internet of Things, *Future Generation Computer Systems*, vol. 89, p. 110–125, 2018, doi: 10.1016/j.future.2018.06.027

25. B. K. Mohanta, D. Jena, U. Satapathy, and S. Patnaik, Survey on IoT security: Challenges and solution using machine learning, artificial intelligence and blockchain technology. *Internet of Things*, vol. 11, 100227, 2020.

26. B. K. Mohanta, D. Jena, U. Satapathy, and S. Patnaik, Survey on IoT security: Challenges and solution using machine learning, artificial intelligence and blockchain technology, *Internet of Things*, vol. 11, 100227, 2020, ISSN 2542-6605, doi; 10.1016/j.iot.2020.100227

27. P. Nowakowski, P. Zorawski, K. Cabaj, and W. Mazurczyk, Detecting network covert channels using machine learning, data mining and hierarchical organisation of frequent sets, *Journal of Wireless Mobile Networks, Ubiquitous Computing, and Dependable Applications (JoWUA)*, vol. 12, no. 1, pp. 20–43, Mar. 2021, doi: 10.22667/JOWUA.2021.03.31.020

28. R. Tanya Bindu and T. Kavitha, A survey on various crypto-steganography techniques for real-time images. In Hemanth, J., Pelusi, D., and Chen, J.I. Z. (eds.), *Intelligent Cyber Physical Systems and Internet of Things. ICoICI 2022. Engineering Cyber-Physical Systems and Critical Infrastructures*, vol. 3. Springer, Cham, 2023, doi: 10.1007/978-3-031-18497-0_28

29. J. Yli-Huumo, D. Ko, S. Choi, S. Park, and K. Smolander, Where is current research on blockchain technology?—A systematic review. *PloS one*, vol. 11, no. 10, e0163477, 2016.

30. V. Scarani, H. Bechmann-Pasquinucci, N. J. Cerf, M. Dušek, N. Lütkenhaus, and M. Peev. Quantum key distribution: A comprehensive review. *Reviews of Modern Physics*, vol. 81, 1301, 2009.

15 Clean-Slate and Cross-Layer IoT Security Design

S. Rajarajeswari and N. Hema
Vellore Institute of Technology, Chennai

15.1 INTRODUCTION: BACKGROUND AND DRIVING FORCES

The Internet of Things (IoT) has become progressively prevalent in our day-to-day lives. The IoT advances web-enabled applications by establishing connections between "everyone" (e.g., people) and "everything" (such as machines, systems, and devices equipment) in a real-world or virtual environment. Using IoT applications and services has become easier because of the exponential expansion of smart devices (such as tablets, smartphones, intelligent circuits, actuators, and sensors). According to predictions, there will be 28.5 billion Internet-connected gadgets by 2022, up from 18 billion in 2017. By 2022, this will translate to an average of 3.6 networked linked devices per person, from 2.3 in 2016. By 2022, it is expected that there will be an average 51% rise in the count of connections and devices per home and per Internet user. By 2022, it is anticipated that this pattern will increase worldwide Internet traffic to 4.8 ZB (zetta-bytes) annually. This represents the IoT's potential scope, in which the network will connect billions of objects. The significance of the information kept, processed, and conveyed increases along with scale, as do the attacks against them. In another way, these forecasts indicate a rise in the quantity and level of threats and attacks against these embedded devices, necessitating stronger security measures. It is therefore a major concern to think about the fundamental security needs while creating an IoT architecture in order to protect confidential information from unauthorized access of users and services. Be aware that for our purposes, a thing can be a single user, a group of devices, a service or program, or another similar entity.

15.2 THE CLEAN-SLATE APPROACH

A system can alter or evolve primarily in one of two ways:

- **Incremental**: with incremental patches, a system is transformed from one stage to another.

- **Clean-Slate**: The system is completely rewritten to provide better efficiency and/or abstractions while maintaining identical functionality based on new fundamental ideas.

The Internet has had tremendous success over the last 30 years utilizing an incremental strategy. But, as an outcome of its popularity, the community has now arrived at a stage where experimentation with the current architecture [1] is either impossible or unaffordable. Perhaps it could be time to consider a fresh start strategy that includes unconventional thoughtful, the construction of alternate network architectures, and research with the architecture in order to assess the concepts, improve them, and give them a chance to be implemented either entirely in a new structure or gradually on/ in the current network.

It is difficult to criticize and change the currently available design principles because they are fundamental to the Internet's current Internet architecture. Yet, as we have shown, given these design principles, it is difficult to handle both the individual and collective difficulties listed above. Advancements in packet optical components and high-speed packet forwarding hardware have contributed to the implementation of faster data transmission and forwarding capabilities; wireless networks, large computational resources, and virtualization techniques have also made new possibilities possible, which raises doubts about certain aspects of the traditional design tenets.

15.2.1 CLEAN-SLATE THINKING

Innovative thinking is required because the community is unsure of what the new architecture will look like. In reality, it's likely that a distinct network design will emerge as an outcome of the capacity to reimagine the network and service architecture. The technological condition during that time, a unique set of design objectives, distinct priorities, and consequently a distinctive arrangement of functionality are a few specific causes. The operations that surround the network are involved in network management; therefore, a new network architecture alone is insufficient to address these issues. In order to encourage researchers to consider alternatives, it is crucial to give them a chance to practice the difficulty of administering such networks firsthand.

15.2.2 EVALUATION OF A CLEAN-SLATE APPROACH

The largest obstacle to using a Clean-Slate method is the requirement for a method to judge when the newly developed architecture is adequate. This must be achievable without understanding what such an architecture may include, which makes it even more difficult. We are aware of numerous strategies for doing this, including "prototypes" and "paperware," both of which fall short. Prototyping is essential since one wants to create a system to test it and persuade others that it is the best option. It is quite difficult to find a novel concept accepted that has not been tested widely and in practical settings. However, it is necessary intelligently because it allows scholars to discover information that may be unfamiliar to them otherwise. Building an experimental facility is another component of the Clean-Slate Design, in addition to research into new network designs.

15.2.3 IMPLEMENTATION OF THE CLEAN-SLATE APPROACH

Even while a Clean-Slate strategy is promoted, enhancing the existing architecture should not come to an end. It is reasonable to assume that a few concepts first put forth as a part of the current Internet have seen the application of an improved architecture, as exemplified by the adoption of IPv6. Also, it makes sense to create intermediary stages to help the existing architecture change into the desired new design after a new one has been identified. Also, the skills and resources gained by running and administering an experimental facility could be extremely beneficial for running the existing Internet and/or private Intranets. Clean-Slate should be considered a design method somewhat a finished product.

15.2.4 EXAMPLES OF SUCCESSFUL IMPLEMENTATION OF CLEAN-SLATE IDEAS

A Clean-Slate approach may end in a variety of success stories. For instance, the cutting-edge services and apps that will be created through the process might become mature enough to be used for commercial purposes on the current Internet. In another way, the research community will develop a novel network architecture [2] that replaces the current one. In yet another way, the process's learnings are adopted by business participants and incorporated into the network architecture of the present. The "conservative" conclusion might be that the existing Internet design is the "best" one that can be made. The most "radical" result is that the testing ground, which permits many sub-system designs and network facilities to coexist, may serve as the model for the future Internet.

15.3 CROSS-LAYER DESIGN IN IoT

An architectural strategy called cross-layer design allows the interplay of many levels to enhance efficiency and reduce energy usage, and design with cross-layers extend the life of the network and offer real-time communications with Quality of Service (QoS). As was previously said, wireless networks and the optimal method of communication between the nodes are the focus of the cross-layer approach [3]. Cross-layer protocols, in particular, can dramatically enhance the energy and bandwidth usage of wireless networks. To maintain and reuse a full-fledged cross-layer stack, smart software architecture is needed.

In the cross-layer, an application-oriented protocol for cyber-physical systems for IoT is the Trustful Space-Time Protocol (TSTP) [4]. The TSTP focuses on providing essential functionality required by systems, rather than maintaining the original protocol interfaces within a layered architecture. It integrates various functionalities such as time synchronization, MAC (Media Access Control), spatial localization, routing, security, and a data-centric API. TSTP achieves this by combining shared data from multiple network services into a single communication architecture, which reduces the overhead associated with control messages. This eliminates the need for redundant information replication across services [5]. The usage of cross-layered design improves the operational effectiveness of Wireless Sensor Networks (WSN) and Wireless Mesh Networks (WMN) in IoT systems [6]. Research endeavors have predominantly concentrated on enhancing WSN. By encapsulating and sharing packet control data with all protocol components, the cross-layer architecture minimizes the amount of control messages and enhances protocol decision-making.

Cross-layer approaches are employed for resource allocation more efficiently and achieve higher throughput while also being adaptable to constantly changing network traffic. When the cross-layer design was introduced to the IoT, the outcomes were better since it placed a greater emphasis on bandwidth and energy use. Networks have good QoS, security, mobility, and performance. The main goals of cross-layer architecture have been to reduce bandwidth and energy use. Complex software design is needed in order to save and reuse a packet. Information hiding was a TCP/IP feature that was causing networks to fail. The necessary communication between layers was never fully established as a result. So, it was challenging to offer remedies and fix such networks. The cross-layer approach is used to tackle this problem. When opportunistic scheduling is employed, the physical and data connection layers split the channel into time slots and perform transmission within Transmission Time Interval (TTI), which speeds up downloads. Data flow and network dependability are worked on by logical link control and MAC.

15.3.1 CROSS-LAYER DESIGN ISSUES AND POSSIBLE SOLUTIONS IN IoT

The major problems with the cross-layer architecture in IoT include deployment of more devices, memory, energy constraints, and processing. Different QoS standards for latency, energy use, and dependability can apply to various things. With battery-operated devices, for instance, saving energy for communication is a key factor. Contrarily, the gadgets linked to a power source do not require this energy restriction. IoT applications make consumers' lives easier, but they can't always guarantee their privacy and security [7]. In order to prevent such problems in the future, several remedies should be offered. A list of IoT application obstacles includes a summary of all IoT protocols as well as issues that they have raised. They also emphasize potential and innovative solutions at various levels. It is tough to guarantee user privacy and security in diverse networks like the IoT. The IoT is a topic of extensive examination, but much more work is required for it to mature. According to a study of the networking protocol layer, some researchers identified the vulnerabilities and threats to WSNs and listed the defense strategies [8]. They also provided an outline of security problems such as key management, cryptography, attack detection, secure data fusion, attack prevention, secure location security, secure routing, and many additional security concerns. In one study, research focused on the privacy necessities for IoT Systems and privacy challenges in IoT. Also, they offered an evaluation of technical Privacy Enhancing Techniques (PETs) in various IoT applications and proposed a new framework that is based on the security and privacy needs of IoT systems and possible privacy concerns with IoT-enabled services. In addition, they examine the whole IoT architecture and suggest appropriate sorts of PETs to tackle potential privacy issues that could already exist in each detected location.

Data confidentiality, data integrity, and data availability are the three key domains that make up security: a renowned model for the creation of security systems. In one of the research works, the authors examined security challenges at each layer and suggested solutions such as routing protocol security, key management, node trust management, and cryptographic methods. A manual and user authentication key agreement technique for diverse ad-hoc WSN employing smart cards was also presented. The assaults included DOS attacks, password change attacks, GWN bypassing attacks, impersonation attacks, and replay attacks. The IoT concept is allowing the distant user to get in touch with a single sensor node directly.

15.3.2 Interoperability of Cross-Layer Design

The cross-layer architecture in IoT faces significant challenges that are presently being studied. One of the key challenges is scaling IoT applications [9] to accommodate a large number of devices, which proves difficult due to limitations in scheduling, memory, processing power, and energy constraints [5]. As millions of devices across various industries, such as retail, smart home automation, smart energy, transportation, and healthcare, become interconnected, a wider explanation of interoperability is now necessary. This more comprehensive definition considers the cross-domain influence of interoperability on the overall performance of interconnected systems. It enables network providers to effortlessly transition between various wireless access networks. However, ensuring end-to-end interoperability poses an additional challenge in the IoT landscape because of the requirement to oversee a substantial quantity of diverse devices belonging to multiple platforms.

While the integration of IoT and cloud computing has the potential to tackle this issue to some extent, interoperability remains crucial in certain situations, such as in Fog Computing. Additionally, many businesses rely on interoperability to establish distinct products that are specifically designed to work seamlessly together [10]. By studying and adopting more standards, and with the participation of enterprise components, the implementation of smart scenarios with diverse devices can be achieved without concerns about interoperability. However, updating a system with a simple layered design becomes challenging, as each layer operates independently, whereas, in a cross-layered design, dependencies create complexities when implementing system improvements.

15.3.3 Movement

In ad-hoc networks, the movement of nodes is frequent, necessitating the identification and resolution of events caused by node mobility, such as route changes and channel switching, to ensure uninterrupted communication. Mobility can also be influenced by factors like channel fading, transmission delays, and high bit error rates, leading to a deterioration in service quality. Wireless technologies are primarily characterized by mobility. The TCP/IP protocol, originally designed for static Internet services, lacks the necessary flexibility to effectively connect wireless devices. As a result, the standard TCP/mobility IP encounters challenges when interacting with IoT networks. Mobility facilitates seamless handoffs and device movement in wireless networks, but traditional architecture often faces mobility management issues due to the separation of network and link layer solutions. To address this, the cross-layer concept has been proposed.

15.3.4 SERVICE QUALITY

To assess service performance and customer satisfaction, the integrated indicator QoS is utilized. The term "QoS" commonly refers to end-to-end QoS, encompassing resource allocation, utilization, and communication across the system layer, network layer, application layer, and middleware layer. In IoT applications, the sensing layer, network layer, and application layer are all employed, and the QoS of each layer has an impact on the overall QoS of any application. The researchers have analyzed the characteristics of IoT application services, transmission networks, and perception to establish QoS criteria. Following that, the researchers put forward a QoS design specifically tailored to the IoT, utilizing a layered structure. The proposed QoS architecture is visually represented in the accompanying graphic.

15.3.5 PROBLEMS WITH CROSS-LAYER ROUTING

There are a few difficulties involved with using cross-layer routing.

- **Difficulty in Redesign**: Enhancing data forwarding performance in IoT's QoS architecture requires dynamic cross-layer designs across different levels. However, most IoT protocols follow a classical design that is ill-suited for wireless devices. Redesigning or modifying these protocols is a challenging task as it involves making changes in multiple subsystems.
- **Lack of Standardization**: The absence of standardized routing protocols in IoT frameworks can lead to QoS and performance issues. Non-standardized protocols create inconsistencies and hinder the overall effectiveness of the system.

The growing utilization of cross-routing protocols has brought about challenges in terms of coexistence with other wireless protocols, making it challenging for multiple protocols to smoothly operate together. These difficulties directly influence the stability and coverage of the network. The distributed algorithms in cross-routing networks make it difficult to tackle the dynamic nature of wireless networking frameworks. Therefore, having numerous incompatible protocols is not suitable for diverse applications.

15.4 XLF: A CROSS-LAYER FRAMEWORK FOR ENHANCING SECURITY IN THE INTERNET OF THINGS

This section introduces the concept of XLF: a cross-layer framework, a comprehensive security framework designed to safeguard IoT systems. Unlike current defenses that often focus on specific applications or vulnerabilities, XLF takes a holistic approach by incorporating multiple functions across each layer. By utilizing

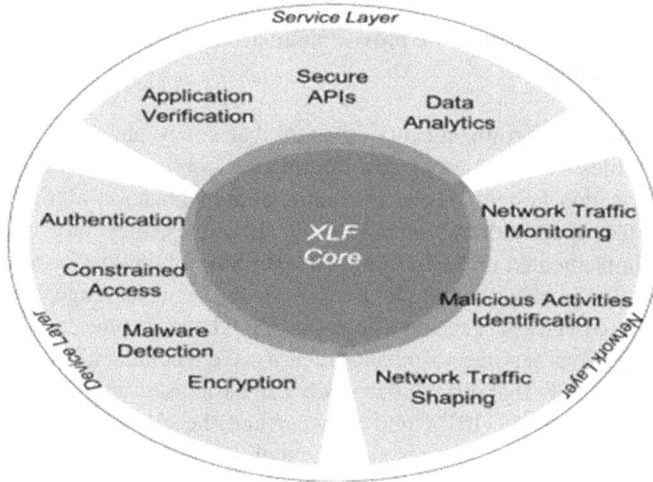

FIGURE 15.1   Design of XLF cross-layer security [5].

a cross-layer strategy, XLF enhances the overall protection capabilities. The development of protective measures at each tier has been guided by our comprehension of the architecture of IoT systems, their capabilities, and the specific vulnerabilities present in the potential points of attack have informed our comprehension associated with each layer. With XLF, we aim to provide robust and complete security for IoT systems beyond the restrictions of application-specific or vulnerability-driven defenses.

Figure 15.1 outlines the fundamental mechanisms of XLF and the recommended features to be incorporated into each layer. We suggest integrating different functionalities in the device layer, service layer, and network layer, which will be further elaborated on in subsequent discussions. Unlike previous approaches where these layers operated independently, XLF promotes inter-layer Interactions made possible by the XLF Core at the center. The XLF Core serves the purpose of not only enabling communication between layers but also gathering raw data and Results of detection obtained from individual layers, conducting in-depth evaluations using advanced techniques like deep learning [5]. By combining the security functions of each system component, this cross-layer design is intended to deliver proactive intrusion protection and extensive anomaly detection. The following parts will explore in detail the specific functions suggested for each layer, including the XLF Core.

15.4.1   SECURITY MEASURES IN THE DEVICE LAYER

We advocate for the implementation of specific security measures derived from an analysis of attack vectors, including authentication, access limitation, malware detection, and encryption. These security features aim to protect IoT devices' firmware, software, and their communications with users, cloud services, and third-party platforms at the device layer. It is important that physical attacks involving hardware components, such as micro probing and reverse engineering, are not currently considered in XLF. In this section, we provide more details about the four primary security functions we propose.

- **Authentication**: Authentication ensures that solely authorized individuals or objects can access private information, relying on a well-designed authentication system. While conventional authentication algorithms and techniques may address some security challenges, they face challenges when implemented in the IoT environment. Simplifying the authentication process for users with multiple accounts is one such challenge. Two-factor authentication (2FA) and single sign-on (SSO) have become common practices in various services to enhance security. We recommend an authentication architecture that differentiates between basic users and knowledgeable users, involving the IoT cloud provider and the cloud service provider. However, scalability and resource limitations pose challenges in implementing these methods on IoT devices. To overcome these challenges, we propose leveraging the functionalities of the XLF Core to handle authentication tasks associated with IoT devices. Delegation can be employed at both the network layer and the device layer, with the delegation proxy serving as a more capable entity compared to the IoT devices. The proxy entity is tasked with various responsibilities, including caching SSO tokens, conducting SSO authentication, validating timestamps, and processing unprocessed data for individuals possessing lower privileges. Furthermore, the XLF Core enables differentiation between LAN and WAN access requests, and the correlation of authentication results can be exchanged and compared between the delegation proxy and the service cloud. This approach improves scalability and fosters collaboration among diverse IoT devices within a unified security framework.
- **Encryption**: Secure transmission of data between IoT devices and cloud providers is crucial to protect user-sensitive information. Traditional encryption techniques offer strong security guarantees, but adapting them to resource-constrained IoT devices is challenging. To tackle this issue, National Institute of Standards and Technology (NIST) has presented a compilation of lightweight cryptography techniques suitable for IoT devices, encompassing hash functions, lightweight block ciphers, message stream ciphers, and authentication codes (MAC). It is crucial for vendors to embrace these lightweight methods in order to guarantee data integrity and data security throughout the entire data transmission process. Encryption alone is often insufficient to protect user privacy. To address this issue, we propose leveraging the network layer's knowledge to enhance encryption techniques, which will be discussed in more detail later.
- **Restricted Access**: This security measure aims to restrict access to specific third-party services and resources, allowing devices to engage without disrupting their normal operations. Network access control (NAC) plays a crucial part in this regard, using predetermined parameters and policies to determine whether to accept or deny network access requests. Implementing

NAC can pose a difficulty due to the complexities of interactions with protocols. Devices or specially configured network devices can enforce NAC. Nevertheless, when devices are hardcoded to connect exclusively to specific corporate domains, they become susceptible to DNS cache poisoning attacks. DNS is crucial to IoT security initiatives, and Domain Name System Security Extensions (DNSSEC) can be employed to protect DNS data. Nevertheless, widespread adoption of DNSSEC by IoT device manufacturers is lacking [11]. Additionally, existing DNS privacy options, such as DNS over HTTPS (proxied), DNS over TLS (DoT), DNS over DTLS, and DNS-Crypt, were not explicitly intended for lightweight IoT devices. We suggest leveraging the computing resources given by the XLF Core serves as a means to bridge the gap between current DNS privacy options and lightweight cryptography-based DNS. By allowing basic encryption algorithms at the network layer and lightweight encryption at the device layer, the XLF Core enhances privacy assurances on the Internet.

- **Detection of Malicious Activities**: This feature offers an additional layer of security in case devices become compromised. IoT botnets and malware pose significant threats, and proactive and reactive techniques are necessary to mitigate these risks. Proactive methods involve inspecting all firmware and software updates using deep packet inspection or fingerprint identification. Reactive systems should possess the capability to identify and prevent malicious activities in real-time. Both techniques require the cooperation and assistance of the network layer. Furthermore, network layer security features should meet the criteria of scalability, resistance to zero-day vulnerabilities, and ease of maintenance and customization. To simplify integration and communication between security functions, we propose incorporating a Platform for Security as a Service within the XLF Core. This platform facilitates the integration of unique security features and ensures seamless collaboration between the network layer and other components. Traffic shaping is also recommended as a minimal yet powerful network function to ensure the privacy of users from internal and external observers. By introducing random delays and redundant packets into network traffic, the proposed design can alter packet transmission rates and make traffic analysis more challenging. Additionally, monitoring network traffic and identifying malicious activity can be achieved by leveraging the XLF Core's machine-learning capabilities. By combining data from the service layer and device layer, patterns of typical IoT device operations can be identified, allowing for the detection of significant deviations and raising alerts when necessary.

15.4.2 SECURITY MECHANISMS IN SERVICE LAYER

As mentioned earlier, cloud-based back-end services play an important role in supporting modern IoT applications by providing essential functions such as content, alerts, and remote administration. However, these cloud endpoints are susceptible to various vulnerabilities, including malfunctioning services, weak authentication, and insecure APIs. The earlier study has identified potential dangers in video surveillance systems, such as false alarms and denial-of-service attacks, resulting from weak authentication, incorrect infrastructure configurations, and unsafe APIs. To address these issues, we propose several security features.

- **Secure APIs**: APIs are vital for communication and information sharing in cloud-based IoT applications. Yet, they are frequently disregarded as a potential threat vector. HTTP APIs and cloud backend APIs are prevalent in the domain of IoT, with SOAP and REST being two common API development methods. While SOAP offers broader security and compliance advantages, REST is favored because of its straightforwardness and compatibility with HTTP. To protect APIs against attacks, proper authorization and authentication mechanisms such as OAuth2 and OpenID Connect should be implemented. Every API call needs to be authenticated using an API token to hinder unauthorized entry to endpoints.
- **Application Validation**: IoT apps are automated programs that gather data acquired from IoT devices and enable their interaction with other devices. However, some design weaknesses in these applications can make them vulnerable to attacks. Security examinations of platforms like Samsung's SmartThings have revealed issues such as excessive privileged access, insecure third-party integration, and inadequate event data protection. While the strategies mentioned earlier can address some of these risks, additional work is needed to identify and rectify problems like excessive privileges and insufficient event data protection. Monitoring and profiling the state transition patterns of IoT applications can help ensure their integrity, and the network layer can play a role in collecting device status information. The XLF Core, employing machine learning methodologies like time series modeling, can monitor and verify the proper functioning of applications. Additionally, users have the ability to generate their personal security programs within the smart gateway to mitigate new security risks.
- **Data Analytics**: The data produced by IoT devices is a valuable resource for security analytics. Automation algorithms define IoT device behaviors, and the collected data should exhibit predictable trends. By integrating data from various domains, multi-dimensional security analytics enable the detection of malicious, suspicious, or unintended anomalies. They also provide contextual insights into the nature of threats, threat vectors, associated business risks, and recommended mitigation strategies. For instance, monitoring the state transitions of a smart lock connected to a smart thermometer can help identify abnormal behavior and potential threats. Integrating a data analytics module for security-related objectives within the cloud architecture enables more accurate threat detection when utilized in conjunction with threat intelligence information.

XLF Core acts as the central component that establishes connections and correlations among security functions spanning multiple layers. It focuses on capabilities such as integrating and analyzing heterogeneous data sources, multi-kernel learning (MKL) for data correlation and classification, and graph-based learning for community discovery. MKL allows the combination of features from multiple sources and simultaneous training of classifiers, enhancing the learning process. Graph-based learning, on the other hand, helps identify correlations and events within user groups or communities in the IoT ecosystem. The implementation of XLF Core can be installed across multiple layers, including the network layer or the service layer, taking advantage of processing capabilities at the edge or in the cloud.

The emergence of edge computing, driven by the growth of IoT and the massive amount of data generated, aligns with the implementation of XLF Core at edge devices like smart IoT gateways. Leveraging this computing paradigm can further enhance the capabilities and benefits of XLF Core in securing IoT systems.

15.5 A BIOMETRIC RECOGNITION SYSTEM ACROSS LAYERS FOR MOBILE IoT DEVICES

In this section, we present a low-complexity cross-layer biometric recognition system shown in Figure 15.2, specifically designed for mobile IoT devices. The system is partitioned into manageable domains, consisting of small, distinct, and interchangeable components within a cross-layer architecture (combining hardware and software). Each layer operates independently but with well-defined interfaces, involving experts from diverse fields including control theory, software engineering, operating system design, and circuit design. The system architecture utilizes information storage and coordination at each layer to improve overall system performance, leading to significant advancements in biometric recognition systems.

Hardware components are employed to accelerate specific application tasks and provide interfaces to the environment, while software components execute on processing cores. This enables the system to leverage advancements in memory technologies and transistors, such as magnetic tunnel junctions or tunnel field-effect transistors. The system's modularity simplifies management, maintenance, and optimization operations, offering enhanced robustness against failures, connectivity loss, security breaches, reduced channel capacity, and communication anomalies. Additionally, the system's service quality can be adjusted based on different application requirements. The hardware components in this heterogeneous design can repurpose the functionality provided by the software components without impacting them. Nevertheless, the system exhibits certain limitations, which encompass suboptimal layering that introduces redundancy, limited accessibility of processing information across layers, and potential performance degradation under strict QoS limitations.

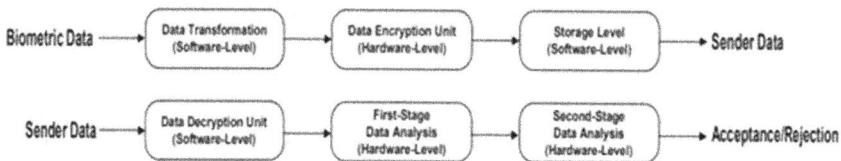

FIGURE 15.2   The architecture of the cross-layer biometric recognition system [3].

The system architecture is depicted in the accompanying diagram. The system initiates with a startup or user enrollment phase, involving the Sender and Receiver sides. During this phase, an individual captures their biometric data (such as fingerprint and/or iris) using a mobile device. The collected data is then segmented into 128-bit sequential plaintext chunks by an embedded software module within the device. These converted data chunks are kept in a plaintext-only queue and subsequently encrypted using a dedicated 128-bit AES hardware module, ensuring their sequential ordering. Following encryption, the data chunks are inserted into a ciphertext queue.

On the recipient's end, the ciphertexts are individually decrypted and stored in a queue. Subsequently, the encrypted data segments are directed to a software-based AES decryption module. Once decrypted, the biometric data is reconstructed and forwarded to a dedicated hardware unit for initial data analysis. This hardware unit compares the biometric data's 12 fundamental statistical measures to their corresponding reference values stored in registers or memory. These statistical measures include the number of average values, maximum values, number of minimum values, median values, standard deviation, mode, number of values below a low threshold, number of values surpassing a high threshold, kurtosis, Manhattan norm, Euclidean norm, and skewness.

The high threshold value is determined by calculating the mean plus three times the standard deviation of the biometric data, while the mean serves as the low threshold value. If 9 out of the 12 statistical measures fall within a selected tolerance boundary, the biometric data proceeds to the subsequent stage of the recognition system. Otherwise, the data is rejected.

In the next stage, the filtered data undergoes noise removal and comparison to identify any fraudulent biometric data. An average filtering technique with a window size of three is employed for noise removal. Next, the filtered data is inputted into a secondary data analysis hardware unit, where it is in contrast to the reference biometric data using four distance measurements: Euclidean distance, Manhattan Hamming distance, city block distance, and Chebyshev distance.

The estimated values of the distance measures for the test biometric data are matched against their reference values. To be accepted, at least three of the four measurements need to be within the designated tolerance boundary. This process is repeated for all incoming data on the receiving side.

15.6 SUMMARY

The IoT has the potential to bring about significant changes in the world. Nonetheless, similar to any emerging technology, it faces various technical and non-technical challenges. One area of focus in addressing these challenges is the development of cross-layer protocols, which have proven to be more effective than traditional network layers. These protocols play an important role in determining the longevity and performance of IoT networks. Thorough research has been carried out regarding cross-layered design and its application in IoT, aiming to identify and address the issues related to the old architecture.

In this chapter, we explore the use of cross-layer design in IoT and delve into the various challenges it presents, along with potential solutions. One example of a cross-layered protocol discussed is the TSTP, which incorporates security, time and space synchronization geographic routing, duty cycling, and an API to ensure the security of IoT services. We provide detailed insights into the communication between the MAC and the router, which covers most of the issues encountered in cross-layer architecture.

REFERENCES

1. J. Roberts. "The clean-slate approach to future Internet design: A survey of research initiatives." *Annals of Telecommunication*, 64, 271–276, 2009. doi: 10.1007/s12243-009-0109-y
2. A. Feldmann. "Internet clean-slate design: What and why?" *ACM SIGCOMM Computer Communication Review* 37, 59–64, July 2007. doi: 10.1145/1273445.1273453
3. S. Taheri and J.-S. Yuan. "A cross-layer biometric recognition system for mobile IoT Devices." *Electronics*, 2018. doi: 10.3390/electronics7020026
4. D. Resner, G. M. de Araujo, and A. A. Fröhlich. "Design and implementation of a cross-layer IoT protocol." *Science of Computer Programming* 165, 24–37, 2018.
5. A. Wang, et al. "XLF: A cross-layer framework to secure the Internet of Things (IoT)." *IEEE 39th International Conference on Distributed Computing Systems (ICDCS)*, pp. 1830–1839, 2019.
6. S. Chintalapudi. "Cross-layer design in Internet of Things (IOT)-issues and possible solutions." Research Gate, 2021.
7. T.-C. Chang, G. Bouloukakis, C.-Y. Hsieh, C.-H. Hsu, N. Venkatasubramanian. "Smart Parcels: cross-Layer IoT planning for smart communities." *IoTDI 2021: 6th ACM/ IEEE International Conference on Internet-of-Things Design and Implementation, Charlottesville (virtual)*, United States, pp. 195–207, May 2021, ⟨hal-03171372⟩. doi: 10.1145/3450268.3453526
8. K. Gupta and S. Shukla. "Internet of Things: Security challenges for next generation networks." *2016 International Conference on Innovation and Challenges in Cyber Security (ICICCS-INBUSH)*, Greater Noida, India, pp. 315–318, 2016. doi: 10.1109/ ICICCS.2016.7542301
9. T. Kavitha, V. Ajantha Devi, S. Neelavathy Pari, and S. Ramanathan. *Internet of Everything: Smart Sensing Technologies*, Nova Science Publishers, Publication Date: June 17, 2022. doi: 10.52305/PNQM1088
10. Y. Zhang, L. Duan, C.-A. Sun, B. Cheng, and J. Chen. "A cross-layer security solution for publish/subscribe-based IoT services communication infrastructure." *2017 IEEE International Conference on Web Services (ICWS)*, Honolulu, HI, USA, pp. 580–587, 2017. doi: 10.1109/ICWS.2017.68
11. S. Pal, M. Hitchens, T. Rabehaja, and S. Mukhopadhyay. "Security requirements for the Internet of Things: A systematic approach." *Sensors* 20, 5897, 2020. doi: 10.3390/ s20205897

16 Edge Computing Technology for Secure IoT

D. Bhuvana Suganthi
BNM Institute of Technology, Visvesvaraya Technological University, Bengaluru, India

D. Indumathy
Rajalakshmi College of Engineering, Anna University, Chennai, India

K. Panimozhi
BMS College of Engineering, Deemed University, Bengaluru, India

P. Kavitha
M A M School of Engineering, Anna University, Thiruchirapalli, India

A. Punitha and S. Saravanan
M A M School of Engineering, Anna University, Thiruchirapalli, India

16.1 INTRODUCTION: EDGE COMPUTING IN IoT

Edge computing has recently become popular in the modern environment. It is becoming one of the major influences in innovative developmental research areas such as the Internet of Things (IoT), secure vehicle-to-vehicle communication, 5G, and artificial intelligence. It is required for fast and compact storage and also for an easily approachable means of communication. The next level of communication moves toward fog computing, followed by cloud computing for a higher level of storage. Using edge computing, we could track and recover the data in a flexible way. A network full of smart devices is called the Internet of Things [1, 2]. The transmission of a huge amount of data from one end to the other needs to be processed and analyzed. The computation of data is collected and the data is permitted to be organized and processed in a nearby location, instead of transmitting the data back to

DOI: 10.1201/9781003477327-16

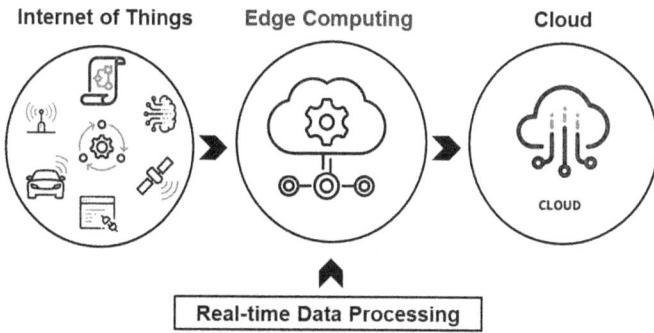

FIGURE 16.1   Real-time data processing using edge computing.

the data center or cloud. Figure 16.1 shows the process of real-time data using edge computing. The buzz about computation toward edge technology is growing, and when something is needed fast (i.e., milliseconds), it must be within the arm's reach of every user. Edge computing is a nearby source of processing and storage for the data needed for IoT devices.

An example of edge computing is nearby public access such as retail outlets and branch offices in which the public can access information easily, compared to public access at head offices or a higher network. 5G cellular is also called edge computing since the deployment of the core station integrates the capability of powerful computing. As per a survey, edge computations would not be accurate [3]. The average website juggles countless microservices to ensure great user experiences, strong security, incremental learning, optimization, and more. Each of these operations adds delays that hamper the user experience.

The edge computing process happens at the nearby physical location either from the user side or the source data side, which benefits the user with improvement in speed and security, which offers a good service and better products. The industrial or company side utilizes this edge computing technology by centralizing the infrastructure by distributing a common pool of resources to meet the requirement of increasing numbers of devices and data. In real time, analyzing the data in the usage of IoT and edge computing together has improved drastically. Especially, the IT environment uses edge technology since cloud computing has become omnipresent. The necessity of edge-level security has become a research area due to the rapid development of edge computing beyond the network. The exponential growth of using mobile and IoT applications due to less delay, super high speed, ease of accessibility, ease of scalability, and high security leads to the generation of high amounts of data and consumption at the edge [4].

16.2   EDGE COMPUTING ARCHITECTURE

In the IoT, most of the diverse measurement data are provided by millions of sensors. End users can serve as human–computer interfaces. Sensors and end users will be interconnected to exchange the data and provide additional services. It is considered a data provider to all the multiple networks of the distributed infrastructure at all levels.

Edge centers provide multiple management and virtualization services. These edge networks deploy multiple edge data centers, which cooperate with each other, and it is connected to the cloud traditionally. Due to this feature, distributive collaborative computing service patterns can be achieved. The edge layer delivers three important characteristics: (1) local data processing, (2) faster decision-making, and (3) filtered data transfer to the cloud.

The cloud network is built with high capacity for computation and storage due to multiple layers of heterogeneous servers, which satisfies the resource requirements for the end devices in a different geographical location. Figure 16.2 shows the architecture of edge computing. It is processed in three layers: IoT Layer, Edge Layer, and Cloud Layer.

The emerging three-tier architecture will consist of three layers, namely, Data Sources, Intelligence, and Actionable Insight [5]. The Data Source tier is the first tier, including anything that can generate data, such as social media feeds, click streams, and machine logs. The Intelligence layer is a second tier and machine learning concepts are involved in this tier. This intelligence layer cuts across the edge layer and public cloud. The Actionable Insight layer is the third tier, which is responsible for taking actions based on the intelligence offered by the previous tier. This tier also acts on behalf of the user [4].

Figure 16.3 shows the architecture of edge computing and its issues, and the nine different challenges are network bandwidth, distributed computing, latency, accessibility, backup, data accumulation, scale, control, and management. In general, bandwidth will be more to data centers and less to the endpoints [6]. Specifically,

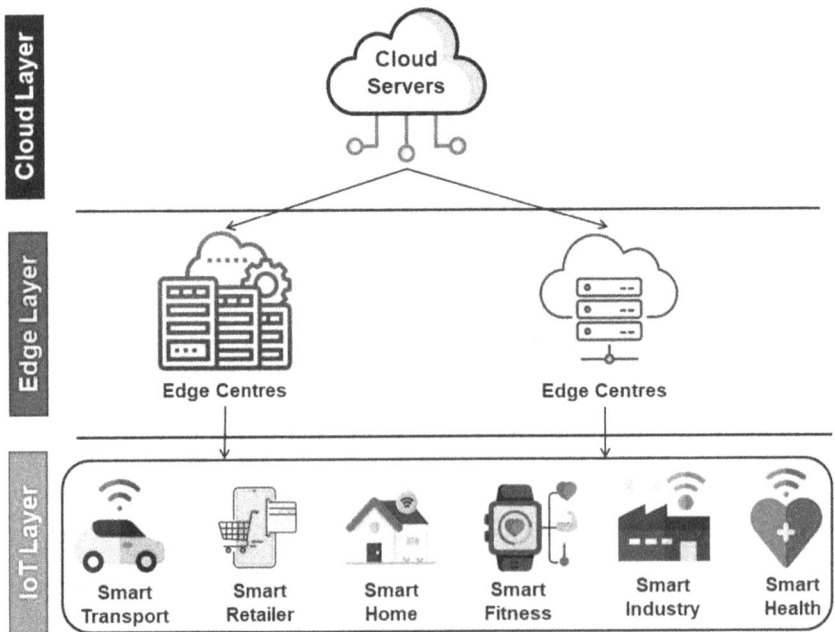

FIGURE 16.2 Edge computing architecture.

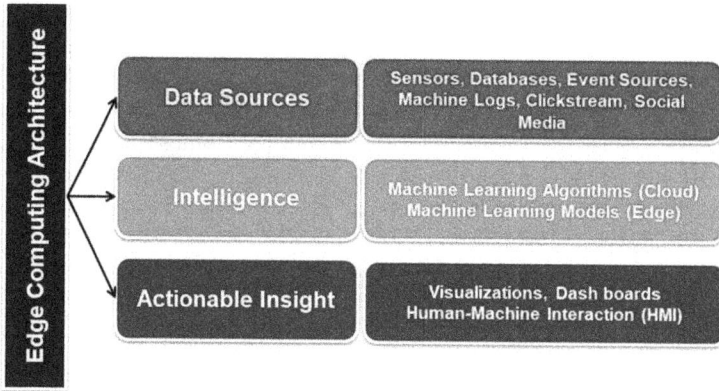

FIGURE 16.3   Edge computing: three-tier architecture.

edge computing requires more bandwidth across the network. It needs to be of the appropriate size and the resources must be in a centralized data center. Application data traverses the network in each direction, dealing with the right access, and sharing the data introduces some delay in the data processing.

16.3   LITERATURE STUDY OF SECURED ALGORITHMS USED IN EDGE COMPUTING

The aim of security level in edge computing should be toward identifying the best edge security at the physical level and keeping application, data security, and network as close to the data center as possible. A recent review by Atos [7] explained that cybersecurity risk is increasing due to many factors like firmware of the device toward edge security, geographical information from multiple devices, storing, processing of data, and when the devices are being procured. It also damages the device and data through some malware's actions like remotely accessing the device to hack data, to access corporate systems, and to sabotage operations, as said by Raj Sharma, founder of CyberPlus consultancy. The security level in the industry or company increases by taking some measures to control physical level security and detect movement and abnormal conditions, if any, by monitoring the environment frequently. Bola Rotibi [2], Research Director of CCS Insight says that challenges toward edge security will increase the capability of data processing. Daniel Paillet [2], Cybersecurity Lead Architect at Schneider Electric's energy management division says that the selection of devices must incorporate adherence to security standards.

The hardware-based root of the trust package needs to be investigated at the time of buying, to prevent identity from being tampered with, and device-level encryption also needs to be in a correct configuration. Paillet says that disabling any inactive devices, abnormal assessment, and patching are initiatives to be taken while selecting a device [8]. Security risks also increase in the transmission of data between edge devices and the cloud. The challenges of securing any network are also due to limited computing capacity. AI-powered tools increase network security and improve entity behavior systems.

Tawnya Lancaster says that parallel computation of security concepts can be performed using tools that augment or supplement and allow the developer to concentrate on their higher-level work. Since data and processing are happening in the outside world, i.e., distributed monitoring and managing of heterogeneous environments, highly dynamic adaptable security control is needed.

The various aspects involved in edge security are secure perimeter, patch cycles, securing applications, managing vulnerability, data storage, backup, password protection, authentication, cloud adoption, and early thread. This secure operation in edge computing takes place using access control, firewalls, apps that work beyond the network layer; early detection of potential breaches; and automatic updates. The cost of an edge facility is minimized by neglecting protection features and treating it as a streamlined data center. This is the highest security risk in edge computing. Just by disconnecting or inserting a hard disk from an edge computing resource, the entire database could be stolen. In such a case, it is impossible to have a storage backup of critical files due to limited local storage options.

To ease the convenience of users and administrators, edge computing may not go through a strong authentication system such as a two-stage authentication. Due to this password authentication issue, security risks increase. Due to limited tools for security at the edge, accessing our own data center also needs support from third parties, in which the security threat again extends to the highest level. Cloud platforms could help edge elements but not to the full level of security due to the change of frameworks and the cost factor.

As a conclusion of the survey on edge security, there are really some serious issues that add to the difficulty in ensuring smooth operations, and its remedies also need a longer discussion. The value of edge technologies should be improved by finding suitable strategies to prevent, anticipate, and overcome all edge security issues.

The conclusion does not give the idea that edge computing is always at risk; that is not the case. It depends on the different platforms it uses. Like M2M, IoT which depends on only one device with limited built-in security is not going to be affected by edge security due to a single device which can easily reduce the vulnerable surface attack of an application. If edge computing uses devices having high built-in security like laptops, desktops, and mobiles, then the cost of security is reduced. The other possible, simple method to reduce security risk is by providing a security layer between the devices and the VPN.

16.4   IoT EDGE SECURITY LEVELS

The physical objects connected to the internet transfer, receive, and analyze data continuously in a feedback loop without human availability at the level of the IoT or smart devices. The final analysis is done by artificial intelligence and machine learning in real time. The devices come with enough storage backup and process data using high-speed computation within milliseconds. However, the IoT also faces security issues due to resource constraints and insufficient designs for security. The recent usage of edge computing with high resources offers a better form of solutions at the security level for IoT applications. The network becomes unprotected due to the highly dynamic network in the edge computing environment.

Multiple devices are connected to the IoT, and a huge array of potential threats are possible. Data provided to the network must be hidden to avoid data breaches. Many algorithms are under research to filter out the personal data from IoT devices connected to the edge network and only raw data is processed in the edge data center. Some frameworks use proxy virtual machines for securing personal information as well as for reducing data traffic [3]. Still, more research work is in the initial stage toward edge-based security for IoT applications. In general, the 5Ps to be considered for edge security are People, Processes, Products, Proofs, and Policies.

People must be trained in cyber security usage since they are the big assets in handling smart devices. They need to organize the process of connecting the hardware to software, and devices to servers. Regular continued testing is mandatory to update edge security. The basic six rules for edge computing security are access control, control edge configuration from central operations, enhanced audit procedures, the highest level of network security, edge as cloud portion which is independent, and monitoring of all activities. Applications and data processes should be controlled centrally with high-quality encryption to prevent failures in critical application components that are migrated to edge facilities. It is critical to handle edge information due to it not being able to conform to the security level of the network connection.

The other way to handle edge security is by providing multifactor authentication or a dongle with full security to be incorporated for all operational access which uses a Wi-Fi network with no access to others' passwords. The completed log details of operations, deployment, access to any higher-order nodes, and configuration changes need to be monitored carefully. Any changes in the network configuration should be brought to notice in advance and in turn can be notified to the management about the escalation procedure if anything unexpected is reported.

For digital transformation, edge technologies are in the next wave of innovation in the category of automated operations, new business model introduction, and experience in rich customer delivery processes. With the advantage of features of software-defined infrastructure and cloud-native application design, millions of IoT structures are into many applications and are integral in making sure that the identity of each device is genuine and secured. Therefore, device manufacturers are facing challenges in providing both integrity and identity of all components such as cloud servers, edge servers, and sensors in cloud-connected systems. This form of safety protection is most basic for highly integrated secure cloud connectivity, secured process management, and zero-touch provisioning [9].

IoT benefits include better computation due to physical devices or existing data sources. A faster response time for IoT devices is possible if the data produced by the IoT is analyzed at the edge. The overall benefits of IoT devices using edge computation are less delay, faster response times, increased efficiency, increased bandwidth, the possibility of working offline in case of network failure, rapid analysis, and decision-making. Devices using edge computing require network connectivity for different uses such as receiving automation instructions, allowing remote management, and forwarding network traffic for analytics.

IoT devices are designed for low power consumption and low cost and are therefore not adaptable to complex technology because of the need to operate them under diverse environmental conditions such as temperature and humidity, dust, or vibration. This demand for diversity can be overcome by using specialized M2M protocols, with sophisticated security features.

16.5   EDGE DATA CENTER

The IoT and edge computing are together used for many use cases such as autonomous vehicles in which real-time traffic, street signs, stop lights, and pedestrian data, which need to be responded to immediately; this could be made possible using edge computing [10]. The service of cloud computing is overtaken by edge computing in such cases to avoid any accident by allowing IoT sensors in vehicles to process the data in real time. Overall, the edge and IoT improve the quality of service.

Edge computing also can be as secure as the central core, based on the responsibility taken by cyber security professionals. The tool called Simplilearn provides a broad view of training on some certification courses and boot camps on cyber security to develop the required skills and knowledge to secure infrastructure of any size or type. There are 15 billion IoT devices connected to the internet, and effective security cannot be provided for all such devices. Hence, the organization should take responsibility by incorporating security agents into edge nodes.

By putting security agents into edge nodes (Figure 16.4), like micro data centers with sufficient processing power, the traffic transported from compromised IoT devices can be detected as corrupt and access to the rest of the network can be denied.

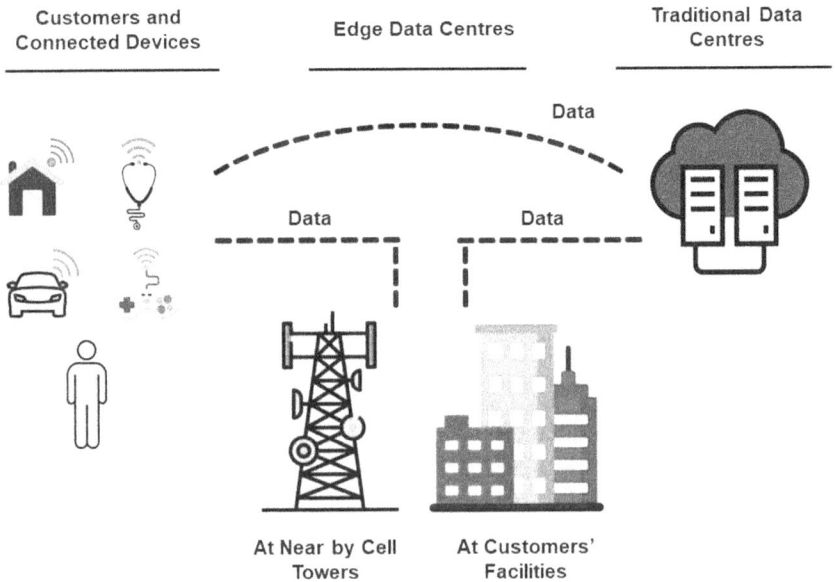

FIGURE 16.4   Edge data center.

Since IoT edge devices are capable of processing the data at the edge of the network rather than centralized on cloud servers, this method has both positive and negative effects on IoT devices and their data security, data minimization, and decentralized infrastructure.

Preprocessing the data before sending it to the cloud helps reduce the amount of data transmission, and due to this, the possibility of data leakage is also reduced. The second level of edge security in the IoT is distributing the data processing to various devices on the edge network; this improves reliability and reduces the occurrence of faults in the data network.

Data centers are needed to process data efficiently. The data centers need more storage, more computing power, faster communication, and higher bandwidth to connect more devices. By deploying an edge data center at the edge network, it brings all resources to end users and devices, which leads to high-speed delivery of information and improves efficiency. Edge computing depends on a distributed data center.

Edge data centers shown in Figure 16.5 can process the cloud computing resources and cache contents to the end devices. The time-sensitive data is processed first at the edge data centers, where the data which is less time-sensitive can be sent to a larger data center called the cloud, for long-term storage.

Cloud data centers are also called hyperscale cloud data centers which are highly loaded with resources that support high capacity and long-term data analysis, but end with high latency. Therefore, edge data centers that are introduced are away from the

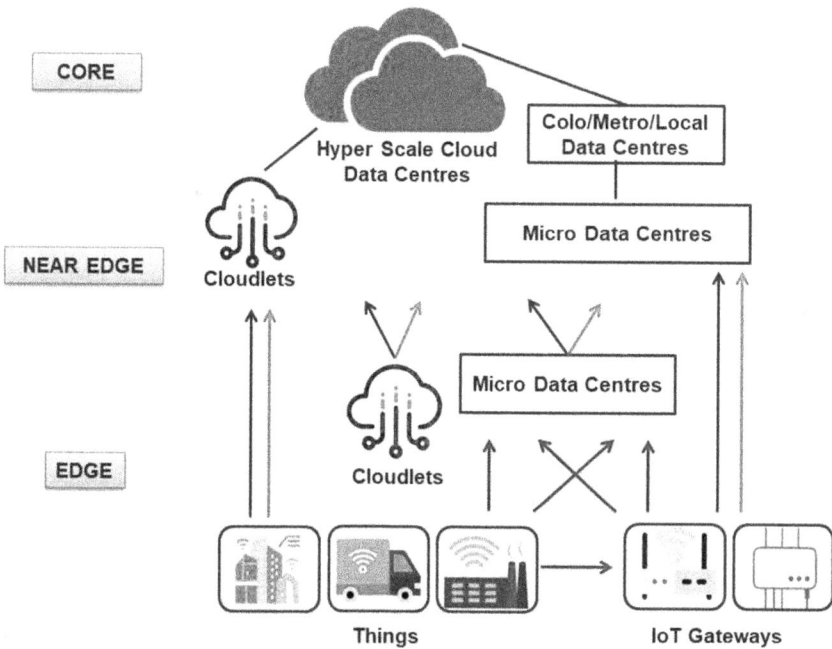

FIGURE 16.5   Edge data center to identify location, speed, and connectivity.

cloud, called micro data centers, which serve the devices remarkably close by. Hence, latency is reduced. This process is possible for small business communication. The overall advantage of moving data centers to the edge leads to reduced latency, increased bandwidth, lower operating costs, and improved security.

When the physical distance between the end device and the data center is reduced, the number of network hops is also reduced, and, hence, the latency is also reduced. The bandwidth is reduced due to local processes at the edge data center and thus the amount of data flowing into the central servers is also reduced. Due to this, users get more network bandwidth, which improves the overall network performance. Also, the cost of data transmission is inherently reduced due to a reduction in the usage of central servers. In general, edge data centers reduce the need for high-cost circuits and interconnection hubs.

Edge data centers enhance security by avoiding the amount of sensitive information, restricting the amount of data stored in any individual location by means of distributed data storage, and decreasing the border network due to the major migration taking place at the border transmission.

High performance is achieved with minimal latency, thereby enhancing the overall user experience.

16.6 DESIGN OF KEY-VALUE STORES FOR IoT EDGE STORAGE

IoT data flows from the IoT sensor and passes through a gateway called the IoT edge gateway which pulls the data from the IoT edge using stream analytics and then streams the data to store it into the database. The IoT gateway connects the network of the sensors and core network to the cloud. This carries data preprocessing to reduce redundancy and the processed data is forwarded from the cloud to the end users.

Every IoT system needs a database to have its own storage position. To pull the data from the IoT sensors, data ingestion techniques such as Kafka, MQTT, and REST are used. The data is classified based on the premise gateway which includes the data involving time-sensitive, locally available data, horizontal scaling, and hot data stored in edge data centers (Figure 16.6). Data is moved to the public cloud using the technology called data lake which makes an issue for cold data analysis. Therefore, it needs horizontal scaling support too due to the increasing number of IoT devices and its data storage passing from the edge to the cloud. This process is called big data analytics.

The most popular databases for IoT apps are InfluxDB, CrateDB, RiakTS, MongoDB, RethinkDB, SQLite, and Apache Cassandra. Depending on the data access method, the right storage for time series and IoT domain use case is selected. Hot database is the best option for real-time data storage, which has interesting features such as querying abilities, flexibility in data formats, messaging and queuing capability, and read and write capabilities with little latency at the lowest cost. This hot database is used for data being frequently queried and updated, whereas a cold database is almost fixed, and cannot be modified once it is stored. The concept of task offloading at the edge is to minimize the total cost in terms of energy consumption and latency by predicting the characteristics in advance and combining them with the task decision to allocate resources dynamically and to improve the quality of service.

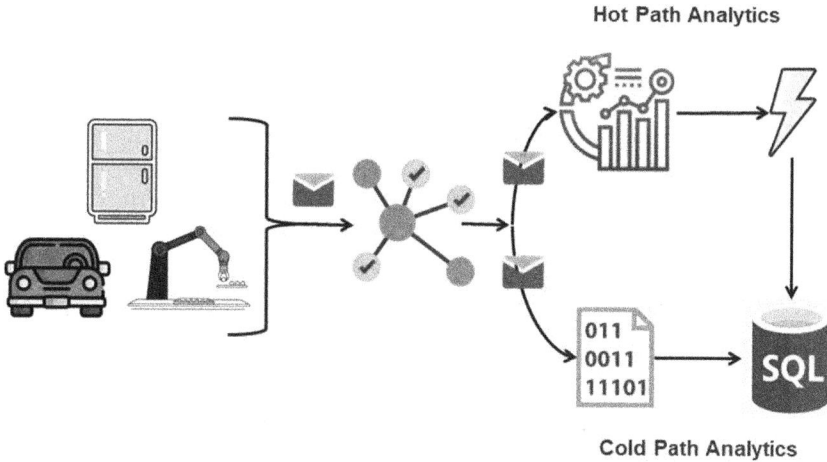

FIGURE 16.6   Edge data storage.

16.7   CHALLENGES AND OPPORTUNITIES

Edge computing has a lot of advantages in terms of processing speed, handling the data sets, and AI-powered functionality but it also has equally high challenges in analyzing and approaching securing devices and networks. Security is good in a few places such as the local area network and hyperscale clouds and for incorporating data centers, which has made organizations hide their information behind screens of security defenses in both online and offline modes. Data is increasingly processed and stored in local data centers, on industrial equipment in remote locations, and on IoT devices. This redistribution of work process will not be suitable for conventional models of IT security which exposes corporate data assets and hold digital information.

The controls required to address the security challenges of edge computing can be divided into two overlapping categories: those that apply to devices and those that concern networks. The way of managing the data and analysis of data in a company is thus transformed in the IoT industry.

The scale level varies with adding more connected devices at the edge and, in turn, the processes in various disciplines such as computing, network, management, storage, security, and licensing are not simple. Edge computing not only deals with more hardware, but it also affects the scale factor toward the software part. The unprecedented scale of data and its complexity have exceeded the capacity of the infrastructure. The major challenges that all industries face are concerned with digital transformation and economic growth related to connectivity and cybersecurity.

It is crucial to implement several types of privacy mechanisms and security levels for edge devices to prevent any malicious attack. The challenges toward security are data confidentiality, data integrity, secure data communication, access control, and privacy preserving [11].

Data collection is the new challenge in edge computing whereas data storage and access are critical. Regardless of the physical location and control, the management must follow similar procedures and protocols consistently.

In secure data communication, the end user's data must be encrypted and outsourced to the edge servers, and it must use a proper search key algorithm to decrypt the data. Unfortunately, basic key algorithms, such as secure ranked keyword search algorithm, attribute-based keyword search algorithm, dynamic search method, and proxy key encryption with keyword search approach, are made to be more complex at the privilege level [12].

In access control, big challenges exist in authentication due to multiple domains operating parallelly, which may have some limited privileges to access the network. This privilege may lead to unauthorized data access. Access rights are increasing every day, changing the security footprint and the need for servers to deal with the network and physical security to reflect traffic patterns. Irrespective of location, overall protection of data must be considered to protect the assets because just backing up the network will not be sufficient for effective protection of data. Therefore, it is to be considered in upcoming research to work on authorized control for the users of the available devices.

In privacy preservation, challenges are due to the remote servers accessing the end user's data and their personal information from edge devices.

16.8   SUMMARY

For IoT, edge computing is a vital part since speed, data, and analytics can be managed effectively and it has high power processing. This combination of IoT and edge has a high potential gain for mobile operators, customers, and partners. At the edge, physical devices use several types of operating systems and apps to manage and monitor remotely.

IoT and edge focus on managing many devices and large volumes of data and guarantee high-speed processing and less complexity of device management. It is less dependent on the cloud. Quantifiable values are produced by edge computing to consumer IoT use cases with zero delay and henceforth businesses on the IoT edge such as mobile networks are developed as the cost of connectivity is reduced.

IoT is increasingly dependent on edge networks. However, security risks in IoT devices make edge security more important than ever before. The same level of importance is handled by both edge computing and IoT. Overall, this combined platform takes some time to recognize as mainstream technologies. Becoming an expert in either edge computing or the IoT domain brings one closer to achieving overall mastery in the respective field. Using IoT edge computing has many advantages in security, interoperability, decreased data exposure, resiliency, cost-effectiveness, consistency, and quick response.

Edge requires fundamental security features such as an open network to administrator, encrypted data, an automated monitoring tool, and limited access to edit the data and resources. The edge to IoT devices resolves the delay and network traffic congestion and also improves the quality of the network connection.

REFERENCES

[1] T. Kavitha, G. Senbagavalli, D. Koundal, Y. Guo, and D. Jain (Eds.), *Convergence of Deep Learning and Internet of Things: Computing and Technology*. IGI Global, 2023, DOI: 10.4018/978-1-6684-6275-1. ISBN: 9781668462751.

[2] T. Kavitha, V. Ajantha Devi, S. Neelavathy Pari, and S. Ramanathan, *Internet of Everything: Smart Sensing Technologies*", Nova Science Publishers, Publication Date: June 17, 2022, DOI: 10.52305/PNQM1088

[3] J. Zhang, et al., "Data Security and Privacy-Preserving in Edge Computing Paradigm: Survey and Open Issues", *IEEE Access*, Volume 6, PP. 18209–18237, March 2018, DOI: 10.1109/ACCESS.2018.2820162

[4] Y. Ren, "Secure Data Storage Based on Blockchain and Coding in Edge Computing", *Mathematical Biosciences and Engineering*, Special Issue: Security and Privacy in Smart Computing, Volume 16, Issue 4, PP. 1874–1892, 2019, DOI: 10.3934/mbe.2019091

[5] M. S. V. Janakiram, "Edge Computing – Redefining the Enterprise Infrastructure", Forbes Newsletter, February 2017.

[6] B. Chen, "Edge Computing in IoT-Based Manufacturing", *IEEE Communication Magazine*, Volume: 56, Issue 9, PP. 103–109, September 2018, DOI: 10.1109/MCOM. 2018.1701231

[7] https://www.atos.net/en/solutions/edge-computing-infrastructure/definition-edge-computing

[8] W. Yu, "A Survey on the Edge Computing for the Internet of Things", *IEEE Mobile Edge Computing*, Volume 6, PP. 6900–6919, November 2017, DOI: 10.1109/ ACCESS.2017.2778504

[9] N. Hassan, "The Role of Edge Computing in Internet of Things", *IEEE Communication Magazine*, Volume 56, Issue 11, PP. 110–115, November 2018, DOI: 10.1109/MCOM. 2018.1700906

[10] Y. Tu, et al., "Task Offloading Based on LSTM Prediction and Deep Reinforcement Learning for Efficient Edge Computing in IoT", *Special Issue Machine Learning for Wireless Communications*, 2022, DOI: 10.3390/fi14020030

[11] https://techmonitor.ai/focus/security-challenges-of-edge-computing

[12] E. T. Michailidis, "Secure UAV-Aided Mobile EDGE Computing for IoT: A Review", *IEEE Access*, Volume 10, PP. 86353–86383, August 2022, DOI: 10.1109/ACCESS. 2022.3199408

17 Cloud Security for the IoT

Prashant Dahiwale
Government Polytechnic, Daman UT, Gujarat Technological University, India

M. M. Raghuwanshi
Symbiosis Institute of Technology, Deemed University, Nagpur, India

Sanjay Mate
Sangam University, Bhilwara, Government Polytechnic, Diu UT, India

17.1 INTRODUCTION

The security of IoT devices has been a growing concern for many years, and with the increasing number of devices and the sophistication of attacks, it has become more critical than ever to ensure that these devices are secured against cyber threats. Cloud-based IoT systems can offer significant advantages over traditional on-premise systems, such as increased agility, easier scalability, and the ability to collect and analyze vast amounts of data [1]. However, they also introduce a new set of risks and challenges that must be addressed.

The security of cloud-based IoT systems is crucial, as these systems often handle sensitive data and control critical infrastructure. In order to protect against attacks, it is essential to understand the unique security risks associated with cloud-based IoT systems and implement appropriate safeguards. This includes understanding the various security measures that can be implemented to mitigate these risks, such as encryption, access control, authentication, and secure communication protocols [2]. Additionally, continuous monitoring and threat detection are necessary to maintain the security of cloud-based IoT systems.

This chapter examines the current state of cloud security in IoT, including threats and attack vectors, and examines various security measures that can be used to mitigate these risks. Additionally, the importance of continuous monitoring and threat management in the security of cloud-based IoT systems is emphasized. By understanding security risks and implementing appropriate protections, organizations can take full advantage of IoT cloud technologies while minimizing the risk of cybercrime.

DOI: 10.1201/9781003477327-17

17.2   ROLE OF CLOUD COMPUTING IN IoT

Cloud computing plays a crucial role in the Internet of Things (IoT) by providing a platform for data storage, management, and processing for IoT devices and applications.

1. Data Storage: With the increasing number of IoT devices and their continuous generation of data, cloud computing provides a centralized data storage platform that can handle large amounts of data. This eliminates the need for local storage and ensures data security and accessibility.
2. Data Management: The cloud allows for the organization and management of large amounts of data generated by IoT devices. It provides tools for data analysis, data visualization, and real-time data processing, which are essential for making informed decisions based on IoT data.
3. Scalability: Cloud computing provides the ability to expand or reduce resources as needed, making it easier for IoT applications to manage changes in data volume and demand.
4. Cost Savings: By using cloud computing, organizations can reduce the cost of maintaining and updating their own data centers and IT infrastructure. This also reduces the need for specialized IT personnel and lowers the cost of data storage and processing.
5. Remote Access: Cloud computing enables remote access to IoT data and applications, making it possible to monitor and manage IoT devices from anywhere in the world with an Internet connection [3].

Overall, the integration of cloud computing and IoT provides organizations with a powerful platform for data management and analysis, enabling them to make informed decisions and drive innovation in the IoT space.

17.3   HOW CLOUD SERVICES BENEFIT THE IoT ECOSYSTEM

Let's understand how these cloud services and platforms can benefit the IoT ecosystem.

1. Scalability: Cloud services allow for easy scalability of computing resources as the number of connected devices and amount of data generated grows.
2. Data Management and Analysis: The cloud provides a centralized platform for storing, processing, and analyzing vast amounts of data generated by IoT devices, making it easier to derive insights and make informed decisions.
3. Cost Savings: By outsourcing data storage and processing to the cloud, organizations can reduce the costs associated with maintaining their own data centers and IT infrastructure.
4. Remote Access and Control: Cloud services enable remote access to IoT devices and data, allowing organizations to monitor and manage their devices from anywhere in the world.

5. Increased Reliability and Security: Cloud service providers invest heavily in infrastructure and security measures, which can provide increased reliability and security for IoT devices and data.

6. Improved Application Development: The cloud provides a platform for developing, testing, and deploying IoT applications, making it easier for organizations to create and launch new IoT solutions.

Overall, cloud services offer a range of benefits to the IoT ecosystem, from scalability and data management to cost savings and improved application development [4]. These benefits make it easier for organizations to realize the full potential of their IoT investments.

17.4 A COMMON ARCHITECTURE FOR INTEGRATING THE IoT WITH CLOUD COMPUTING

A common strategy for integrating the IoT with cloud computing often includes the following.

1. IoT Devices: IoT devices collect data from physical sensors, machines, and other sources, and transmit it to the cloud.

2. Edge Computing: Edge computing provides processing and analysis capabilities at the edge of the network, close to IoT devices, to reduce data transfer to the cloud and improve response time.

3. Gateway: The gateway acts as a bridge between IoT devices and the cloud, providing connectivity, security, and data management capabilities. The gateway can also preorder and collect data before sending it to the cloud.

4. Cloud Platform: The cloud platform provides a centralized repository for storing, processing, and analyzing data from IoT devices. It also provides services for application development, management, and deployment.

5. Data Management and Analysis: The cloud platform provides tools for organizing, storing, and analyzing large amounts of data from IoT devices, allowing organizations to derive insights and make informed decisions.

6. Applications: Applications built on the cloud platform can provide a range of services, such as monitoring and control, predictive maintenance, and real-time analytics. These applications can be accessed through a web interface, mobile app, or other channels.

7. Security: Security is a key concern in an IoT-cloud architecture and should be incorporated at multiple levels, including device security, data encryption, and access control.

   This architecture provides a scalable and flexible platform for integrating IoT devices and cloud computing, enabling organizations to realize the benefits of IoT and cloud computing [5], such as data management and analysis, cost savings, and improved application development.

17.5   SECURITY ISSUES IN CLOUD

Security is a critical concern in cloud computing, as sensitive data and applications are stored and processed on remote servers. Some of the common security issues in the cloud include the following.

1. Data Breach: A data breach occurs when an unauthorized person gains access to sensitive data stored in the cloud. This may be due to hacking, malware, or human error.
2. Insider Threats: Insider threats refer to security incidents caused by employees or contractors who have access to cloud systems. This can include accidental or intentional data breaches or theft of intellectual property.
3. Account Hijacking: Account hijacking refers to unauthorized access to a cloud account by an individual who obtains or guesses a user's login credentials.
4. Lack of Visibility and Control: Organizations may lack visibility and control over their data in the cloud, making it difficult to monitor their security status and provide personal information.
5. Compliance Issues: Many employment laws regulate the handling of sensitive data. Cloud service providers may not always comply with these regulations, which can lead to compliance issues and possible fines.
6. Dependence on Service Providers: Organizations that use cloud services are dependent on their cloud service providers for security as shown in Figure 17.1 [6], which can be a concern if the service provider experiences security breaches or fails to keep up with evolving security threats.

To address these security issues, organizations need to implement effective security measures such as access, multi-factor authentication, and access control, and conduct audits, security, and regular audits [7]. They must also carefully review and analyze cloud service providers' security practices and negotiate service-level agreements that clearly define security operations.

FIGURE 17.1   Security in cloud [6].

17.6   CLOUD SECURITY CHALLENGES

Cloud security poses several challenges that organizations need to overcome to ensure the safety of their data and applications in the cloud.

1. Data Privacy: Ensuring sensitive data is protected and accessible only to authorized users is a major challenge in cloud computing.
2. Compliance: Many business laws govern the handling of sensitive data, and organizations must ensure that their cloud service providers comply with these laws.
3. Data Sovereignty: Data sovereignty refers to the laws and regulations governing data storage and processing in various countries and regions. Organizations must ensure that their data is stored and processed in accordance with appropriate laws and regulations.
4. Multi-cloud Security: Organizations that use multiple cloud service providers may struggle to ensure consistent security across their cloud environments.
5. Responsibility Model: The role model in the cloud involves shared responsibility between cloud service providers and customers, which can lead to conflicts over who is responsible for the security of many aspects of the cloud environment.

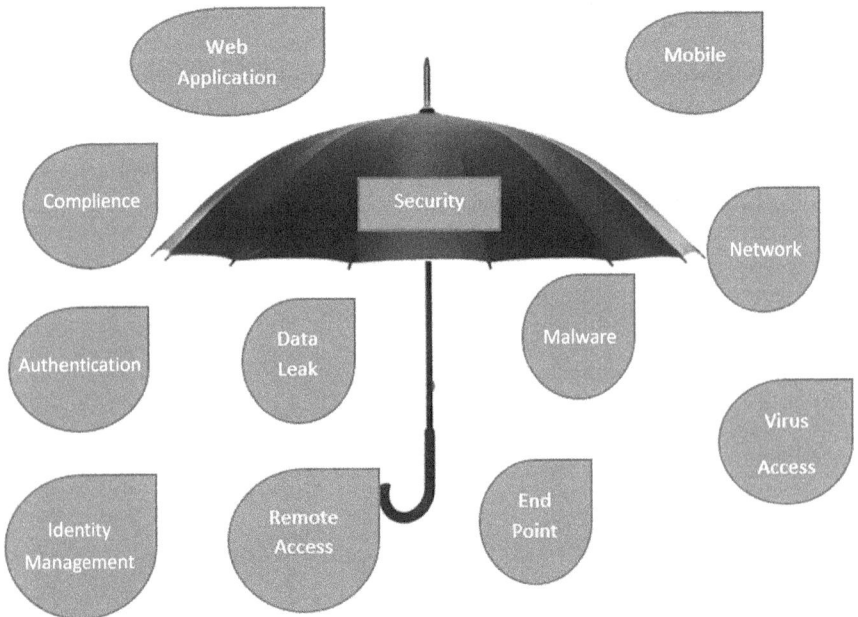

FIGURE 17.2   Challenges in IoT and cloud [6].

6. Complex Security Architectures: The dynamic and scalable nature of cloud computing can lead to complex security architectures that are difficult to manage and monitor.
7. Legacy Security Tools: Organizations may struggle to use their existing security tools and processes in the cloud, leading to gaps in security coverage.

To address these security challenges, organizations need to implement a comprehensive security strategy that covers all aspects of their cloud environment as shown in Figure 17.2 [6]. This may include implementing encryption, multi-factor authentication, access controls, and other security measures, as well as regularly reviewing and updating their security practices [8].

Additionally, organizations should seek to educate their employees about cloud security best practices and ensure that their cloud service providers have robust security programs in place.

The Pros and Cons of cloud security mechanisms are presented in Table 17.1.

TABLE 17.1
Pros and Cons of Cloud Security Mechanism

Cloud Security Mechanism	Pros	Cons
Encryption	1. Provides confidentiality and integrity of data 2. Protects data even if the network is compromised 3. Encryption algorithms are widely available and easy to implement	1. Encryption and decryption can cause performance issues 2. Key management can be complex.
Authentication	1. Ensures that only authorized users or devices access data 2. Prevents unauthorized access to sensitive data 3. Can be implemented using different authentication methods, such as biometric, token-based, or password-based	1. Can be vulnerable to attacks, such as phishing or password guessing 2. Authentication systems can be complex to manage
Authorization	1. Controls access to resources based on user or device privileges 2. Helps prevent unauthorized access to sensitive data. 3. Can be used in conjunction with authentication to provide an additional layer of security	1. Can be difficult to implement and manage 2. Can be vulnerable to attacks, such as privilege escalation or backdoor access
Virtual Private Networks (VPNs)	1. Provides a secure connection between devices and the cloud 2. Encrypts data in transit 3. Can be used to connect remote devices to the cloud securely	1. Can add latency and reduce performance 2. VPNs can be complex to configure and manage

(Continued)

TABLE 17.1   (CONTINUED)

Cloud Security Mechanism	Pros	Cons
Firewalls	1. Protects devices from unauthorized access 2. Monitors and controls incoming and outgoing traffic 3. Can be used to block traffic from known malicious IP addresses	1. Firewalls can be complex to configure and manage 2. May not protect against new or unknown threats 3. Can add latency and reduce performance
Intrusion Detection and Prevention Systems (IDS/IPS)	1. Monitors network traffic for suspicious activity 2. Can detect and prevent attacks 3. Can be used to block traffic from known malicious IP addresses	1. IDS/IPS can be complex to configure and manage 2. May generate false positives or false negatives 3. Can add latency and reduce performance
Security Information and Event Management (SIEM)	1. Centralizes security event monitoring 2. Can detect and respond to security incidents quickly 3. Provides a holistic view of security events across devices and the cloud	1. SIEM can be complex to configure and manage 2. Can generate a large volume of alerts 3. Can be costly to implement

17.7   SECURITY ECOSYSTEM: HOW CAN WE PROTECT OURSELVES?

Various cloud security issues as shown in Figure 17.3 [9] are threats to cloud security ecosystems, a security ecosystem is a comprehensive approach to protecting an organization's assets and data in the cloud. To secure themselves in the cloud, organizations can follow the following best practices.

1. Adopt a Risk-Based Approach: Organizations should prioritize their security efforts based on the level of risk associated with each asset and data set.
2. Encrypt Sensitive Data: Encrypting sensitive data at rest and in transit helps protect it from unauthorized access.
3. Use Multi-factor Authentication: Multi-factor authentication (MFA) requires users to provide multiple types of authentication, such as passwords and security tokens, to access the cloud.
4. Implement Access Control: Access control determines who can access data and resources in the cloud and how they can work. Organizations need to implement access control responsibilities to ensure that users can only access the information and resources they need to do their jobs.
5. Monitor and Audit Cloud Usage: Regularly monitoring and auditing cloud usage helps organizations identify and address security incidents and potential threats.
6. Implement Security-Focused Software Development Practices: Organizations should implement secure software development practices, such as threat modeling and secure coding, to help ensure that their cloud applications are secure.

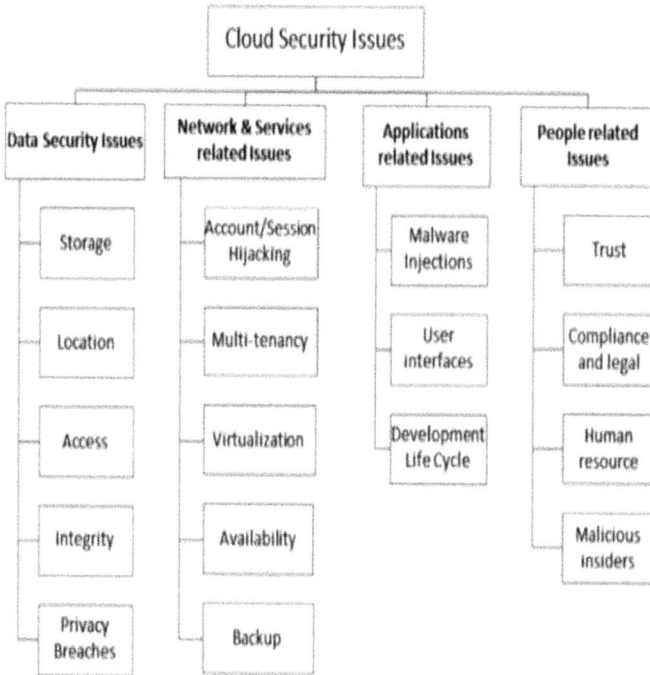

FIGURE 17.3 Cloud security issues [9].

7. Work with a Trusted Cloud Service Provider: Organizations should carefully evaluate the security practices of their cloud service providers and choose one that has a strong security program and a track record of security excellence.

8. Educate Employees on Cloud Security: Employee awareness and education are key components of the security ecosystem. Organizations should provide regular training to their employees to help them understand the importance of cloud security and how to protect data and applications in the cloud.

By following these best practices and implementing a comprehensive security ecosystem, organizations can help protect their data and applications in the cloud and minimize their risk of security incidents [10].

17.8 IoT CLOUD SECURITY SOLUTIONS

IoT cloud security solutions are measures taken to secure connected devices and the data they generate, process, and transmit. These solutions include the following:

1. Encryption: protecting data in transit and at rest using strong encryption algorithms.

2. Authentication: verifying the identity of devices and users accessing the IoT network.

3. Authorization: controlling access to data and resources based on predefined roles and permissions.
4. Firewall: A network security system that monitors and controls network access.
5. Threat detection and response: monitoring the network for potential security threats and responding to them promptly.
6. Software Updates: regular updating of devices' software to fix vulnerabilities and improve security.
7. Virtual Private Network (VPN): a secure communication tunnel between IoT devices and the cloud.
8. Identity and Access Management (IAM): managing identities, roles, and permissions of users and devices within the IoT network.

Implementing these solutions can help organizations secure their IoT systems and prevent unauthorized access, data breaches, and other security threats.

17.8.1 ENCRYPTION OF DATA AT REST

Data at rest encryption refers to the process of converting data into a secure, unreadable form to protect information from unauthorized access and theft while it is stored. Encrypted data can be decrypted and accessed only using a key or password. During this process, raw data is converted into ciphertext using an encryption algorithm. Encryption algorithms create a unique key for each data set, which must be kept secret to ensure the security of the data. Still, data can be stored on a variety of devices such as hard drives, flash drives, and cloud storage. Data encryption at rest helps protect sensitive data such as financial information, personal information, or confidential business information from theft or unauthorized access [11]. Examples of encryption algorithms used to encrypt data at rest include AES (Advanced Encryption Standard), RSA, and Blowfish. It is important to remember that encryption alone is not enough to maintain data security and must be combined with other security measures such as key management, access control, and regular security monitoring.

17.8.2 ENCRYPTION OF DATA IN TRANSIT

Data encryption in transmission refers to the process of converting data into a secure, unreadable form when it is transferred from one device to another. Encrypted data can be decrypted and accessed only using a key or password. During this process, raw data is converted into ciphertext using an encryption algorithm. Encryption algorithms create a unique key for each data set, which must be kept secret to ensure the security of the data. Data encryption in transit is important to protect sensitive information, such as financial information, personal information, or confidential business information, from illegal access or theft as it travels across the network [12]. Examples of encryption algorithms used to encrypt data in transit include SSL (Secure Sockets Layer), TLS (Transport Layer Security), and IPSec (Internet Protocol Security). It is worth noting that encryption alone is not enough to protect data in transit because it must be combined with other security measures such as key management, access control, and regular security monitoring.

17.8.3   Device Identity

Device identity refers to a unique identifier assigned to a device connected to an IoT network. The device identity allows the system to differentiate between devices and recognize them when they connect to the network. having a unique identity for each device is essential for secure communication and management within an IoT network [6]. It enables the system to perform authentication and authorization processes, ensuring that only authorized devices can access the network and its resources.

Device identity can be established using various methods, such as the following.

1. Unique Device Identifiers (UDID): A fixed, unique identifier assigned to a device during the manufacturing process.
2. Digital Certificates: A certificate issued by a trusted certificate authority (CA) that verifies the device's identity and helps establish a secure connection.
3. Public Key Infrastructure (PKI): A system that uses public and private keys for authentication and encryption. The public key is used for encrypting messages and verifying digital signatures, while the private key is used for decryption.
4. Token-Based Authentication: A process where a unique token is assigned to each device, allowing it to authenticate and access the network.

Having a secure and unique device identity is critical for ensuring the security of an IoT network and protecting it from unauthorized access and potential security threats.

17.8.4   Device Authentication Using OAuth 2.0

OAuth 2.0 is a standard authentication tool that provides a secure way for devices to access protections on behalf of the user. It is widely used for IoT device authentication. In OAuth 2.0, a device first requests access to protected resources from an authorization server. The authorization server then asks the user to grant the device permission to access the resource. If the user gives permission, the authorization server offers access to the device, which the user uses to access protected resources [13].

Here are the steps to authenticate a device using OAuth 2.0.

1. Device Registration: Before the device can receive a user ID and secret, it must first register with an authorization server.
2. Device Authorization: The device sends an authorization request to the authorization server, including the user ID and required access.
3. User Authorization: The user will be asked to allow or deny access to the device.
4. Granting an Access Token: If the user allows access, the server allows access to the device.
5. Data Access: Devices use tokens to access protected data.

OAuth 2.0 provides a secure and flexible way to authenticate devices, allowing devices to access security without having to store and manage the user credentials used. It also supports various authorization types such as authorization code and implicit user credentials, allowing it to be used in various IoT situations.

17.8.5   USER ROLE AND POLICY

User responsibilities and rights refer to the permissions and restrictions assigned to users within an organization or system. User roles define the activities and roles that users are allowed to perform on the system, such as accessing certain resources, modifying files, or performing administrative tasks. User roles are typically assigned based on the user's job title or level of responsibility. Policy defines the rules and guidelines that govern access to resources, information, and systems. Indicates the conditions under which the user is authorized to perform certain actions, such as reading, changing, or deleting information. Policies can be designed to fit business needs, such as data privacy laws, or to meet an organization's security policy [13, 14]. The combination of user roles and policy controls ensures that users have the permissions they need to perform their jobs while limiting their access to sensitive data and resources. This helps reduce the risk of unauthorized access and protects the system from security threats. It's important to regularly review and update user roles and policies to ensure they remain relevant to your organization's changing needs and security.

17.8.6   CERTIFICATE-BASED AUTHENTICATION

Certificate-based authentication is a type of authentication that uses digital certificates to identify a user or device. This authentication method is widely used for secure communications and access to protected Internet and Intranet environments. In certificate-based authentication, each user or device is issued a digital certificate signed by a trusted CA. The certificate contains information regarding the identity of the user or device, such as the CA's name, public key, and digital signature. When a user or device attempts to access a protected resource, the client device verifies the certificate by analyzing the digital signature and comparing it to the CAs list [14]. If the certificate is valid and signed by a trusted CA, the user or device is allowed to access the resource.

Certificate-based authentication provides several benefits, including the following.

1. Strong Security: Digital certificates are more secure than passwords and provide a higher level of assurance that the user or device attempting to access the resource is who they claim to be.
2. Non-repudiation: The use of digital signatures ensures that the identity of the user or device cannot be denied.
3. Scalability: Certificate-based authentication can easily be scaled to support large numbers of users or devices.

It's important to note that certificate-based authentication requires the proper management of digital certificates, including regular updates and revocations, to ensure their security and validity.

17.9 CONCLUSION

Securing the IoT cloud is a crucial aspect of protecting the vast network of connected devices and the data they generate. Ensuring the security of the IoT cloud requires a multi-layered approach that covers all aspects of the device-to-cloud connection, including data encryption, device identity and authentication, user role and policy management, and certificate-based authentication. IoT cloud security solutions must also be designed to keep pace with the rapid evolution of technology and the increasing number of connected devices. This requires regular updates, testing, and continuous monitoring to identify and mitigate new security threats.

Implementing effective IoT cloud security measures can help organizations protect their data, maintain privacy, and comply with industry regulations. It also ensures the reliable operation of connected devices, builds trust with customers and partners, and supports the growth of the IoT industry.

REFERENCES

1. Waqas Ahmed, Abdul Rahman Javed, Thar Baker, and Zunera Jalil, "Cyber Security in IoT-Based Cloud Computing: A Comprehensive Survey", *Electronics*, Vol. 11, No. 16, 2022, DOI:10.3390/electronics11010016
2. Irfan Mohiuddin and Ahmad Almogren, "Security Challenges and Strategies for the IoT in Cloud Computing", *2020 IEEE 11th International Conference on Information and Communication Systems (ICICS)*, 07–09 April 2020, Irbid, Jordan, DOI:10.1109/ICICS49469.2020.239563
3. M. Mamun-Ibn-Abdullah and M. Humayun Kabir, "A Multilayer Security Framework for Cloud Computing in Internet of Things (IoT) Domain", *Journal of Computer and Communications*, Vol. 9, No. 7, July 2021.
4. S. Liu, T. Zhao, X. Liu, Y. Li, and P. Wang, "Proactive Resilient Dayahead Unit Commitment with Cloud Computing Data Centers", *IEEE Transactions on Industry Applications* (Early Access Article), 2022.
5. A. Mohiyuddin, A. R. Javed, C. Chakraborty, M. Rizwan, M. Shabbir, and J. Nebhen, "Secure Cloud Storage for Medical IoT Data Using Adaptive Neuro-Fuzzy Inference System", *International Journal of Fuzzy Systems*, 1–13, 2021.
6. Pokuri Rajani and Parupally Anuja Reddy, "Security Ecosystem in IoT & Cloud", *International Journal of Engineering and Computer Science*, Vol. 5, No. 03, 15966–15970, March, 2016, ISSN 2319-7242.
7. Aderemi A. Atayero, Olusegun A. Ilori, and Michael O. Adedokun, "Cloud Security and the Internet of Things: Impact on the Virtual Learning Environment", *Proceedings of EDULEARN15 Conference, 6th-8th* July 2015, Barcelona, Spain, ISBN: 978-84-606-8243-1.
8. X. Li, et al., "Enhancing Cloud-Based IoT Security through Trustworthy Cloud Service", *IEEE Access*, Vol. 7, 2019. DOI:10.1109/ACCESS.2018.2890432
9. W. Ahmad, A. Rasool, A. R. Javed, T. Baker, and Z. Jalil, "Cyber Security in IoT-Based Cloud Computing: A Comprehensive Survey", *Electronics*, Vol. 11, 16. DOI:10.3390/electronics11010016
10. J. Singh, et al., "Twenty Security Considerations for Cloud-Supported Internet of Things", *IEEE Internet of Things Journal*, Vol. 3, No. 3, 269–284, 2016.
11. L. Wang and R. Ranjan, "Processing Distributed Internet of Things Data in Clouds", *IEEE Cloud Computing*, Vol. 2, No. 1, 76–80, 2015.

12. Security in IoT. https://sist.sathyabama.ac.in/sist_coursematerial/uploads/SECA7021.pdf
13. Importance of Cloud Computing for Large-Scale IoT Solutions. https://www.einfochips.com/blog/importance-of-cloud-computing-for-large-scale-iot-solutions/
14. IoT Cloud Security and IoT Application Security. https://www.embitel.com/blog/embedded-blog/iot-cloud-and-application-security

18 IoT Security Using Blockchain

P. Sankar
Hindustan Institute of Technology and Science,
Chennai, India

Anitha Kumari
GITAM University (Deemed), Bengaluru, India

18.1 INTRODUCTION

The IoT industry has the opportunity to benefit greatly from blockchain technology. The IoT has been shown to be a powerful tool for businesses looking to get an edge in their respective markets. The user's privacy and security are put at risk with each new piece of information gathered, as it is used to trigger or request an action without human participation. The term "Internet of Things" (IoT) refers to a system in which embedded systems, sensors, software, and artificial intelligence are utilized to collect data from the internet and then apply it to various intelligent applications (Kavitha and Saraswathi, 2017). By using blockchain technology, it is possible to prevent double expenditure on a single transaction. One of the problems that arose as a result of the IoT's rapid growth was its susceptibility to cyberattacks. Blockchain's decentralized digital phase makes it well-suited for the cryptocurrency transactional environment, and this fact has led to its inevitable adoption in the IoT (Zheng et al., 2017). Using blockchain technology to increase IoT security is one option.

Researchers have done a number of studies on how blockchain systems can be used with IoT. Oscar Novo (2019) studied a proof-of-concept architecture in which blockchain was used to manage access to the different systems to controls in a distributed way. However, blockchain technology can be used to make this IoT more secure. Data privacy is compromised by the centralized client–server architecture of the IoT. Data security is severely compromised by the server's centralized design (Khan and Salah, 2018). That data integrity, confidentiality, and security are compromised in the core infrastructure of the IoT is a main roadblock to the technology's mainstream adoption (Garg et al., 2021).

Sensor data measurements can be tracked and protected from being duplicated by malicious parties using blockchain technology. IoT device deployments can be complicated, but a distributed ledger can facilitate easy identification, authentication, and secure data sharing across nodes. An ecosystem of systems is necessary to understand the full potential of the IoT. This requires integrated solutions and protection. The goal of this chapter is to acquaint the reader with the most recent developments in IoT

DOI: 10.1201/9781003477327-18

217

security using blockchain technology. This paper examines the IoT environment's issues and potential, as well as how blockchain integration can mitigate some of them.

18.2   IoT SECURITY

The IoT connects individuals, locations, and objects, thus creating the potential for value creation and capture. Embedded in physical objects are sophisticated electronics, sensors, and actuators that transfer data to the IoT network. This data is utilized by the IoT's analytical capabilities to translate insights into action, which impacts corporate operations and leads to new methods of working. However, some technical and security issues still need to be addressed (Deloitte, 2022). IoT devices can continue to connect to the internet as long as there is a telecommunications infrastructure and hardware that makes this possible. This platform integrates real-world and virtual worlds in addition to facilitating the collection of data from computers, connecting them to backend systems, and enabling the development of IoT applications. Each intelligent gadget utilizes data in real time and incorporates predictive analytics and large amounts of data. In addition, automated learning provides context and independently develops actions (Chakray, 2022a). IoT increases the amount of data-gathering gadgets in businesses and homes (Alamri et al., 2019).

Security concerns have hampered IoT deployment. Due to shortcomings in security, IoT devices are frequently attacked by Distributed Denial of Service (DDoS) assaults. In DDoS assaults, numerous infected computers bombard a target with a large number of simultaneous data requests, depriving users of the targeted system of service. Recent DDoS assaults have disrupted businesses and individuals. Cybercriminals can use unreliable IoT devices' inadequate security to execute DDoS assaults. Scalability is another IoT challenge. This could make blockchain more centralized as it grows and needs to keep track of records, which is bad for its future (Chandel et al., 2020). The necessity for centralized systems to verify, authorize, and link nodes will become a bottleneck as the number of IoT devices rises. This would necessitate large investments in servers capable of handling large amounts of data transmission. If a server goes down, the entire network can fall down. Gartner predicts IoT endpoints will expand by 32% annually from 2016 to 2021, reaching around 25 billion units. As IoT devices are projected to become so important in the next few years, companies must invest in tackling security and scalability issues (Deloitte, 2022). IoT security is based on technology that protects both connected equipment and the IoT itself. Data is automatically sent across the network using unique identifiers. Precautions have been taken to ensure network and device security. Frameworks have been released by the Global System for Mobile Communications (GSMA), the Industrial Consortium, the IoT Security Foundation, and others. Also, the General Data Protection Regulation (GDPR) of the European Union protects privacy and is used by IoT devices (Chakray, 2022a).

18.3   BLOCKCHAIN IN IoT SERVICES

Despite the great potential of emerging technologies like blockchain and the IoT, businesses are reluctant to use them due to technical and security issues. Blockchain and IoT will continue to evolve into global standards (Eberendu and Chinebu, 2021). Since blockchain technology is the backbone of crypto coin mining, its popularity has grown in tandem with the industry. But today, IoT is also a significant factor. Extremely large data sets are processed by IoT devices, which are linked in a chain and hence vulnerable to cyberattacks. In this situation, it becomes feasible to use blockchain technology to standardize, verify, and protect the approval of data processed by the devices. To ensure the integrity of the IoT, blockchain can track the data gathered by sensors without allowing any false information to be added to the chain. Using Blockchain technology, sensors may share information without involving a central authority. IoT security and scalability could be helped by blockchain or distributed ledger technology.

Blockchain's unique powers and benefits make it an "information game changer." A distributed ledger technology that is shared by participants over the internet is the foundation of a blockchain system. A transaction or event cannot be altered or withdrawn from the ledger after it has been verified and recorded. It lets people record and exchange information. In this community, certain people keep a copy of the ledger and use a consensus process to confirm new transactions. Each blockchain transaction is substantiated by an ensemble of users., increasing the system's resistance against fraud. IoT sensors that rely on batteries, are small, have limited processing power, and lack adequate storage space incur an additional burden in the form of security (Dasgupta et al., 2019). Blockchain and IoT devices are addressable and capable of listing the history of connected equipment, forming a bank to help solve difficulties in the future (Chakray, 2022b). Some uses of blockchain in IoT scenarios include secure device identity verification, data integrity assurance, and supply chain transparency. However, low computational power in IoT devices can be a challenge for joining blockchain networks as it requires efficient consensus mechanisms and lightweight cryptography to minimize resource consumption.

18.4   BUSINESS AND SOCIAL IMPACTS ON IoT SECURITY

Blockchain can significantly impact IoT security by enhancing data integrity, providing transparency, enabling decentralization, fostering trust, reducing costs, and promoting interoperability. These impacts benefit businesses through improved cybersecurity, operational efficiency, and market competitiveness, while also benefiting society by safeguarding privacy, ensuring transparency in supply chains, and empowering individuals and communities.

- **Enhanced Security**: Blockchain's immutable ledger helps protect sensitive IoT data from tampering and unauthorized access, boosting consumer trust and IoT adoption. It ensures privacy and safety, preventing breaches and harm from compromised IoT devices.
- **Data Integrity**: Blockchain assures data integrity, crucial in industries like healthcare and finance, reducing risks and compliance costs. It improved data integrity and safeguards personal info, medical records, and financial data, preserving trust in IoT tech.

- **Decentralization**: Decentralized blockchains promote innovation and competition in IoT, reducing reliance on central authorities. It empowers individuals and creates a more equitable IoT landscape.
- **Interoperability**: Blockchain fosters interoperability among IoT devices, cutting development costs and expanding market reach. Enhanced interoperability leads to user-friendly IoT solutions, benefiting diverse individuals and preferences.
- **Supply Chain Transparency**: Blockchain ensures supply chain visibility, reducing fraud and guaranteeing product authenticity. Transparent supply chains promote fair trade, sustainability, and safer products.
- **Trust and Accountability**: Blockchain fosters trust with transparent, tamper-proof IoT records. Improved trust encourages IoT adoption and societal benefits.
- **Cost Reduction**: Blockchain streamlines IoT management, reducing operational costs. Lower deployment costs can lead to more affordable IoT services, benefiting a broader range of users.

Security issues in IoT networks are a major concern due to the proliferation of connected devices. Blockchain technology offers solutions to several security issues in IoT networks by providing a secure, tamper-resistant, and decentralized infrastructure. However, it's important to carefully design and implement blockchain solutions to ensure they effectively address these security concerns while considering factors like scalability, interoperability, and energy efficiency.

18.5 IoT–BLOCKCHAIN IMPLEMENTATION

IoT networks are able to process data transactions originating from a wide variety of connected devices, each of which may be owned and operated by a distinct organization. This makes it hard to find the source of any data leaks if cybercriminals attack. Also, since the IoT creates a lot of data and since there are many parties involved, it's not always clear who owns the data (Deloitte, 2022). In the following ways, Blockchain can help ease IoT's security and scalability problems:

- Blockchain's tamper-proof distributed ledger eliminates the need for trust. No single organization controls all IoT data.
- Blockchain storage of IoT data will increase network security. It is nearly impossible to delete data on blockchains because of their strong encryption.
- A blockchain allows anyone with network access to track prior transactions. This can help identify a data leak's source and take prompt action.
- Chains of transactions enable quick transaction processing and device coordination.
- As the number of devices that can talk to each other grows, distributed ledger technology makes it possible to handle a huge number of transactions.
- By eliminating IoT gateway processing overheads, blockchain can help IoT enterprises decrease expenses.

Smart contracts, which are agreements between two parties that are stored on a distributed ledger, might enable the implementation of contractual arrangements between parties if specific conditions are satisfied. An illustration of a distributed ledger system is smart contracts. A smart contract, for instance, can immediately allow payments without requiring human involvement when all conditions for providing the service have been completed.

18.5.1 BLOCKCHAIN IN AN IoT NETWORK

Blockchain can be implemented in IoT through various approaches and use cases. Implementing blockchain in IoT requires careful consideration of factors such as scalability, energy efficiency, and the choice of blockchain platform. It's essential to tailor the solution to specific IoT use cases and ensure that the benefits of blockchain, such as security and transparency, outweigh the associated complexities and costs. Figure 18.1 illustrates blockchain–IoT networks' interactions with data. Building a blockchain-based IoT infrastructure requires first thinking about how devices will communicate with one another (Safonov, 2022).

- **IoT–IoT**: The simplest way to integrate blockchain technology into the IoT network is to use a shared register to store IoT data. There will be different routing strategies used for data transfer outside of the blockchain. This will result in faster transaction times and fewer delays. This strategy enables devices to operate offline. It is a straightforward approach to deployment since it does not require major changes to the workflow of IoT devices. The blockchain can be used instead of a cloud or server to send, store, and extract data.
- **IoT–Blockchain**: In this method, the blockchain serves as a cloud for conventional IoT networks and mediates interactions between IoT devices. Tracing, communication security, process automation, and increased capacity are all

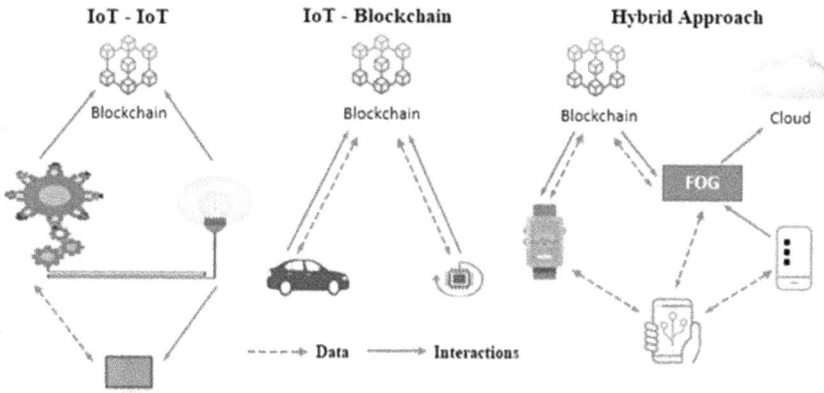

FIGURE 18.1 Blockchain–IoT interactions.

positive outcomes that can be anticipated. It will be much more complicated and time-consuming if the blockchain is too slow. Due to the extensive
modifications to both the operation of IoT devices and the development of
the blockchain, integrating this technology into IoT networks is challenging.
Furthermore, a blockchain that has improved throughput, capacity, and no
transaction fees should be used. This may be a blockchain based on IOTA,
Modum.io, Riddle, or Code.

- **Hybrid Approach**: The majority of data and interactions are shared across
  IoT devices, and only a subset of data types is retained by the blockchain.
  It has many benefits, but IoT devices are unlikely to operate with minimal
  latency and high speed in real time due to this issue. This strategy facilitates fog computing to compensate for blockchain and IoT limitations. In
  addition, using peripheral devices instead of cloud computing could reduce
  operating costs when extracting, storing, and analyzing private data.

18.5.2 MAJOR GAPS IN IoT

Figure 18.2 shows how the IoT is interconnected. IoT is a concept that describes a
system in which smart things interact as humans do via the internet. Sensors, wearables, and complicated mechanisms can connect to the internet and other devices
with the help of routers and gates. They exchange information and commands.
Furthermore, devices with user interfaces, such as personal computers, laptops, tablets, and so on, can be used in IoT networks. This technology is utilized for a variety
of reasons, including data transfer between meteorological and hydrometeorological
stations and the development of infrastructure for smart homes and cities. The difficulties, methodologies, and applications of IoT are constantly evolving. So do their
security and scaling requirements. According to Gartner, these two issues will be the
most difficult for IoT development in the coming years, with blockchain technology

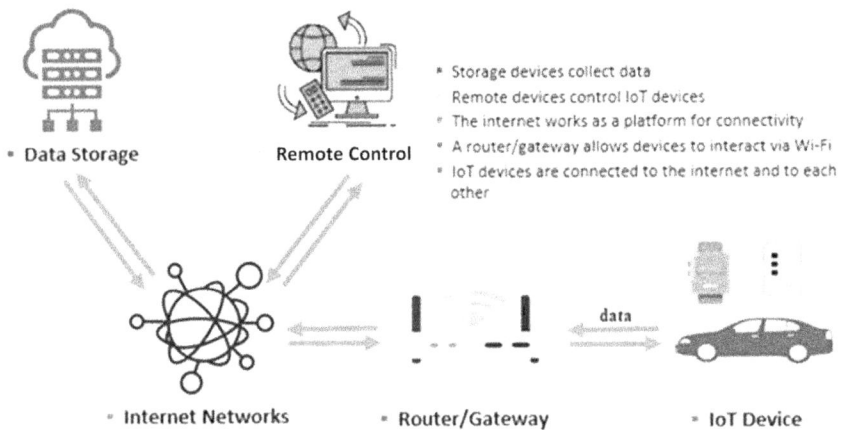

* Data Storage   Remote Control   ▪ Storage devices collect data
                                           Remote devices control IoT devices
                                           ▪ The internet works as a platform for connectivity
                                           ▪ A router/gateway allows devices to interact via Wi-Fi
                                           ▪ IoT devices are connected to the internet and to each
                                             other

data

▪ Internet Networks   ▪ Router/Gateway   ▪ IoT Device

FIGURE 18.2   Internet of Things.

providing the greatest solution. These are the most significant problems that blockchain technology can solve.

- **Obsolete Firmware**: Only a few providers update their IoT systems regularly, and fewer consumers update the firmware when new versions are released. Computers with outdated firmware may have security holes and weak spots, making them harder to hack.
- **Inadequate Authentication**: Simple authentication for applications with pre-set, strong passwords is used by most IoT systems. This form of security can be readily circumvented by simply locating the correct password.
- **Insecure Connection**: It measures data transfers between IoT devices and the cloud, or when data is stored on a device or in the cloud.
- **Physical Intrusion**: Hackers may change the configuration of an IoT device in order to overhear conversations, record video, or launch DDoS attacks.

Major gaps in IoT issues with blockchain theoretically involve leveraging blockchain's core features to provide solutions to common IoT challenges. The theoretical application of blockchain to address IoT gaps demonstrates its potential to provide innovative solutions to the most pressing challenges in IoT deployments. However, practical implementation considerations, including scalability, energy efficiency, and the choice of blockchain platform, must be carefully evaluated to achieve the desired outcomes.

18.6 CHALLENGES IN IoT–BLOCKCHAIN

Some are merging them in order to determine the possibilities of reducing security and other associated business risks. There may be difficulties along the way, but more firms are investing in blockchain-based IoT solutions (Eberendu and Chinebu, 2021). Security is a major issue with the current generation of IoT systems, which are based on a client–server model that is administered by a centralized authority. Blockchain solves this issue by moving the decision-making to a distributed network of computers based on a system of shared consensus.

Consensus protocol choice is difficult. To choose the best blockchain solution for IoT, consider that many mechanisms cannot be employed owing to high computational power needs, scaling challenges, excessive fees, etc. Each transaction might cost $2 to $20 using the Bitcoin or Ethereum blockchain. IoT interacts with dozens or millions of transactions per day, which requires a lot of money. With IoT devices having limited capabilities and computing power, blockchains are careful about validators. To work with the blockchain consensus algorithm and the IoT, it must meet the following requirements:

- Low transaction fees
- Low computational costs
- Low communication complexity
- High fault tolerance
- IoT-focused validation methodologies

- Resilience to Sybil assaults
- Resistance to denial-of-service attacks

There are currently just a few consensus algorithms that are capable of meeting these requirements, and the majority of them are founded on the Byzantine Fault Tolerance (BFT) protocol. The following is a comparison of the most popular processes for reaching a consensus.

IoT Infrastructure Limitations: The next issue is that IoT devices have limited CPU capabilities and small memory. No one has yet found a solution to this problem, but IBM appears to be very close. They pioneered the idea of flexible and localized blockchains. The network nodes are split into three categories: Simple P2P nodes that save address and balance, basic P2P nodes that store recent transactions, and P2P exchangers that copy the blockchain and analyze data.

Insufficient Encryption: Encryption is used in many modern apps, programs, and systems. IoT encryption security can be boosted with entropy encryption, which is based on quantum random number generation. In this method, quantum physics data is used instead of classical physics data. The generation of quantum random numbers improves the encryption of IoT networks.

Scaling Difficulties: The issue is that IoT networks continue to expand rapidly, necessitating the processing of additional smart devices, transactions, and data. All of this makes scaling challenging, particularly for blockchains with slow processing speeds. Blockchain architecture for the IoT should be capable of processing thousands of transactions per second. This can be accomplished in several ways.

Parallel Computing: This improves the efficiency of the transaction while simultaneously processing a limited number of transactions at once. This method of computing can be utilized for the purposes of data collection and analysis in the IoT, for the analysis of huge volumes of data, and for the processing of side projects that require a significant amount of computing power.

Other Blockchain Solutions: They may be able to boost the work rate by using affiliate networks, parental chains, and root chains. By including smart contracts that are able to communicate with the main blockchain, these protocols may be able to accelerate work even further.

No protocol Communication: Devices in the traditional IoT network are usually connected to the internet securely and quickly using wired or wireless technologies such as DSL/ADSL, Wi-Fi, 4G, and LTE. Low-power and low-bandwidth protocols are commonly used to connect the smart devices used in blockchain and IoT integration to the internet. However, blockchains are not compatible with these protocols.

Blockchain will open up new avenues for implementing IoT systems. Despite these few shortcomings, blockchain offers several significant advantages, but its acceptance in the sector is still in its early stages (Eberendu and Chinebu, 2021). To solve this problem, developers must construct unique blockchain and IoT protocols. Time and money are needed. IoT devices overload smart contracts. This chapter explored the IoT's benefits for smart contracts. It is still challenging and needs third-party data. It can be compromised in this manner, hence IoT needs strong authentication, security, and trust. Decentralized smart contracts demand a lot of computer power, a problem for the IoT. Reduce overloading while increasing device, transaction, and bandwidth productivity (Safonov, 2022). The safety of IoT devices is as follows:

- Encrypted passwords, which must be strong even if changed frequently.
- Advanced encryption, because resource constraints are common.
- Assets that were never intended for networking
- Adopting a single protective framework improves security and interoperability.
- 5G networks handle millions of linked communications devices.
- Temporary delays, high speed, and low power usage are goals.
- The government uses IoT for lighting, security, and traffic.
- IoT creates a hyper-connected world that requires constant control and security. It uses programming codes to solve this problem.

18.7 OPPORTUNITIES

In recent years, blockchain has evolved into a technology with several capabilities that will help the IoT expand in the next few years (Christidis and Devetsikiotis, 2016). The decentralized nature of both blockchain and the IoT bodes well for their eventual merging. However, owing to the vast disparity in processing power and network connectivity across IoT devices, there will be inconsistencies in the blocks such devices generate. However, gaining access to massive devices will lead to a dramatic growth of blockchain data (Xu et al., 2021). IoT necessitates new ways of conducting business and, consequently, new ways of generating revenue. IoT offers enterprises real-time sensor data and information services. It automates business and production processes, remotely controls operations, and optimizes supply chains. IoT enhances staff productivity and effectiveness by automating regular operations and accelerating decision-making and communications. IoT can improve customer experiences. In addition to being functional, its products and services are appealing and customizable. Blockchain technology solves scalability, privacy, and IoT security confidence challenges (Chakray, 2022b).

- The blocks are public, but the transaction content is safeguarded by private keys.
- It is decentralized, and there is enough trust. Network members agree on transactions.
- The database grows while keeping records. Changing old records would be expensive.
- Blockchain identities are used in high-capacity technology such as authentication systems. These want to become laws because they protect business and personal information.

- It also enables users to collaborate on the use of several files simultaneously. It ensures resource robustness and stability, reducing traffic and delays. The network is secure and user identities are confidential.

As this area of technology has made a lot of progress, sensors and smart chips are becoming more portable and able to interact with blockchain ledgers in real time. When applied to the IoT, blockchain technology with IoT can facilitate the development of a marketplace where devices can trade services with one another, and businesses can get value from the information they collect. Already, the proliferation of new blockchain protocols, alliances, and IoT device suppliers suggests that blockchain technology could be a strong fit for the IoT market (Deloitte, 2022).

18.8   SUMMARY

In summary, blockchain technology has the potential to greatly enhance both business and social aspects of IoT security by providing robust, transparent, and decentralized solutions that promote trust, data integrity, and interoperability while reducing costs and enhancing accountability. Both blockchain and IoT are promising new technologies that have yet to achieve widespread adoption due to lingering concerns about their reliability and safety. Many companies in the industry are working on applications that leverage both technologies together. This is because the two technologies can be used together to reduce security risks and the business risks that come with them. IoT networks can be made safer and more productive with Blockchain technology. It can also make them decentralized and give them the ability to use smart contracts. However, using this technology comes with a number of problems, such as limited IoT resources, weak encryption, scaling issues, and communication protocols that focus on both blockchain networks and IoT devices. The process can be done through custom development, which should only be done by a reliable technical partner. Companies should start thinking about how to use blockchain and IoT to solve their business problems. The goal of this innovation hub is to inspire and guide businesses as they develop their own IoT and blockchain initiatives.

REFERENCES

Alamri, M., Jhanjhi, N.Z. and Humayun, M. (2019). Blockchain for Internet of Things (IoT) Research Issues Challenges & Future Directions: A Review, *International Journal of Computer Science and Network Security*, 19(5): 244–258.

Chakray (2022a). What Is the Internet of Things (IoT) and What Challenges Does It Pose? *Chakray*.

Chakray (2022b). Blockchain and IoT Security: Everything You Need to Know, *Chakray*.

Chandel, S., Zhang, S. and Wu, H. (2020). Using Blockchain in IoT: Is It a Smooth Road Ahead for Real? *Advanced in Information and Communication*, pp. 159–171.

Christidis, K. and Devetsikiotis, M. (2016). Blockchains and Smart Contracts for Internet of Things, *IEEE Access*, 4: 2292–2303.

Dasgupta, D., Shrein, J.M. and Gupta, K.D. (2019). A Survey of Blockchain from Security Perspective. *Journal of Banking and Financial Technology*, 3: 1–17.

Deloitte (2022). Perspectives: Can Blockchain Accelerate Internet of Things (IoT) Adoption? *Deloitte*.

Eberendu, A.C. and Chinebu, R. I. (2021). Can Blockchain Be a Solution to IoT Technical and Security Issues, *International Journal of Network Security & Its Applications*, 13(6): 123–132, doi: 10.5121/ijnsa.2021.13609

Garg, R., Gupta, P. and Kaur, A. (2021). Secure IoT via Blockchain. *IOP Conference Series: Materials Science and Engineering (ICCRDA 2020)*, 012048, doi: 10.1088/1757-899X/1022/1/012048

Kavitha, T. and Saraswathi, S. (July, 2017). New Sensing Technologies or/and Devices for Emergency Response and Disaster Management, Book Chapter, IGI Global International Publisher, pp. 1–40, doi: 10.4018/978-1-5225-2575-2, Ch001. ISBN: 9781522525752.

Khan, M.A. and Salah, K. (2018). IoT Security: Review, Blockchain Solutions, and Open Challenges, *Future Generation Computer Systems*, 82: 395–411.

Oscar Nova, E. (2019). Scalable Access Management in IoT Using Blockchain: A Performance Evaluation, *IEEE Internet of Things Journal*, 6(3): 4694–4701.

Safonov, A. (2022). Blockchain: How to Implement Blockchain in IOT? *Merehead.*

Xu, L.D., Lu, Y. and Li, L. (2021). Embedding Blockchain Technology into IoT for Security: A Survey, *IEEE Internet of Things Journal*, 8(13): 10452–10473.

Zheng, Z., Xie, S., Dai, H., Chen, X. and Wang, H. (2017). An Overview of Blockchain Technology: Architecture, Consensus, and Future Trends, *IEEE 6th International Congress on Big Data*, pp. 557–564.

19 IoT Ecosystem-Level Security Auditing, Analysis, and Recovery

Anita Shukla, Ankit Jain, Puspraj Singh Chauhan, and Raghvendra Singh

Pranveer Singh Institute of Technology, APJ Abdul Kalam Technical University, Lucknow, India

19.1 INTRODUCTION

In order to build cyber-physical systems, the Internet of Things (IoT) connects "things," or individually programmable objects, with physical sensing and/or actuation capabilities. These gadgets primarily gather information from the outside world and act on it, with the potential to collaborate with processing services [1]. There are many advantages that are provided, and they are presently used in various cases, including healthcare [2, 3], fitness [4], the agriculture business [5, 6], and manufacturing [7, 8]. Self-driving cars [9] and smart cities [10] are examples of ideas that were once thought to be futuristic and far-fetched. Healthcare [11], IoT devices [12], smart homes [13], transportation systems [14], and recently workplace accident reduction [15] are all key applications of these technologies. Studies of IoT systems, security assessments, and recovery procedures all have a big impact on the IoT ecosystem. The IoT is a collection of connected devices [16]. Furthermore, big data and cloud computing are intimately tied to it. In order to detect and secure the expanding number of networked "things," IoT security was created as an on-demand cloud service with a subscription model. IoT security auditing is the process of inspecting IoT devices to find security weaknesses in both the hardware and software. We live in a world where the idea of "Smart Everything" and the IoT are realities. As a result, more and more IoT systems are being deployed in the real world to try to take advantage of the various opportunities and benefits they provide. As a result of the network's interconnection with billions of frequently insecure devices and the absence of clear security architecture for the creation of IoT systems and platforms, these systems now have a larger attack surface and are therefore more open to assault by hostile actors.

The majority of such devices, though, are not immediately tracked or secured. Due to the low-power, resource-efficient, and autonomous character of IoT devices, achieving a high level of security is unfortunately not an easy task [11, 17–19]. Additionally, IoT devices are only intended for a limited range of functions, and the

DOI: 10.1201/9781003477327-19

current surge in demand for IoT devices has resulted in the creation of unreliable gadgets. In addition, because the IoT is still relatively new, certain companies that make IoT devices lack experience. The majority of businesses, or 80%, prioritize performance over vulnerability testing for IoT systems, according to IBM [20]. IoT devices make it challenging to apply software fixes and updates. The sophistication required to offer real security to IoT devices is still far from being attained. One problem is the IoT devices' variety, which makes finding a single solution all but impossible. Therefore, the new problems brought forth by the IoT cannot be solved by traditional security methods. Despite efforts in the area of IoT security, no genuine framework has been developed that can serve as a baseline for appropriate security. Furthermore, there is still much to learn about the auditing of security in the IoT. IoT security auditing is being done by several security frameworks. Even if they work, the solutions for every IoT device are either too (application) specialized [17, 21] or too network-oriented [13, 22, 23].

In the present work, pertinent research has been tried using case studies to look at an IoT-based weather monitoring system for irrigation. Along with keeping an eye on these factors, the data will be processed utilizing the Blynk servers and original hardware circuits on a permanent PCB. These parameters are temperature, humidity, light intensity, pressure, rainfall detection, soil moisture, and water management. Node MCU and IoT technology are used to show and analyze the data. In addition, security audits have been handled, which find potential security holes in all varieties of connected devices, analysis of linked devices with systems, and enable successful data recovery from anywhere in the globe.

19.2 LEVEL SECURITY AUDITING FOR IoT ECOSYSTEMS

The IoT is a complex ecosystem with a number of significant components that one should be aware of. In terms of cyber security, each of these elements brings new security risks, resulting in a larger attack facade. Here, a coordinated process for breaking down complicated IoT ecosystems into more manageable sections is provided in the subsequent subsection. Further, key security practices and controls are highlighted that must be implemented in each of these components to increase the ecosystem's overall security. Some features of security auditing, analysis, and recovery mechanisms pertaining to the IoT ecosystem have been tabulated further.

19.2.1 A TYPICAL IoT ECOSYSTEM'S ASSET TAXONOMY

Despite the diversity of IoT ecosystems presently set up in the real world, in any ecosystem that is chosen for analysis, there are some elements that are essentially the same. By recognizing these elements, a mental model of a typical IoT ecosystem has been created and generalizes its many resources and capabilities. According to an asset taxonomy presented in Figure 19.1, a typical IoT ecosystem has nine different basic assets, each of which plays a unique function in the ecosystem, is connected to a different stakeholder, and poses unique security threats. There is little doubt that the asset taxonomy technique is not new.

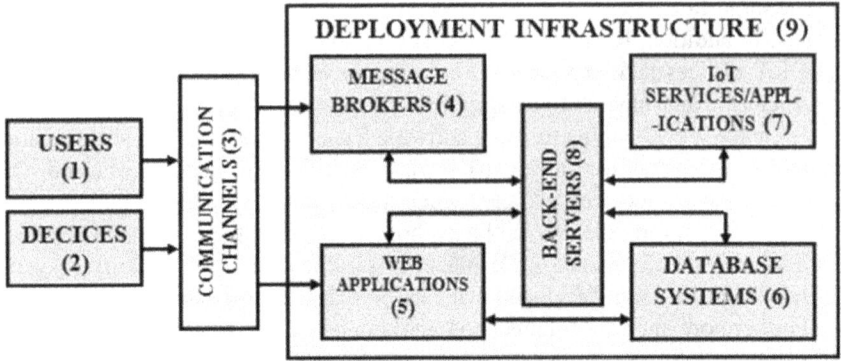

FIGURE 19.1   Asset taxonomy of a typical IoT ecosystem.

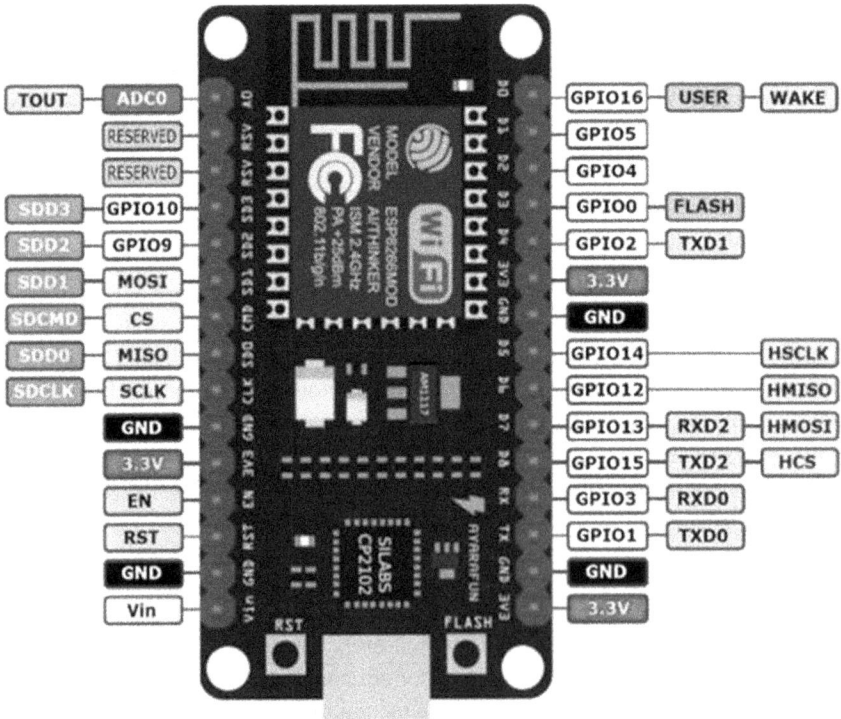

FIGURE 19.2   Node MCU ESP8266 Wi-Fi Module.

TABLE 19.1

Features of IoT Security Auditing, Analysis, and Recovery Mechanisms

Confidentiality	The method through which only authorized objects or users have access to the on-air and stored information's secrets and confidentiality.
Integrity	A method is used to ensure correctness and there is no data alteration.
Non-repudiation	The process through which an IoT system confirms the authenticity and source of an event.
Availability	Ensuring that services are available to those who require them even in the event of a power outage or other malfunction.
Privacy	The process through which an IoT system accesses sensitive data by abiding by rules and procedures.
Auditability	The method through which an IoT system keeps track of its behavior.
Accountability	The system via which IoT system users will be held accountable for their behavior.
Trustworthiness	The process through which an IoT system may confirm a person's identity and build confidence with a third party.

However, our taxonomy differentiates between the following asset types by taking into account the human factor and further breaking down the platform aspect for a more systematic assessment of its parts.

1. **Users**: The entities that make up the ecosystem's end users and who gain from its use. Individuals, groups, businesses, and even entire nations can be users. This category also includes participants who are not employed in development or marketing, in addition to profit-makers.
2. **Device**: This asset category includes smart watches, smart home equipment, sensors [24], security cameras, and other usually low-resource internet-connected gadgets. Both detecting and actuating abilities can be included in the gadgets.
3. **Communication Channels**: People and devices connecting and communicating with remote storage and compute areas, sometimes known as the cloud are made possible via communication channels, which are intangible things.
4. **Message Brokers**: The primary points of entry for data arriving from IoT devices are the message brokers, which are companies that are a part of the ecosystem's platform. To encourage interoperability and enable communication with other IoT devices, the message brokers often offer a wide range of application protocols.
5. **Web Applications**: The web applications, or standard web interfaces, are the users' access points into the IoT platform and allow them to log in and carry out the numerous operations that each platform offers. Adding or removing devices, managing data and using logic, or exporting particular data from the platform are some examples.
6. **Database Systems**: Data storage is handled via database systems. They are essential elements for numerous methods, including authentication

and authorization, data management, and the furnishing of data to IoT applications. Depending on the platform, they can be relational or non-relational.

7. **IoT Services/Applications**: Applications that process data from databases or IoT devices logically and produce results that are useful to users are referred to as this asset. The majority of platforms offer a number of pre-defined services, and a few of them also allow the development of unique applications that are typically run in virtualized environments.

8. **Back-end Servers**: The back-end coordinates IoT platform functionality. The databases, message brokers, web applications, and other IoT services are carefully taken care of, along with the logical operation of the various assets.

9. **Deployment Infrastructure**: IoT systems at the edge or in the cloud are built on top of this layer. It refers to the actual servers on which the platforms are hosted as well as the network configuration (routing, DNS, etc.) that enables user and device interaction with remote platform services. Additionally, it gives platforms networking capabilities and bandwidth management.

19.3 CHALLENGES FACED IN IoT ECOSYSTEMS ANALYSIS AND RECOVERY

Analysis of IoT ecosystems differs fundamentally from that of traditional ecosystems and developments. It presents a difficulty in and of itself because it blends hardware devices with software development. However, data collection, transport, and analysis are also part of the IoT development process. Analysis and recovery of IoT ecosystems encounter a variety of issues.

19.3.1 IoT Security Challenges

Lack of Encryption: Despite being an excellent method for preventing data theft, encryption presents one of the major IoT security challenges. Identical to those found on a traditional computer, drives offer similar processing and storage capacities. The number of attacks where hackers can simply alter the security algorithms has increased as a result.

Inadequate Upgrading and Testing: As the number of IoT devices increases, device manufacturers are more driven to create and market their products as soon as possible without giving security any care. Most devices do not receive adequate testing or updates, leaving them open to security dangers like hackers.

Risks Associated with Default Passwords and Brute Force: Due to faulty credentials and login information, all IoT devices are vulnerable to password hacking and brute force attacks. Any business that uses default factory passwords exposes not only its own assets but also the sensitive data of its clients to the risk of a brute-force attack.

IoT Ransomware and IoT Malware: Ransomware uses encryption to successfully lock out users from a range of devices and platforms while maintaining access to the user's crucial data and information. An example of this is when a hacker takes pictures on a computer camera. By using malware access points, the hackers might demand money to unlock the device and return the data.

An IoT Botnet Targeting Crypto Currencies: Employees of IoT botnets have the ability to change data privacy, which poses serious risks for a market for open cryptocurrencies. Threats to the precise value and construction of the cryptocurrency code come from malicious hackers. The companies that use blockchain are working to increase security. Even if the actual blockchain technology itself is not particularly insecure, the process of developing apps is.

19.3.2 PROBLEM WITH IoT DESIGN

A Drawback Is Battery Life: Packaging and integrating tiny chips with low weight and power consumption are more challenging issues, although no size restrictions exist on display in the mobile.

Increased Cost and Time to Market: Cost has just a little impact on embedded systems. To manage cost modeling/cost-effective use of digital electrical components, better design techniques are needed for IoT devices. To commercialize the embedded device, designers must find a solution to design time.

System Security: In order to be designed and put into use, systems must be secure with cryptographic algorithms and security protocols. It uses a number of techniques to secure every element of embedded systems, from prototype to finished product.

19.3.3 ISSUES WITH IoT DEPLOYMENT

Connectivity: When gadgets, programmers, and cloud platforms are combined, it is the main problem. The value of connected gadgets that provide useful information and front is quite high. However, a challenge arises when limited connectivity is required for IoT sensors to monitor process data and deliver insights.

Cross-Platform Capability: Future technological improvements must be taken into account while developing IoT applications. Its development requires a delicate balance between hardware and software operations. IoT application developers face a problem in maintaining the best possible performance of the device and IoT platform drivers despite rising device rates and repairs.

Data Collecting and Processing: The development of the IoT depends on data. In this case, the way the data was processed or how useful it was more crucial. Along with security and privacy, development teams must make sure they carefully consider how data is gathered, kept, or processed within an environment.

Inadequate Skill Set: All of the aforementioned development issues will only be solved if a qualified resource is working on the IoT application development. When creating IoT apps, having the right talent will always help you get beyond the toughest challenges.

19.4 IoT-BASED WEATHER MONITORING SYSTEM FOR GREEN HOUSE

In the present work, an IoT-based weather monitoring system for greenhouse/irrigation using case studies and relevant research has been examined. The parameters like temperature, humidity, Light intensity, pressure, rainfall detection, soil moisture, and water management are monitored. Further, these parameters are displayed and processed. The roof of the greenhouse can be opened and closed manually as well as automatically according to the intensity of light. The data will be displayed and interpreted using Node MCU and IoT technology. Also, care should be taken in security auditing, which identifies potential security flaws in all types of connected devices, analysis of connected devices with the system, and successful recovery of the data from anywhere in the world.

The main components which are used in this system are listed below:

- Node MCU
- I2C LCD 16 × 2
- Temperature and humidity sensor (DHT11)
- Light sensor (LDR sensor module)
- Pressure sensor (BMP18)
- Rain sensor
- Ultrasonic sensor (HC-SR04)
- Servo motor

The circuit diagram is shown in Figure 19.4; the system is built with a Node MCU microcontroller board. A greenhouse is generally used for sophisticated crops or for certain crops which are used for research purposes in the field of irrigation. The I2C protocol 16 × 2 LCD and BMP180 pressure sensors are interfaced using SDA and SCL pin of Node MCU D2 and D1, respectively. DHT11 sensor is employed to measure the temperature and humidity. If the temperature range goes below the threshold level, it will automatically switch on the fan using a relay for cooling purposes. A digital light sensor is employed to measure the light intensity; it will automatically switch on the artificial light if the light intensity goes below a certain threshold level. This circuitry has one push button to open or close the roof of the greenhouse using a servo motor manually as well as automatically according to the intensity of light through the Blynk server.

A buzzer is also provided for the alarm. A rain sensor is used to detect the rainfall, it works as a switch. It consists of two parts: a sensor pad while the one is the sensor module. When water drops hit the surface of the sensing pad, the switch closes from its normal open position and gets information about rainfall. A soil moisture sensor is employed to detect moisture levels in soil. An ultrasonic sensor is employed to

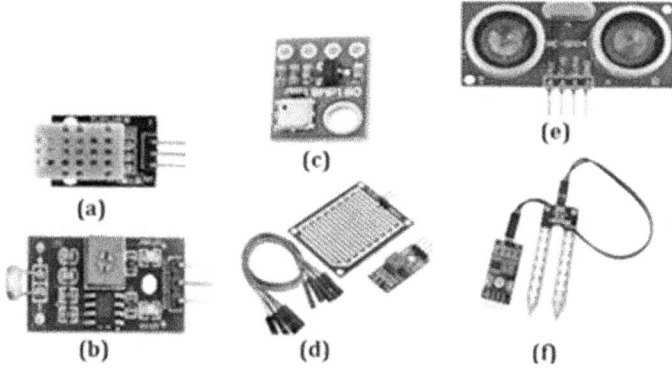

FIGURE 19.3 (a) DHT11 sensor, (b) LDR module, (c) pressure sensor, (d) rain sensor, (e) ultrasonic sensor, and (f) soil moisture sensor.

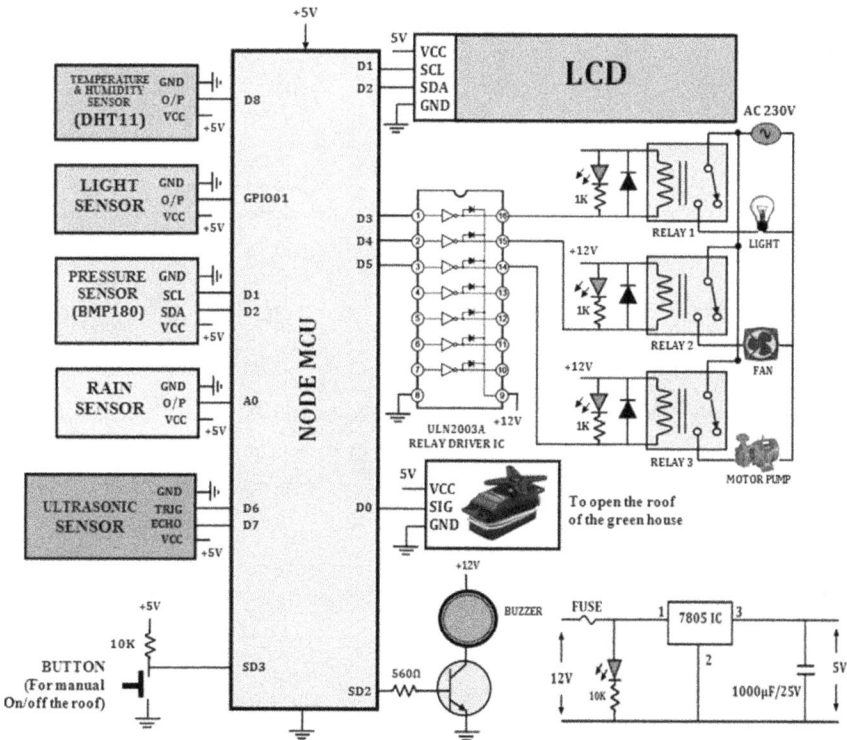

FIGURE 19.4 Circuit diagram of IoT-based weather monitoring system for greenhouse.

detect the water level in the tank using the concept of ultrasonic-like radar. It detects the water level and switches on the water pump when the tank is empty. Node MCU is not able to provide enough current which is required to operate the relay, which is why ULN2003A is used as a relay driver module. All the sensor data will be displayed over the Blynk server using IoT and the range is all over the world.

19.5   SETTING UP BLYNK APPLICATION

Just go the https://blynk.io and click on start free as shown in Figure 19.5a. After that, register with Gmail ID. Enter your email address and password to join up for the Blynk IoT cloud server (Figure 19.5b), and then hit the icon of the new project to assign the project name.

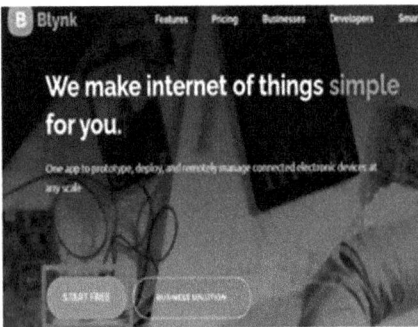

(a)

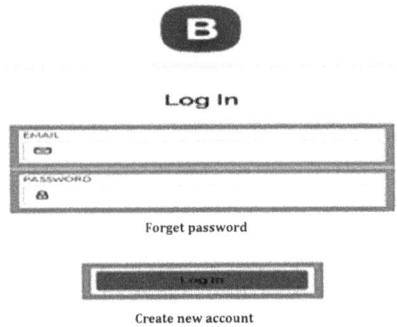

(b)

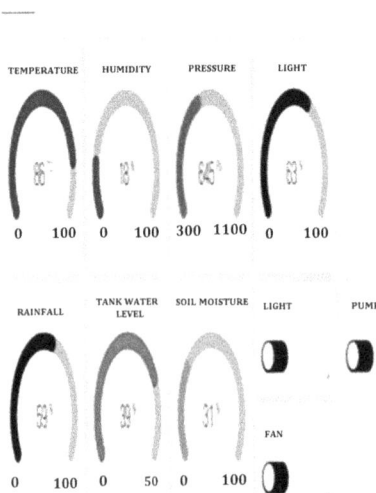

(c)

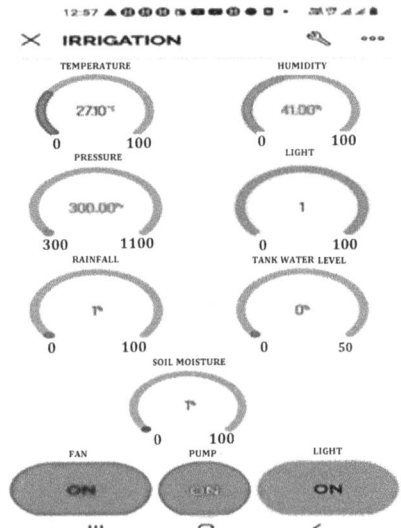

(d)

FIGURE 19.5   (a) Blynk cloud user interface, (b) log in with registered mail ID, (c) Blynk desktop dashboard, and (d) Blynk mobile dashboard.

Choose the Node MCU board, and then pick Wi-Fi as the connection type. Finally, press the Create button as displayed in Figures 19.5c and 19.5d. The same information can also be displayed on LCD and Blynk desktop and mobile dashboard.

19.6 RESULTS AND DISCUSSION

In this work, we have presented a case study in the form of a greenhouse in order to understand IoT ecosystem-level security auditing, analysis, and recovery by testing in real-time situations. Here, an IoT-based weather monitoring system for irrigation has been examined. It is well-known that a greenhouse is generally used for sophisticated crops or for certain crops that are used for research purposes in the field of irrigation. The proposed system was put to the test in real time following its development, and the outcomes were good and appropriate. It is essential to often check the temperature and humidity of the system and switch on and off the fan to maintain the temperature level of the greenhouse in manual as well as in automatic mode. As we know sunlight is necessary for the plants used in a greenhouse, according to the intensity of light during day time, our greenhouse roof will be opened with the help of a manual switch or by using the server as well by operating a servo motor. During the dark, an artificial light will automatically be switched on. The pressure and moisture level of the soil will give proper readings; whenever the soil needs water, the water pump switches on automatically. The water level on the tank should be measured using an ultrasonic sensor and if found empty, the water pump is switched on. Figures 19.6a and 19.6b show the original hardware of the proposed system and greenhouse, respectively, on which the result has been tested.

The system was found to work successfully and all sensors also worked satisfactorily. The user would be able to see all sensor data on LCD and transfer the data through IoT to the Blynk server. By careful observation and real-time testing, we conclude that our system is secure with a password. The deployed IoT system is able to detect possible security flaws in both the hardware and software and there are also

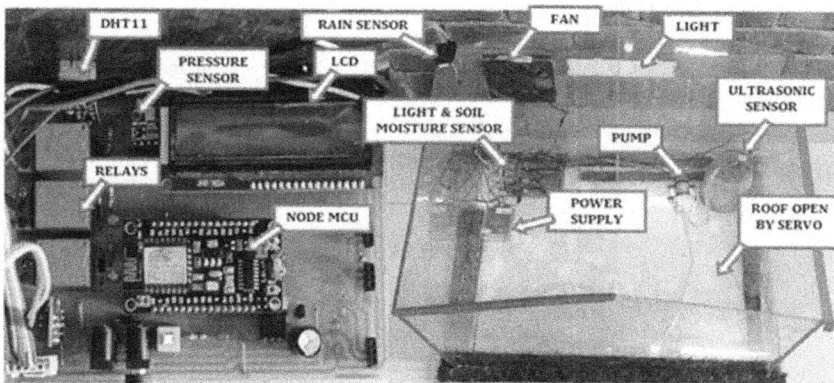

FIGURE 19.6 Original picture of the proposed system: (a) PCB hardware and (b) greenhouse system.

provisions to recover the system from detected flaws. Hence, this presented system covers the full domain of IoT ecosystem security audit.

19.7 SUMMARY

Security audits, IoT system analysis, and recovery procedures all play significant roles in the IoT ecosystem. The practice of examining IoT devices to identify security flaws in both the hardware and software is known as IoT security auditing. By careful observation and using real-time testing, we conclude that our system is secure with a password. We are able to measure and display all readings of sensors on LCD as well as that on the Blynk server successfully. If any security flaw is detected, then our system is able to detect and recover it automatically using the typical IoT ecosystem's asset taxonomy discussed above in detail. All the sensors work on low power, so energy saving is also taken care of. If potential security flaws are identified in any type of connected device or if any sensor gives a false reading, then our deployment infrastructure tries to recover the data successfully from anywhere in the world. The challenges faced in IoT ecosystems has been taken into account.

REFERENCES

[1] A. B. Chebudie, R. Minerva, and D. Rotondi. 2015. Towards a Definition of the Internet of Things (IoT). Ph.D. Dissertation.

[2] B. Farahani, F. Firouzi, and K. Chakrabarty. 2020. Healthcare IoT. 515–545. https://doi.org/10.1007/978-3-030-30367-9_11

[3] G. Kaur and M. Sohal. 2018. IOT Survey: The Phase Changer in Healthcare Industry. *International Journal of Scientific Research in Network Security and Communication* 6 (04), 34–39. https://doi.org/10.26438/ijsrnsc/v6i2.3439

[4] H. Qiu, X. Wang, and F. Xie. 2017. A Survey on Smart Wearables in the Application of Fitness. 303–307. https://doi.org/10.1109/DASC-PICom-DataCom-CyberSciTec.2017.64

[5] M. S. Mekala and V. Perumal. 2017. A Survey: Smart Agriculture IoT with Cloud Computing. 1–7. https://doi.org/10.1109/ICMDCS.2017.8211551

[6] J. Ruan, H. Jiang, C. Zhu, X. Hu, Y. Shi, T. Liu, W. Rao, and F Chan. 2019. Agriculture IoT: Emerging Trends, Cooperation Networks, and Outlook. *IEEE Wireless Communications* 26 (12), 56–63. https://doi.org/10.1109/MWC.001.1900096

[7] H. Xu, W. Yu, D. Griffith, and N. Golmie. 2018. A Survey on Industrial Internet of Things: A Cyber-Physical Systems Perspective. *IEEE Access* 6 (2018), 78238–78259. https://doi.org/10.1109/ACCESS.2018.2884906

[8] L. Xu, W. He, and S. Li. 2014. Internet of Things in Industries: A Survey. *IEEE Transactions on Industrial Informatics* 10 (11), 2233–2243. https://doi.org/10.1109/TII.2014.2300753

[9] M. Dikmen and C. Burns. 2017. Trust in Autonomous Vehicles: The Case of Tesla Autopilot and Summon. In *2017 IEEE International Conference on Systems, Man, and Cybernetics (SMC)*. 1093–1098.

[10] S. Mohanty. 2016. Everything You Wanted to Know about Smart Cities. *IEEE Consumer Electronics Magazine* 5, 60–70. https://doi.org/10.1109/MCE.2016.2556879

[11] PwC. 2017. The Wearable Life 2.0: Connected Living in a Wearable World. https://www.slideshare.net/PwC_Spain/the-wearable-life-20

[12] J. A. Stankovic. 2014. Research Directions for the Internet of Things. *IEEE Internet of Things Journal* 1 (1), 3–9.

[13] M. Vučinič, B. Tourancheau, F. Rousseau, A. Duda, L. Damon, and R. Guizzetti. 2015. Oscar: Object Security Architecture for the Internet of Things. *Ad Hoc Networks* 32, 3–16.

[14] L. Atzori, A. Iera, and G. Morabito. 2010. The Internet of Things: A Survey. *Computer Networks*, 54 (15), 2787–2805.

[15] J. Twentyman. 2016. Wearable Devices Aim to Reduce Workplace Accidents. *Financial Times*, June 2016.

[16] T. Kavitha, V. Ajantha Devi, S. Neelavathy Pari, and S. Ramanathan. 2022. *Internet of Everything: Smart Sensing Technologies*. Nova Science Publishers, Publication Date: June 17, 2022. https://doi.org/10.52305/PNQM1088

[17] A. Costin and J. Zaddach. 2018. *Iot Malware: Comprehensive Survey, Analysis Framework and Case Studies*. BlackHat USA.

[18] A. Cui and S. J. Stolfo. 2010. A Quantitative Analysis of the Insecurity of Embedded Network Devices: Results of a Wide-Area Scan. *Proceedings of the 26th Annual Computer Security Applications Conference*. ACM, pp. 97–106.

[19] M. Conti, A. Dehghantanha, K. Franke, and S. Watson. 2018. Internet of Things Security and Forensics: Challenges and Opportunities. *Future Generation Computer Systems 78*. https://doi.org10.1016/j.future.2017.07.060

[20] P. I. LLC., 2017. Study on Mobile and Internet of Things Application Security. https://www.ponemon.org/blog/2017-studyon-mobile-and-internet-of-things-application-security

[21] E. Fernandes, J. Jung, and A. Prakash. 2016. Security Analysis of Emerging Smart Home Applications. *2016 IEEE Symposium on Security and Privacy (SP)*. IEEE, pp. 636–654.

[22] S. Babar, A. Stango, N. Prasad, J. Sen, and R. Prasad. 2011. Proposed Embedded Security Framework for Internet of Things (IoT). *2011, 2nd International Conference on Wireless Communication, Vehicular Technology, Information Theory and Aerospace & Electronic Systems Technology (Wireless VITAE)*. IEEE, pp. 1–5.

[23] J. Frahim, C. Pignataro, J. Apcar, and M. Morrow. 2015. Securing the Internet of Things: A Proposed Framework. Cisco White Paper.

[24] T. Kavitha and S. Saraswathi. 2017. *New Sensing Technologies or/and Devices for Emergency Response and Disaster Management*. Book Chapter, IGI Global International Publisher, pp. 1–40, July 2017. https://doi.org/10.4018/978-1-5225-2575-2, Ch001. ISBN:9781522525752.

20 Risk Assessment and Vulnerability Analysis in the IoT

Vanajaroselin Chirchi
The Oxford College of Engineering, Visvesvaraya
Technological University, Bengaluru, India

S. Nirmala
AMC Engineering College, Visvesvaraya Technological
University, Bengaluru, India

Emmanvelraj M. Chirchi
Don Bosco Institute of Technology, Visvesvaraya
Technological University, Bengaluru, India

S. L. Karthik Raj
The Oxford College of Engineering, Visvesvaraya
Technological University, Bengaluru, India

E. C. Khushi
Dayanand Sagar University, Bengaluru, India

20.1 INTRODUCTION

The Internet of Things (IoT) has emerged as a powerful technology for connecting physical devices and objects to the Internet, enabling the creation of smart environments and systems. However, the deployment of IoT devices and systems also brings new security and privacy challenges, as these devices are often vulnerable to attacks and exploits. The basic architecture of IoT includes three major layers: physical layer, network layer, and application layer. Although there are a number of different layered architectures of IoT, they are created based on the common three-layer architectures by improving or modifying different aspects, shown in Figure 20.1.

Attacks can occur in any of these three layers. As seen in Table 20.1, the attacks occur at these three layers. The effects of the attacks are more severe in the IoT environment compared to conventional IoT devices due to three important IoT characteristics:

DOI: 10.1201/9781003477327-20

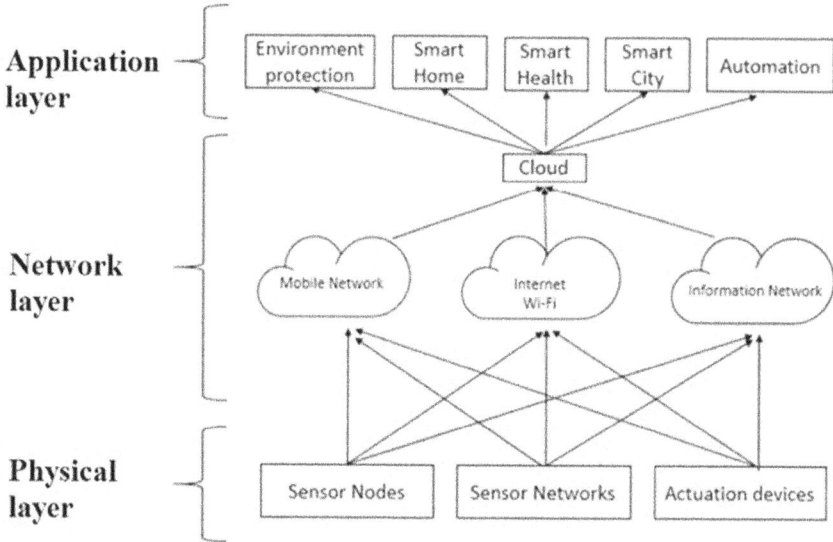

FIGURE 20.1   Three-layer IoT architecture.

TABLE 20.1
Attacks at Different Layers

Layer	Attacks
Application layer	Data spoofing, SQL injection, DOS or DDOS, replay attacks, resource exemption, and reversal attacks
Network layer	Traffic flooding, man-in-the-middle, misrouting, packet sniffing, resource exemption
Physical layer	Impersonating attacks, jamming attacks, device tampering

---

1. Though IoT devices interact with the physical world similar to conventional IT devices, their operational requirements in terms of performance, reliability, resilience, and safety are different from conventional IT devices.
2. Many IoT devices cannot be accessed, managed, or monitored just like conventional IT devices.
3. The capabilities such as availability, efficiency, and effectiveness of cybersecurity are different for IoT devices than conventional devices [1, 2].

To mitigate these security risks, it is important to perform risk assessment and vulnerability analysis to identify potential threats and vulnerabilities and develop effective mitigation strategies.

20.2   RISK ASSESSMENT IN IoT

Risk assessment includes the analysis, prioritization, and identification of security risks, including the identification of potential threats and the estimation of their impact on the system. Analysis of vulnerability involves the assessment and identification of potential weaknesses in the system, including the identification of potential attack vectors and the estimation of their exploitability. The taxonomy of common risks and vulnerabilities in IoT systems is given in Figure 20.2.

Assessment involves detecting the presence or probable occurrence of vulnerabilities, listed in Figure 20.2, in an IoT system. Both risk assessment and vulnerability analysis are critical steps in the evolution of effective safety strategies for IoT systems and are essential for ensuring the security and dependability of these systems.

Recently, in the area of risk assessment and vulnerability analysis in IoT, there has been a significant amount of increase in interest, as researchers and practitioners seek to develop effective strategies for mitigating the security threat associated with the systems. This has led to the publication of numerous papers, surveys, and reviews on the topic, which provides valuable information about the current state of the art and the key challenges and opportunities in this field.

20.2.1   RISK ASSESSMENT CHALLENGES IN IoT

In IoT systems, risk assessment and vulnerability analysis are complicated by the sheer scale and diversity of the devices. IoT systems often involve a large number

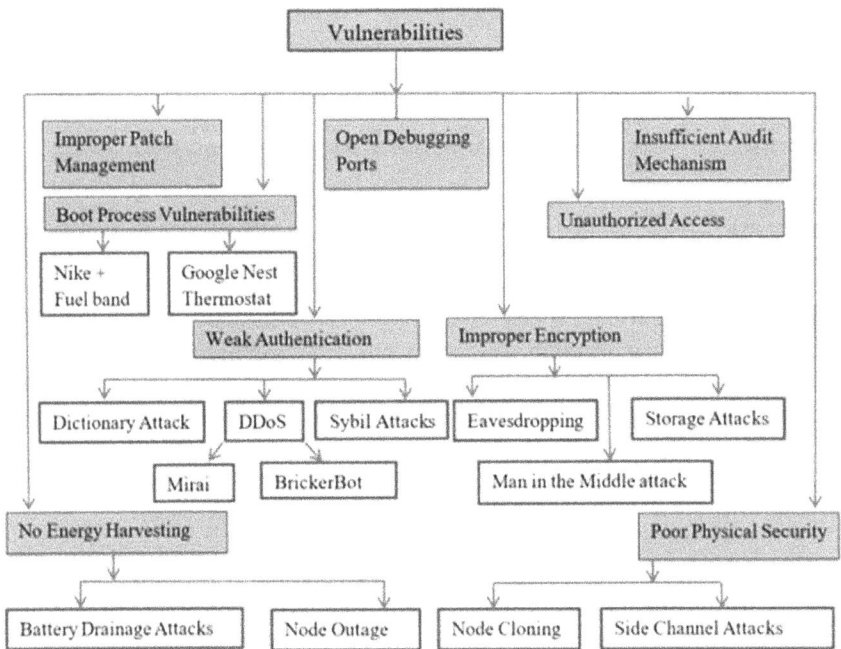

FIGURE 20.2   Risks and vulnerabilities in IoT.

of heterogeneous devices and systems, each with its own unique security and privacy requirements, making it hard to develop a unified security strategy that covers all systems and devices. In addition, IoT devices and systems are often resource-constrained, with limited processing power, memory, and battery life. This makes it difficult to implement robust security measures and to perform risk assessment and vulnerability analysis, as these tasks often require significant computational resources. Furthermore, IoT devices are often deployed in remote or inaccessible locations, making it challenging to update or replace devices that have been compromised or that are no longer secure. A summary of the emerging challenges in the risk assessment of IoT systems is shown in Figure 20.3.

Despite these challenges, risk assessment and vulnerability analysis are critical components of any security strategy for IoT systems. By performing these tasks, organizations can identify potential threats and vulnerabilities, prioritize risks based on their potential impact, and develop effective mitigation strategies to reduce the likelihood of security incidents and to minimize the impact of security breaches. It is also important to note that the threat landscape in IoT is constantly evolving, with new security threats emerging as new devices and technologies are developed and deployed. This requires organizations to regularly perform risk assessment and vulnerability analysis to ensure that their security strategies remain effective and up-to-date. Additionally, organizations must also consider the privacy implications of IoT systems, as these systems often involve the collection and processing of personal data. This requires organizations to perform privacy risk assessments and develop privacy-enhancing technologies that can protect the personal data of users while still enabling the use of IoT systems. Another important consideration in the area of risk assessment and vulnerability analysis in IoT is the development of standards and best practices. As the deployment of IoT systems continues to grow, it is important to

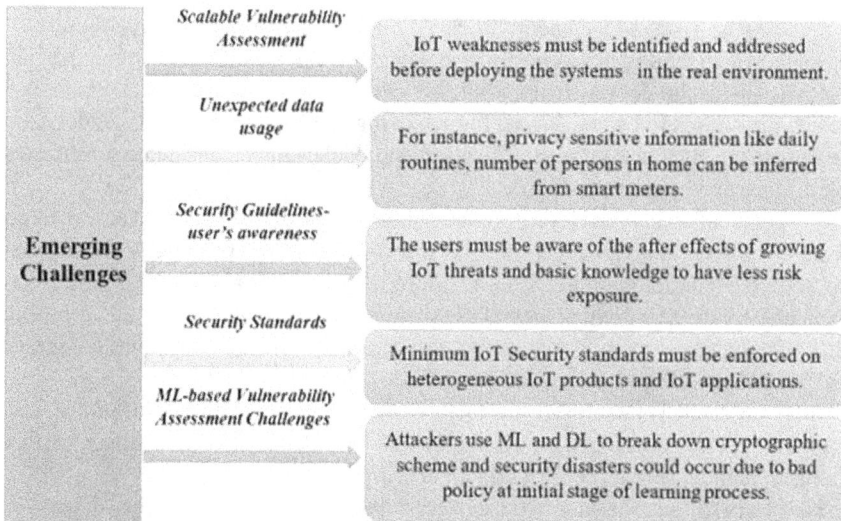

FIGURE 20.3 Emerging risk assessment challenges in IoT.

develop standardized approaches to risk assessment and vulnerability analysis that can be applied across different industries and domains. This will enable organizations to more effectively manage the security risks associated with IoT systems and to develop more secure and reliable systems.

20.2.2 RECENT TRENDS IN IoT RISK ASSESSMENT

One emerging trend in the area of risk assessment and vulnerability analysis in IoT is the development of automated tools and methods. These tools and methods are designed to simplify and streamline the process of risk assessment and vulnerability analysis, making it easier and more efficient for organizations to manage the security risks associated with IoT systems.

Another trend is the integration of risk assessment and vulnerability analysis into the development process of IoT systems. This involves incorporating security considerations of IoT devices and systems on outline and growth making it easier to identify and address potential security risks and vulnerabilities at a prior stage in the developmental process.

Moreover, it's also important to note that organizations must also consider the legal and regulatory requirements associated with IoT systems. This includes compliance with privacy and data protection regulations, as well as with security standards and best practices. Organizations must ensure that their risk assessment and vulnerability analysis processes take these requirements into account and that they are in compliance with all relevant regulations and standards.

20.2.3 RISK ASSESSMENT AND VULNERABILITY ANALYSIS

Risk assessment and vulnerability analysis are critical components of a comprehensive security strategy for IoT systems. By providing a comprehensive understanding of the security risks and vulnerabilities associated with these systems, these tasks can help organizations develop effective security strategies and minimize the risks associated with the deployment of IoT devices and systems. This work attempts to present the factors to be considered for achieving such a comprehensive understanding of security risks and vulnerabilities and approaches to be considered for effective risk mitigation design.

They critically analyzed four frameworks: IoT hazards are discussed and analyzed through rigorous analysis regarding the IoT risk category and impacted industries. The study concentrated on IoT systems utilized in financial and medical technology. The authors offered a distinct way of ranking and quantifying IoT risk and a cutting-edge computational method for calculating the cyber risk for IoT systems used in the healthcare field.

Andrade et al. [4] analyzed the factors of IoT systems contributing to security risks. The authors developed a formula for risk based on four factors: attack surface, interdependence, susceptibility, and vulnerability.

Ma et al. [5] put forward a framework related to graph theory that detects the vulnerability risk in IoT systems. The authors built a correlation model between devices using the hidden Markov model. Potential vulnerabilities in the form of an attack graph are found using this correlation model.

Nurse et al. [7] presented the challenge in extending existing IT risk assessment methodologies for IoT systems. The existing IT risk assessment methodologies could not address the threats related to the high degree of coupling or connectivity of digital, cyber-physical, and social systems in IoT systems. The authors stressed the need for new methodologies to assess risks in this context considering the dynamics and uniqueness of IoT.

Bouveret et al. [8] proposed a risk assessment framework to assess risk quantitatively for cyber–physical systems for financial sectors. This framework assesses the stability risk in financial cyber–physical systems. The authors addressed the challenges in measuring cyber risks for financial systems.

An integrated methodology for managing cybersecurity risks that proactively assesses and manages the risks was proposed by Kure et al. [9]. The concept is supported by current risk management practices. The stakeholder model, cyber and physical system components, as well as their interdependencies, are considered when calculating risks. The framework can identify risky situations, appraise them, and develop mitigation strategies.

20.2.4 Mitigation Strategies

From the survey, the two most important factors of comprehensive understanding of security risks and vulnerabilities and implementing effective mitigation strategies are policies to prevent risk and risk management.

20.2.4.1 Policies

IoT security policy is fundamental to implementing any long-term vulnerability management framework. The policies are written for people and things. The policy written for people should influence the IoT which is allowed in the organization network. The devices that are used by people such as mobiles, laptops, and desktops are at the very least at risk because they have already developed security configurations which reduce the risk greatly. For an effective policy design, understanding the taxonomy of the IoT vulnerability is important.

The taxonomy was developed using a framework that takes into account the security impact, attacks, remediation methods, and situation awareness capabilities of IoT vulnerabilities. The "Layers" category investigates various IoT elements, such as device-based, network-based, and software-based sensitivities. In the "Security Impact" category, vulnerabilities are rated according to how much they compromise key security goals including confidentiality, integrity, and availability. While the "Countermeasures" category explains the methods that can be used to mitigate these vulnerabilities, the "Attacks" category describes the security weaknesses and how they can be exploited.

The remediation techniques are split into three broad classes: Authentication and Access Controls, Software Assurance, and Security Protocols. The "Authentication and Access Controls" class encompasses a variety of security measures designed to ensure that access to IoT devices and data is granted only to authorized users. This category covers context-aware permissions, biometric-based models, algorithms and authentication schemes, and firewalls.

The "Software Assurance" class highlights the remaining potential to ensure the solidarity of IoT devices and their data. This class is critical in mitigating the impact of software vulnerabilities, which are a major source of security breaches in the IoT. The "Software Assurance" class includes measures such as code signing, software fingerprinting, and other integrity-checking methods.

The "Security Protocols" class encompasses lightweight security schemes aimed at remediating identified vulnerabilities in the IoT. This class includes a range of security protocols, such as secure transport protocols, secure data storage protocols, and security protocols designed specifically for the IoT.

The "Situation Awareness Capabilities" class includes methods such as Intrusion detection, vulnerability assessment, network discovery, and honeypots. Vulnerability assessment involves the use of testbeds, attack simulation methods, and fuzzing techniques to identify vulnerabilities in IoT devices. Honeypots aim to capture and analyze malicious activities in the IoT for further investigation. Network discovery involves identifying vulnerable and compromised IoT devices across the Internet, while intrusion detection is concerned with detecting and characterizing malicious activities in the IoT.

With regard to the security of the various layers of the IoT architecture, it is important to acknowledge that vulnerabilities exist in all three layers, which may put the core security goals of the system at risk. In particular, vulnerabilities present in each layer must be delved.

Starting with the first layer, device-based vulnerabilities, we must consider that many IoT devices operate in an unmanned manner, meaning that there are limited or no tamper-resistant policies and methodologies in place. This leaves the devices open to exploitation through physical access. In this case, an attacker could have the ability to cause significant damage to the devices or gain infinite access to the data stored in the memory. Testing of consumer IoT devices indicated this type of vulnerability, which might allow an adversary to change boot parameters, extract the root password, and access other private or sensitive data if they had physical access to the hardware.

Therefore, it is crucial that measures are put in place to mitigate these types of device-based vulnerabilities in the IoT, to prevent attackers from compromising the systems safety. The safety of IoT devices has been a growing concern for both consumers and organizations alike. A number of studies [5, 6, 10] have been conducted to shed light on the various vulnerabilities that exist within the IoT ecosystem, with a particular focus on the device layer, exploring the security implications of the IoT devices of the consumers and the impact of the physical approach on the security of these devices.

Protecting privacy and gaining confidentiality can be achieved by securing the Internet of Things. The utilization of the physical layer to enhance confidentiality in IoT devices has been suggested as a potential opportunity for strengthening security. With reference to network-based vulnerabilities, various research works have focused on addressing the security challenges posed by network and protocol weaknesses in IoT systems. An example which is designed for low-power and low-rate wireless networks is the ZigBee. It uses symmetric keys for establishing secure communications [11–16], with the level of key sharing among nodes being dependent on the chosen security mode.

20.3 RISK MANAGEMENT

Risk management of IoT is a crucial aspect of ensuring the safe and secure use of IoT technologies. It is a comprehensive approach that considers all potential risks associated with the deployment and use of IoT devices, systems, and networks. Identification, prioritization, and assessment are included in the risk management of IoT, followed by the development and implementation of strategies to reduce their impact or risk mitigation. The ultimate goal is to help organizations effectively leverage the benefits of IoT while reducing the potential harm caused by security threats, privacy breaches, and system failures.

Risks affecting organizations can have broad ramifications in terms of financial performance, professional reputation, as well as environmental, safety, and societal results, according to the ISO 31000 standards for risk management. The success and long-term viability of organizations operating in today's fast-changing and uncertain environment therefore depend on their capacity to manage risks effectively. IoT risk management provides a structured and systematic approach to addressing these challenges, enabling organizations to effectively use emerging technologies while minimizing the inherent risks associated with their deployment.

20.3.1 FACTORS TO BE CONSIDERED FOR RISK MANAGEMENT DESIGN

In the field of risk assessment and vulnerability analysis in IoT, there are several challenges that need to be addressed, and a number of approaches are being developed to design effective risk management strategies. Some of them are as follows.

Complexity: IoT systems are often complex, comprising a large number of interconnected devices and systems. This complexity makes it difficult to accurately assess the security risks and vulnerabilities associated with these systems.

Scalability: As the number of IoT devices increase and systems expand, it gets harder to evaluate risk assessment and vulnerability analysis procedures, to keep pace with the growth.

Heterogeneity: IoT systems are often highly heterogeneous, comprising devices and systems from a wide range of manufacturers and using a variety of communication protocols and technologies. This heterogeneity makes it difficult to accurately assess the security risks and vulnerabilities associated with these systems.

Rapid change: The IoT space is rapidly evolving, with new devices, systems, and technologies emerging on a regular basis. This rapid change makes it difficult to keep up with the latest security risks and vulnerabilities and to ensure that risk assessment and vulnerability analysis processes are up-to-date.

20.3.2 APPROACHES FOR RISK MANAGEMENT DESIGN

The following approaches can be considered in the design of effective risk management strategies.

Automated tools and methods: Automated tools and methods are being developed to simplify and streamline the process of risk assessment and vulnerability analysis. These tools and methods can help to reduce the complexity of the process and make it more efficient.

Integration into the development process: Another approach is to integrate risk assessment and vulnerability analysis into the development process of IoT systems. This involves incorporating security considerations of IoT devices into development, design, and systems, making it easier to identify and address potential security risks and vulnerabilities at an early stage in the development process.

Collaboration and information sharing: Collaboration and information sharing between organizations, researchers, and government agencies can help to address the broader security and privacy implications of IoT and ensure that these systems are deployed in a secure and responsible manner.

Compliance with legal and regulatory requirements: Organizations must also consider the legal and regulatory requirements associated with IoT systems. This includes compliance with privacy and data protection regulations, as well as with security standards and best practices. Organizations must ensure that their risk assessment and vulnerability analysis processes take these requirements into account.

20.4 RISK MITIGATION

Risk mitigation helps to avert access and interference with data information during transmission or at rest which would release very important data or may tamper or disrupt the operations of IoT devices.

Risk mitigation in IoT is possible if strategies such as secure connections, identification of high-risk features, and use of secondary networks are considered, which focus on the quality, device updation, understanding of the given guidelines, certification, and purchase of secure smart-enabled devices. The avoidance strategy involves taking steps to prevent a risk from occurring. This could involve making changes to resources or processes to reduce the risk of harm. For instance, if there is a risk of not being able to complete a task due to a lack of specialists, you could hire additional specialists to ensure the risk is avoided. However, it is important to consider the cost of taking these measures and to balance the resources available.

The reduction strategy focuses on reducing the likelihood of a risk occurring or the impact should it occur. For example, if the budget for a project is limited and there is a risk of not being able to complete the project, steps could be taken to proactively manage the costs and minimize the risk.

The third party such as an insurance company or contractor uses the transference strategy which involves passing on the risk consequences. For example, if a project is delayed due to an external contractor, the contractor might face retribution for any loss of revenue incurred by the business.

Finally, there is the acceptance strategy, where a risk is accepted as is. This might occur when the probability of the risk is small or the negative impact is minor, or when the potential reward outweighs the risk. In such cases, it is important to monitor the risk closely and regularly assess whether it continues to be the best approach.

20.4.1 RISK ANALYSIS STRUCTURES

A number of standards, guidelines, and best practices are included in the industrial environment which are available to understand and to help in mitigating them. They are as follows.

IEC 62443: It is the most commonly used standard across the industry. It has the following parts.

- 62443-3-2 for risk assessments
- 62443-3-3 for foundational requirements

The assessment frameworks are classified into two categories:

- OCTAVE (Operationally Critical Threat, Asset, and Vulnerability Evaluation) from the Software Engineering Institute at the University of Carnegie Mellon.
- FAIR (Factor Analysis of Information Risk) from Open Group.

20.4.1.1 OCTAVE

There are numerous standard guidelines and best practices available in the industrial setting to help one identify risks and minimize them. The following standards are part of IEC 63443 and include risk assessments with standards 62443-3-2 and 62443-3-3 [3] for basic security measures used to protect the industrial environment, mostly from networking and communications. In this chapter, the focus is on frameworks for evaluating risks such as OCTAVE and FAIR, and mitigating risk [3].

A methodology for identifying, managing, and evaluating information security threats is called OCTAVE. This methodology aids an organization in (a) developing qualitative risk evaluation criteria that specify the operational risk tolerances of the organization, (b) determine and assess the potential repercussions for the organization if threats are realized, (c) identify assets that are critical to the organization's mission, (d) identify vulnerabilities and threats to those assets, and (e) initiate continuous improvement measures to minimize risks.

The OCTAVE methodology is primarily intended for those in charge of controlling the operational risks faced by an organization. This can include employees in the business units of an organization, those responsible for information security or conformance within an organization, risk managers, the information technology department, and every employee taking part in OCTAVE method risk assessment activities.

OCTAVE Allegro is a lightweight and easy-to-implement process. When it comes to Operational Technology (OT), the methodologies and presumptions are sound; however, security-focused assets are absent.

Step 1: Establish Drivers – Establish risk measurement criteria: It provides a fairly simple means for emphasis on impact, value, and measurement.
Step 2: Profile Assets – Information asset development profile: The profile is constructed based on assets based on priority, asset attributes, technology assets, and security requirements. It is a multi-stage process.
Step 3: Profile Assets – Integrity information asset containers: It is the range of locations and transports where the information is stored. It is not on the asset level and the emphasis is on the container level.
Step 4: Identify Threats – Identify concern area: This step involves mapping the security-related attributes to business-related use cases. In this step, analysts refer to risk profiles and dig into previous risk analyses.
Step 5: Identify Threats – Identify the scenario of threat: Threats are potential undesirable events. Threats are malicious or accidental. The precise identification of outcomes and actors is a valuable factor.
Step 6: Identify and Mitigate Risk – Identify risk: In this step, identify the risk possibility of the undesired outcome. Identify and localize the potential impact on the organization.
Step 7: Identify and Mitigate Risk – Analyse risk: Efforts are placed on the qualitative evaluation of the impacts of the risk.
Step 8: Identify and Mitigate risk – Mitigation process: Accept the danger and take no action other than to record the circumstance and the justification. It is necessary to reduce the risk based on control measures. Risk is ultimately neither acknowledged nor reduced.

20.4.1.2 FAIR

It is a technical definition that the open group has put forward. It places value on a clear understanding of risk and its associated characteristics. The frequency of threat events is applied to vulnerability.

FAIR emphasizes how temporary costs are required when information security is the goal. Operational effectiveness places a high value on the OT [3].

Six types of losses are outlined by FAIR. Two losses are internally centered, while the other two have an external focus.

FAIR is designed to quantify risks and define the chances of those risks becoming serious threats. It also helps industries minimize all possible chances of risks by identifying the factors contributing to them. It is a standard measurable model for information reliability and functioning threat. It helps companies minimize all possible chances of risks by identifying the factors contributing to them. It is very essential to quantify the level of risk to an organization's cybersecurity infrastructure and initiate a swift and appropriate mitigating response.

20.5 SUMMARY

Risk assessment and vulnerability analysis in IoT are critical factors that need to be looked at in order to ensure the security and privacy of these systems. The progress of IoT systems and the increasing number of devices and systems that are interconnected have created new security risks and vulnerabilities that need to be addressed. To address these challenges, a number of approaches are being developed, including the development of automated tools and methods, integration into the development process, collaboration and information sharing, and compliance with legal and regulatory requirements.

Effective risk assessment and vulnerability analysis are essential to understand that IoT systems are deployed in a secure and responsible way, and to secure the privacy and safety of individuals and organizations. As IoT continues to grow and evolve, it is important that the field of risk assessment and vulnerability analysis continues to evolve and adapt to keep pace with the latest security risks and vulnerabilities. This requires continued investment in research and development, as well as collaboration between organizations, researchers, and government agencies.

REFERENCES

1. M. Yu, J. Zhuge, M. Cao, et al., "A Survey of Security Vulnerability Analysis, Discovery, Detection, and Mitigation on IoT Devices", *Future Internet*, Vol. 12, 2019, p. 27, doi: 10.3390/fi12020027
2. B. Northern, T. Burks, M. Hatcher, et al., "VERCASM-CPS: Vulnerability Analysis and Cyber Risk Assessment for Cyber-Physical Systems", *Information*, Vol. 12, No. 10, p. 408, doi:10.3390/info12100408
3. K. Kandasamy, S. Srinivas, K. Achuthan, et al., "IoT Cyber Risk: A Holistic Analysis of Cyber Risk Assessment Frameworks, Risk Vectors, and Risk Ranking Process", *EURASIP Journal on Information Security*, 2020, p. 8.
4. R. Andrade, I. Ortiz-Garcés, X. Tintin & G. Llumiquinga, 'Factors of Risk Analysis for IoT Systems", *Risks*, Vol. 10, 2022, p. 162.
5. Y. Ma, Y. Wu, D. Yu, L. Ding & Y Chen, "Vulnerability Association Evaluation of Internet of Thing Devices Based on Attack Graph", *International Journal of Distributed Sensor Networks*, Vol. 18, No. 5, 2022. DOI: 10.1177/15501329221097817
6. H. L. Hassani & A. Bahnasse, "Vulnerability and Security Risk Assessment in a IIoT Environment in Compliance with Standard IEC 62443", *Procedia Computer Science*, Vol. 191, 2021, pp. 33–40.
7. J. R. C. Nurse, S. Creese & D. De Roure, "Security Risk Assessment in Internet of Things Systems", *IT Professional*, Vol. 19, No. 5, 2017, pp. 20–26.
8. A. Bouveret, *Cyber Risk for the Financial Sector: A Framework for Quantitative Assessment*, 2018 International Monetary Fund.
9. H. I. Kure et al., "An Integrated Cyber Security Risk Management Approach for a Cyber-Physical System", *Applied Sciences*, Vol. 8, 2018, p. 898.
10. P. Oser, R. W. van der Heijden, S. Lüders & F. Kargl, "Risk Prediction of IoT Devices Based on Vulnerability Analysis", *ACM Transactions on Privacy and Security*, Vol. 25, No. 2, Article 14, May 2022, p. 36.

11. T. Kavitha & R. Kaliyaperumal, "Energy Efficient Hierarchical Key Management Protocol," *2019 5th International Conference on Advanced Computing & Communication Systems (ICACCS)*, Sri Eshwar College of Engineering Coimbatore, India, 2019, pp. 53–60. doi:10.1109/ICACCS.2019.8728343

12. T. Kavitha, S. J. S. Priya & D. Sridharan, "Design of Deterministic Key Pre Distribution Using Number Theory," *Proceedings of 3rd International Conference on Electronics Computer Technology (ICECT)*, IEEE, 8–10 April 2011, vol. 5, pp. 134–137.

13. T. Kavitha, S. J. S. Priya & D. Sridharan, "Design of Deterministic Key Pre Distribution Using Number Theory," *Proceedings of 3rd International Conference on Electronics Computer Technology (ICECT)*, IEEE, 8–10 April 2011, vol. 5, pp. 134–137.

14. T. Kavitha & D. Sridharan, "Optimal Resource Key Management Protocol for Clustered Heterogeneous Wireless Sensor Networks", *Malaysian Journal of Computer Science* (0127-9084), Univ Malaya, Vol. 26, No. 3, 2013, pp. 211–231.

15. T. Kavitha & D. Sridharan, "Key Distribution Scheme Using Modulo Operation for WSN", Information (P-1343-4500), International Information Inst Publisher, Japan, Vol. 16, No. 11, Nov 2013, pp. 8213–8228.

16. K. Rajadurai, T. Kavitha & V. J. Subashini, "Application of Modulo Key-Predistribution Protocol", *Research Journal of Applied Sciences, Engineering and Technology* (P-2040-7459), Vol. 11, No. 7, Nov 2015, pp. 780–787.

21 Threat Models and Attack Strategies in the Internet of Things

V. M. Sivagami, K. Kiruthika Devi, and V. Vidhya
Sri Venkateswara College of Engineering, Anna University, Chennai, India

S. Swarna Parvathi
(BAEG)-Biological and Agricultural Engineering, University of Arkansas, Fayetteville, AR, USA

21.1 INTRODUCTION TO THREAT MODELS AND ATTACK STRATEGIES IN THE INTERNET OF THINGS

Internet of Things (IoT) systems are composed of connected devices, networks, and applications. Each component of an IoT system is vulnerable to attack and must be secured to protect the system as a whole. Attackers may target any component of an IoT system to gain access to sensitive data or resources. To better understand the threats posed by an IoT system, it is important to first identify the attack surface. This includes all the connected devices, networks, and applications that an attacker may target. It is also important to identify the system's assets and the data or resources that an attacker may be interested in. Knowing the attack surface and the system's assets can help security professionals prioritize security measures and focus on the most vulnerable areas. Once the attack surfaces and assets are identified, it is important to understand potential attack strategies. Common attack strategies in the IoT include hacking into connected devices, exploiting firmware vulnerabilities, and spoofing device identifiers. Attackers may also use social engineering techniques to access sensitive information or manipulate user behavior. When designing an IoT system, it is important to consider security measures that can help mitigate these threats [1, 2]. These measures may include regular patching, monitoring, and encryption.

Threat models are used to evaluate the security of IoT systems by identifying potential threats and vulnerabilities. Additionally, attackers may use social engineering techniques to access sensitive information or manipulate user behavior. Security measures such as regular patching, monitoring, and encryption can help mitigate these threats.

DOI: 10.1201/9781003477327-21

21.2 THREAT MODELING

Threat modeling is a systematic process employed to identify, assess, and quantify potential security risks within an IoT system [2–4]. By gaining an understanding of the various types of threats that may exist and the potential consequences of these threats, organizations can take proactive measures to safeguard their IoT systems against potential attacks. This article delves into the core principles of threat modeling for IoT, encompassing the different categories of threats, the stages involved in threat modeling, and the tools and methodologies used to pinpoint and mitigate these threats.

Within the context of IoT, threats can broadly be classified into two main categories: physical attacks and cyber-attacks. Physical attacks entail the direct manipulation or interference with the physical components of an IoT system, whereas cyber-attacks revolve around exploiting software vulnerabilities or manipulating data through network channels. Physical attacks usually target the disruption or incapacitation of the physical components of an IoT system [5, 6]. Examples of physical threats include tampering with the device physically, jamming or interfering with signals, and physically damaging the device itself. On the other hand, cyber-attacks predominantly focus on data manipulation or theft. Common instances encompass malware, distributed denial-of-service (DDoS) attacks, man-in-the-middle attacks, and social engineering tactics. Cyber threats can also encompass attacks on data stored in cloud environments or on third-party servers, as well as attacks on the underlying network infrastructure [2, 7].

21.2.1 STAGES OF THREAT MODELING

Threat modeling involves several steps and processes, which can be divided into four broad stages as discussed in [5, 8] are as follows:

1. **Identify Assets**: The first step in threat modeling is to identify the assets that are of value to an organization and need to be protected. In the context of the IoT, this includes physical devices, such as sensors and actuators, as well as any data stored on them, or transmitted over the network.
2. **Identify Threats**: The next step is to identify the threats that could potentially target the identified assets. This includes identifying the potential attack vectors and the actors that could be involved in any attack.
3. **Analyze Risks**: Once the threats have been identified, the next step is to analyze the risks posed by each threat. This involves assessing the likelihood of the threat occurring, the potential impact of the threat, and the potential countermeasures that can be implemented to mitigate the risk.
4. **Mitigate Risks**: The final step is to implement measures to mitigate the identified risks. This includes implementing technical measures such as firewalls, encryption, and authentication, as well as other measures such as user awareness training and physical security measures.
5. **Tools and Techniques**: Common tools and techniques used in threat modeling include attack trees, risk assessment matrices, and security design reviews. Finally, security design reviews are used to assess the security of

a system from the design phase onward. The review process involves analyzing the system design, analyzing the system code, and examining the system architecture for potential weaknesses [5, 7, 9].

21.3   TYPES OF THREATS

The IoT represents a swiftly expanding network comprising interconnected physical objects equipped with software, sensors, and various technologies that empower them to gather and share data. With the continuous expansion of the IoT, the concurrent rise of novel and evolving threats poses a growing risk to the system's security. These various types of threats within the IoT landscape are detailed in Figure 21.1.

1. **Malware**: It is malicious software that can be used to gain unauthorized access to a computer system or network. In the IoT, malware can be used to gain access to a device or network, allowing an attacker to steal or manipulate data, or cause disruption. Common types of malware used in the IoT include botnets, ransomware, and spyware.
2. **Botnets**: A network of connected systems known as a botnet is used to remotely manipulate a victim's system and spread malware. Cybercriminals use command-and-control servers to manage botnets that they use to launch DDoS and phishing attacks, steal sensitive information, and gather online banking information. Botnets can be used by cybercriminals to attack IoT devices that are connected to a variety of other devices, including laptops, desktop computers, and smartphones [8, 10].

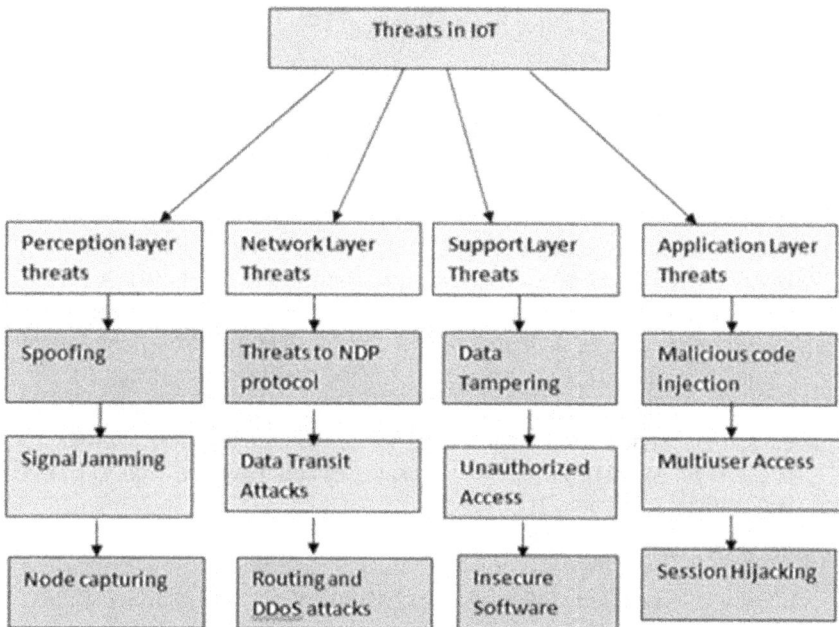

FIGURE 21.1   Types of security threats in IoT.

3. **Ransomware**: Ransomware attacks have emerged as one of the most widely recognized online threats. In this type of attack, a hacker employs malware to encrypt data that is crucial for the functioning of businesses. The hacker then demands a ransom payment in exchange for decrypting this vital data. Studies have shown that hackers can employ this method not only to disrupt operations but also to exert control, refusing to revert to normalcy until the ransom is paid. This approach can be applied similarly to target Industrial Internet of Things (IIoT) systems and smart home technology. For example, a hacker could potentially target a smart home and demand a ransom from the owner to regain control [11, 12].

4. **Spyware**: Spyware represents a security breach typically occurring when it infiltrates a computer or laptop inadvertently, often through a user's inadvertent click on an unfamiliar link or the opening of an attachment. This results in the simultaneous download of spyware alongside the attachment. It is advisable to exercise caution when selecting the sources from which to obtain files for one's system. Spyware, as a type of software, is employed unethically to clandestinely seize a user's personal or business-related information, which is then transmitted to a third party without the user's awareness or consent. Spyware may stealthily enter a computer or laptop as a concealed component, often through free or shared software. These malicious programs possess the capability to surreptitiously monitor a user's activities, gain unauthorized access to data, or even instigate system crashes.

5. **Data Breaches**: It occurs when sensitive or confidential information is exposed without authorization. In the IoT, data breaches can occur when insecure or outdated IoT devices are connected to the network. Without adequate security measures in place, attackers can gain access to the device and steal or manipulate data.

6. **Identity and Data Theft**: Attackers can carry out more intricate and sophisticated identity theft by gathering such data. IoT devices with connections to other IoT devices and enterprise systems are also susceptible to attack. For instance, hackers can break into an organization and access its corporate network by targeting a weak IoT sensor. Attackers can access numerous enterprise systems in this way and steal valuable company information.

7. **Distributed Denial of Service**: As discussed in [2, 3, 13, 14], A DDoS attack is when multiple computers are used to send many requests to a single server or system, overwhelming it and preventing it from functioning properly. In the IoT, a DDoS attack can be used to disrupt the communication between devices or take down an entire network. Zombified IoT devices and botnets have made DDoS attacks easier than before. It is when a device is made unavailable to the user due to an immense traffic flow. DDoS is illustrated in Figure 21.2.

8. **Insecure Interfaces**: Insecure interfaces are the most common type of threat in the IoT. These occur when a device or system has an interface that is not secure or is vulnerable to attack. Attackers can exploit insecure interfaces to gain access to the device or system and steal or manipulate data or cause disruption.

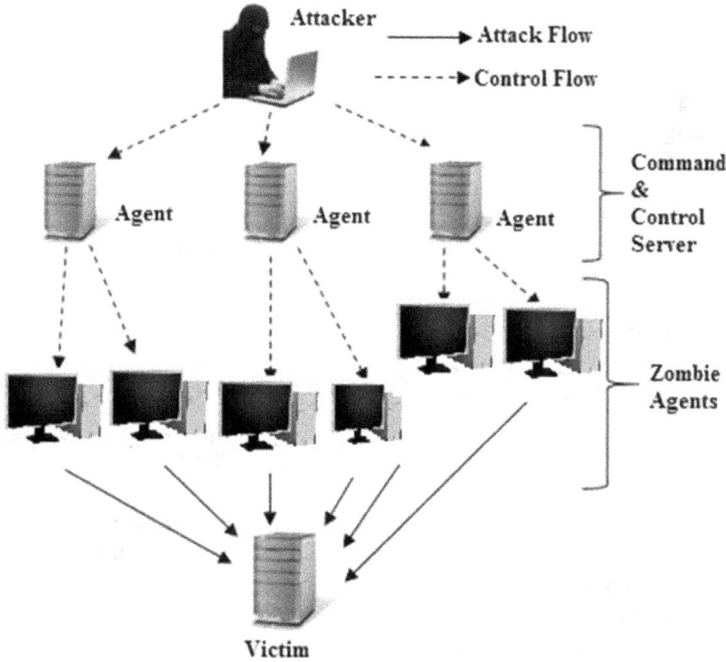

FIGURE 21.2 Distributed denial-of-service attack.

9. **Unencrypted Communications**: Unencrypted communications is another common threat in the IoT. Unencrypted communications are communications that are sent or received in plain text, making them vulnerable to interception and manipulation by attackers. If a device or system is transmitting unencrypted data, an attacker can intercept and manipulate it without detection.

10. **Unauthorized Access**: Unauthorized access is when an attacker gains access to a device or system without permission. In the IoT, unauthorized access can be used to steal or manipulate data or cause disruption. Preventing unauthorized access requires strong authentication measures, such as two-factor authentication, and an effective security system.

11. **Phishing**: Phishing is when an attacker attempts to gain access to a device or system by sending malicious emails or links to unsuspecting users. Figure 21.3 shows how phishing can be used to gain access to a device or system, allowing an attacker to steal or manipulate data, or cause disruption [10, 15–17].

12. **Insecure Software**: Insecure software is software that is not properly secured or updated. In the IoT, insecure software can be used to gain access to a device or system, allowing an attacker to steal or manipulate data, or cause disruption. Keeping software updated and patched is essential to prevent insecure software from being exploited by attackers.

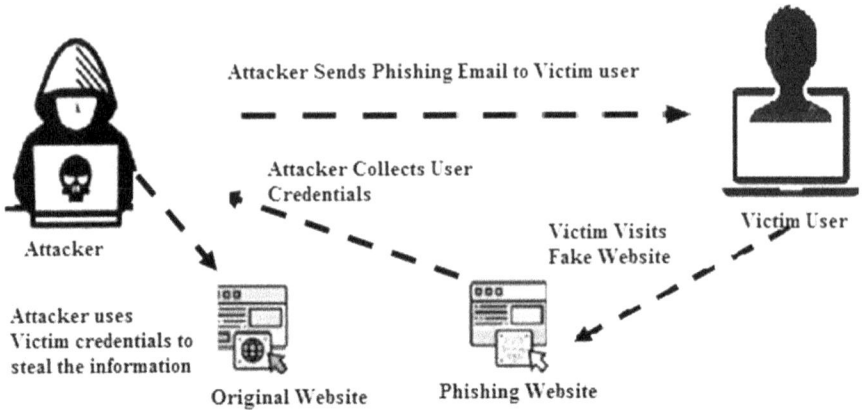

FIGURE 21.3 Phishing attack.

13. **Unsecured Networks**: Unsecured networks are networks that do not have adequate security measures in place. In the IoT, unsecured networks can be used to gain access to a device or system, allowing an attacker to steal or manipulate data, or cause disruption. Implementing strong authentication measures and an effective security system is essential to prevent unsecured networks from being exploited.

14. **Advanced Persistent Threats**: Many organizations have serious security concerns about advanced persistent threats (APTs). Targeted cyber attacks in which an attacker gains unauthorized access to a network and remains undetected for a long time are known as APTs. APTs are used by attackers to keep an eye on network activities and steal important data. It is challenging to stop, find, or mitigate these cyber attacks.

21.3.1 Types of Attacks

Physical attacks are when an attacker physically accesses a device or system, either in person or remotely. In the IoT, physical attacks can be used to gain access to a device or system, allowing an attacker to steal or manipulate data, or cause disruption. Implementing physical security measures, such as locks and alarms, is essential to prevent physical attacks as discussed in [10, 17].

- **Physical Tampering**: Hackers can access the physical location of the devices and easily steal data from them. In addition, they can install malware on the device or break into the network by accessing the ports and inner circuits of the device.
- **Eavesdropping**: The attacker can use a weak connection between the server and an IoT device. They can intercept network traffic and gain access to sensitive data. Using an eavesdropping attack, the intruder can also spy on your conversations using the data of the microphone and camera IoT device [3, 7, 14].
- **Brute-Force Password Attacks**: Cybercriminals can break into your system by trying different combinations of common words to crack the

password. Since IoT devices are made without security concerns in mind, they have the simplest password to crack.

- **Privilege Escalation**: Attackers can gain access to an IoT device by exploiting vulnerabilities, such as an operating system oversight, unpatched vulnerabilities, or a bug in the device. They can break into the system and crawl up to the admin level by further exploiting vulnerabilities and gaining access to the data that can be helpful for them.
- **Man-in-the-Middle Attack**: By exploiting insecure networks, cybercriminals can access the confidential data being passed by the device to the server. The attacker can modify these packets to disrupt communication [7, 13, 15] (Figure 21.4).
- **Malicious Code Injection**: Cybercriminals can exploit an input validation flaw and add malicious code to that place. The application can run the code and make unwanted changes to the program [7, 14] (Figure 21.5).
- **Social Engineering Attack**: Hackers utilize social engineering to trick people into disclosing their private information, including passwords and bank account information. Alternatively, thieves may gain access to a system through social engineering to covertly install harmful software. Typically, social engineering assaults use phishing emails to trick people, thus an attacker must create convincing emails to do this. But with IoT devices, social engineering assaults might be easier to carry out [5, 8, 9].

21.3.2 SECURITY GUIDELINES FOR PROTECTING IOT DEVICES

- **All data being gathered, and information being stored should be accounted for**. Every piece of information being gathered and stored needs to be verified. Every bit of data and information that moves around an IoT system should be mapped appropriately [3, 9].

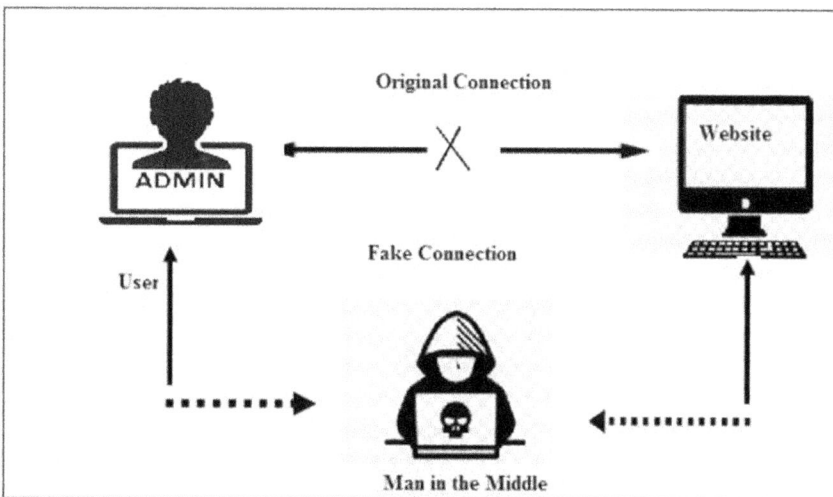

FIGURE 21.4   Man-in-the-middle attack.

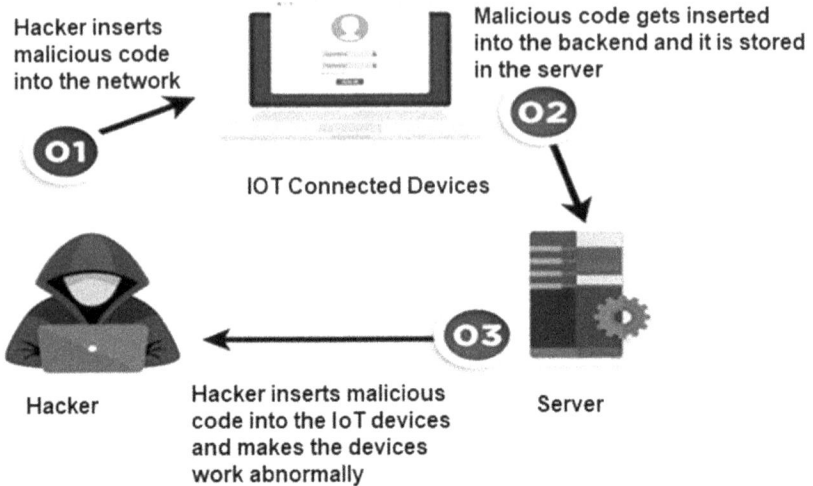

FIGURE 21.5   Malicious code injection.

- **Each device that connects to the network should be configured with security in mind**. This applies not only to the data collected by the sensors and devices deployed in the environment but also to any potential credentials in automation servers or other IoT applications. Before connecting a device to the network, make sure the settings are secure. Encryption, multi-factor authentication, and the usage of strong username and password combinations are all examples [4, 18].
- **The security plan for the organization should be based on the premise that it has been compromised**. Although preventing breaches and compromise is crucial, understanding that there is no foolproof protection against constantly changing threats can aid in the development of mitigation processes that can greatly confine and lessen the effects of an effective attack.
- **Every device needs to be physically protected**. It is crucial to consider how easily people can access IoT devices physically. An IoT device should be kept in a secure location or protected with the proper locks or other tools if it doesn't have any physical protection against tampering. For instance, IP cameras can be directly tampered with by a cybercriminal. They might have malicious hardware or software installed that spreads malware or causes system failures [12, 19].

21.3.3   END-TO-END SECURITY SOLUTIONS

- **Creation of Secure IoT Devices**: In order to guarantee the security of IoT devices, numerous layers of security measures must be implemented across the entire product development process. Employing a secure by design approach, the products are safeguarded from their inception. As an integral part of the security workflow, security is seamlessly integrated by design and Vulnerability Assessment and Penetration Testing (VAPT) into the product development lifecycle. Ultimately, this approach empowers clients to launch secure products into the market, shielding them from potential risks associated with IoT security [4, 20].

- **Data Security Solutions**: By encrypting and safeguarding data, these solutions can assist in protecting it from unauthorized access [9, 20].
- **Data Loss Prevention (DLP)**: By detecting and preventing unauthorized access, copying, and transmission, DLP systems can aid in preventing data loss.
- **Data Governance Solutions**: By ensuring that their data is only accessible to authorized individuals, kept in a secure environment, and shielded from unauthorized alterations, these solutions can assist organizations in managing their data safely.

21.4   ADVANTAGES OF THREAT MODELING

In the realm of IoT, threat modeling assumes a paramount role as a fundamental security practice aimed at safeguarding devices, networks, and data against malevolent entities. This practice entails a comprehensive analysis of the diverse elements constituting an IoT system, pinpointing possible security vulnerabilities and threats, and formulating effective measures to counteract these threats. By engaging in this systematic procedure, organizations can confidently ascertain the security and compliance of their IoT systems with prevailing industry standards and regulatory requirements.

Advantages of threat modeling in IoT environments

1. **Risk Identification**: Threat modeling enables organizations to proactively identify potential risks and vulnerabilities in their IoT systems. By understanding these threats, they can take preemptive measures to mitigate them before they are exploited by Malicious attackers.
2. **Resource Allocation**: It helps organizations allocate their security resources more efficiently. By prioritizing the most critical threats, they can focus their efforts and investments on the areas of greatest concern, optimizing their security posture.
3. **Compliance and Regulations**: IoT systems often must adhere to specific industry regulations and standards. Threat modeling aids in ensuring that an organization's IoT implementation aligns with these compliance requirements, reducing the risk of non-compliance and associated penalties.
4. **Reduced Security Costs**: Identifying and addressing security issues early in the development or deployment of IoT systems can be significantly more cost-effective than dealing with breaches or vulnerabilities after they have been exploited.
5. **Enhanced Security Awareness**: Threat modeling fosters a culture of security awareness within an organization. It encourages stakeholders to consider security aspects at every stage of IoT system development, from design to deployment.
6. **Improved Communication**: It facilitates communication among various teams and stakeholders involved in the IoT project. By visualizing threats

and vulnerabilities, threat models provide a common language for discussing security concerns and solutions.

7. **Customized Security**: Threat modeling allows organizations to tailor security measures to their specific IoT system. This customization ensures that security controls are relevant and effective in addressing the unique risks associated with the system.

8. **Resilience**: It helps in designing IoT systems that are more resilient to attacks. By anticipating threats and designing countermeasures, organizations can build systems that can withstand various security challenges.

9. **Safeguarding Reputation**: A security breach in IoT could tarnish an organization's reputation. Threat modeling aids in protecting the brand and maintaining customer trust by reducing the likelihood of security incidents.

10. **Regulatory Alignment**: Many industries have specific security requirements. Threat modeling helps ensure that IoT systems align with these requirements, making it easier to obtain regulatory approvals and certifications.

In summary, threat modeling in IoT provides a structured approach to identifying and mitigating security risks, promoting efficient resource allocation, and enhancing overall security posture. It is an invaluable practice for organizations aiming to secure their IoT deployments effectively. Finally, threat modeling in IoT helps organizations build trust with their customers. By understanding the potential threats and vulnerabilities of their system, organizations can develop strategies to protect their customers and create a secure environment. This helps organizations build trust and loyalty with their customers and ensures that their customers' data is secure. Figure 21.6 depicts how to protect IoT devices from all these types of attacks.

21.5 CHALLENGES AND FUTURE DIRECTIONS

Threat modeling is a security technique used to identify, quantify, and mitigate potential threats to a system. This chapter will discuss the challenges and future directions of threat modeling in IoT.

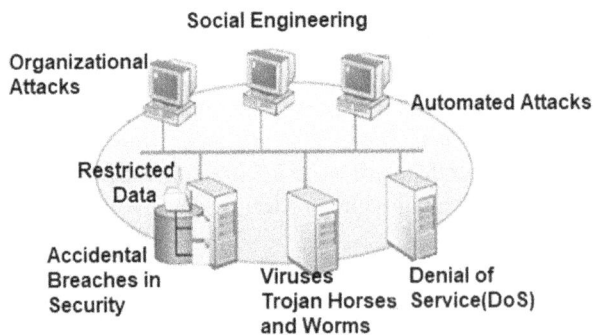

FIGURE 21.6   Protecting IoT devices from all these types of threats.

21.5.1 Challenges

There are several challenges associated with threat modeling in IoT. One of the biggest challenges is the sheer number of IoT devices and their complexity. With the proliferation of IoT devices, it is difficult to keep track of all the potential threats and how they could affect each device. Additionally, due to the variety of IoT applications and the different ways that they interact with each other, it is hard to create a comprehensive threat model that covers all potential threats. Another challenge is the fact that IoT devices typically have limited computing power and memory, which makes it difficult to implement sophisticated security measures. Additionally, many IoT devices have limited connectivity, which makes it difficult to update security measures in a timely manner. Furthermore, there are often multiple stakeholders with different security requirements, which makes it difficult to create a unified security policy. Finally, IoT devices are often connected to the internet, which exposes them to a wide range of threats, including malware, data leakage, and denial-of-service (DoS) attacks. Additionally, these threats can often be difficult to detect and mitigate due to the lack of visibility into the operation of the device.

21.5.2 Future Directions

Despite the complexities associated with conducting threat modeling in the IoT realm, there exist several measures to enhance security. One of the foremost steps involves ensuring the proper configuration and maintenance of devices. This encompasses keeping devices up to date with the latest security patches and configuring them in adherence to established best practices. Furthermore, implementing robust authentication mechanisms, such as two-factor authentication, becomes crucial to restrict access exclusively to authorized users.

Another pivotal measure involves the adoption of secure communication protocols like TLS or DTLS to encrypt data transmissions. This safeguards data exchanges against interception or tampering by potential attackers. It is equally important to employ secure communication protocols, such as MQTT, which not only support encryption but also provide authentication capabilities.

Finally, the implementation of a comprehensive security monitoring system emerges as a vital component of IoT security. Such a system should possess the capability to detect and promptly alert against potential threats. Moreover, it should furnish detailed information regarding the nature of the attack. Additionally, this monitoring system should maintain meticulous logs of all activities related to the device, which can serve as valuable resources for detecting anomalies and conducting investigations into potential security threats.

21.6 SUMMARY

In conclusion, threat models and attack strategies are essential components of any IoT system. By understanding the different types of threats that can be encountered, it is possible to develop an effective strategy to mitigate those threats. This includes ensuring the system is securely configured, monitoring the system for suspicious activity, and implementing a variety of attack strategies. Additionally, it is important to have an incident response plan in place in the event of a successful attack. By following these steps, it is possible to ensure the security of an IoT system and protect it from Malicious attackers.

REFERENCES

1. Omotosho, A., Ayemlo Haruna, B., and Olaniyi, M. O. "Threat modeling of internet of things health devices." *Journal of Applied Security Research*, vol. 14, no. (1) (2019): 106–121.
2. Whitehouse, O. "Security of things: An implementers' guide to cyber-security for internet of things devices and beyond." *NCC Group* (2014).
3. Ramazanzadeh, M. A., Barzegar, B., and Motameni, H. "Automatic generation of threat paths in internet of things-based systems." *IET Communications* (2022).
4. Simonjan, J., Taurer, S., and Dieber, B. "A generalized threat model for visual sensor networks." *Sensors*, vol. 20, no. 13 (2020): 3629.
5. Griffioen, P. and Sinopoli, B. "Assessing risks and modeling threats in the Internet of Things." (2021). arXiv preprint arXiv:2110.07771.
6. Bradbury, M., Jhumka, A., Watson, T., Flores, D., Burton, J., and Butler, M. "Threat-modeling-guided trust-based task offloading for resource-constrained Internet of Things." *ACM Transactions on Sensor Networks (TOSN)*, vol. 18, no. 2 (2022): 1–41.
7. Al Asif, Md R., Fida Hasan, K., Islam, Md Zahidul, and Khondoker, R. "STRIDE-based cyber security threat modeling for IoT-enabled Precision Agriculture Systems." In *2021 3rd International Conference on Sustainable Technologies for Industry 4.0 (STI)*, pp. 1–6. IEEE, 2021.
8. Firdous, S. N., Baig, Z., Valli, C., and Ibrahim, A. "Modelling and evaluation of malicious attacks against the IoT MQTT protocol." In *2017 IEEE International Conference on Internet of Things (iThings) and IEEE Green Computing and Communications (GreenCom) and IEEE Cyber, Physical and Social Computing (CPSCom) and IEEE Smart Data (SmartData)*, pp. 748–755. IEEE, 2017.
9. Brown, M. L., et al. *The IoT Revolution and our Digital Security: Principles for IoT Security*. Wiley Rein, LLP. U.S. Chamber of Commerce: Washington D.C., 2017.
10. Sequeiros, J. A. B. F., Chimuco, F. T., Samaila, M. G., Freire, M. M., and Inacio, P. R. M. "Attack and system modeling applied to IoT, cloud, and mobile ecosystems: Embedding security by design." *ACM Computing Surveys*, vol. 53, no. 2 (Mar 2020). [Online]. doi: 10.1145/3376123
11. Hossain, E., Khan, I., Un-Noor, F., Sikander, S. S., and Sunny, M. S. H. "Application of big data and machine learning in smart grid, and associated security concerns: A review." *IEEE Access*, vol. 7, 2019: 13 960–13 988.
12. Williams, R., McMahon, E., Samtani, S., Patton, M., and Chen, H. "Identifying vulnerabilities of consumer Internet of Things (IoT) devices: A scalable approach." In *2017 IEEE International Conference on Intelligence and Security Informatics (ISI)*, pp. 179–181. IEEE, 2017.
13. Aufner, P. "The IoT security gap: A look down into the valley between threat models and their implementation." *International Journal of Information Security*, vol. 19, no. 1 (2020): 3–14.
14. Rizvi, S., Pipetti, R., & McIntyre, N., and Todd, J. "Threat model for securing Internet of Things (IoT) network at device-level." *Internet of Things*, vol. 11 (2020): 100240. doi: 10.1016/j.iot.2020.100240

15. Isabar, D. "Threat modeling and penetration testing of a Yanzi IoT-system: A survey on the security of the system's RF communication." (2021).

16. Rizvi, S., Pipetti, R., McIntyre, N., Todd, J., and Williams, I. "Threat model for securing Internet of Things (IoT) network at device-level." *Internet of Things*, vol. 11, 2020: 100240.

17. Abbas, S. G., Zahid, S., Hussain, F., Shah, G. A., and Husnain, M. "A threat modelling approach to analyze and mitigate botnet attacks in smart home use case." In *2020 IEEE 14th International Conference on Big Data Science and Engineering (BigDataSE)*, pp. 122–129. IEEE, 2020.

18. Ferrag, M. A., Maglaras, L., and Derhab, A. (2019). "Authentication and authorization for mobile IoT devices using biofeatures: Recent advances and future trends." *Security and Communication Networks* (2019).

19. Ghazanfar, S., Hussain, F., Rehman, A. U., Fayyaz, U. U., Shahzad, F., and Shah, G. A. "IoT-flock: An open-source framework for IoT traffic generation." In *2020 International Conference on Emerging Trends in Smart Technologies (ICETST)*, pp. 1–6. IEEE, 2020.

20. Baker, S. "Cybersecurity and the Internet of Things." *Applied Cybersecurity Strategy for Managers*. Essec Business School: Clergy, France, July 1, 2016.

21. https://owasp.org/www-project-top-ten/, OWASP top 10 Vulnerabilities.

22 Security and Privacy Issues in IoT

K. P. K. Devan, B. S. Liya, and P. Indumathy
Easwari Engineering College, Anna University,
Chennai, India

22.1 INTRODUCTION

The Internet of Things (IoT) is a revolutionary technology paradigm that has gained significant prominence in recent years. It refers to the interconnected network of physical objects, devices, and sensors that can collect, exchange, and process data over the internet, often without human intervention. While IoT promises to bring about numerous benefits, such as increased efficiency, automation, and convenience in various aspects of our lives, it also presents a host of security challenges that must be addressed to fully realize its potential.

Proliferation of Devices: One of the fundamental characteristics of IoT is the sheer number of devices involved. With billions of interconnected devices worldwide, ranging from smart thermostats and wearable fitness trackers to industrial machines and autonomous vehicles, the attack surface for Malicious attackers expands exponentially. Each device represents a potential entry point for cyberattacks.

Diverse Ecosystem: IoT encompasses a wide range of devices and technologies, each with its own unique vulnerabilities and security requirements. This diversity makes it challenging to develop standardized security measures, as one-size-fits-all solutions are rarely applicable. Consequently, securing IoT devices becomes a complex and multifaceted endeavor.

Limited Computing Resources: Many IoT devices are constrained by limited processing power, memory, and energy resources. These constraints often hinder the implementation of robust security measures, making IoT devices susceptible to attacks that leverage their inherent weaknesses.

Data Privacy Concerns [1]: IoT devices collect and transmit vast amounts of data, often of a highly sensitive nature, such as personal health information, location data, and home security details. Ensuring the privacy and confidentiality of this data is paramount, as breaches can have severe consequences for individuals and organizations.

Inadequate Authentication and Authorization: Weak or nonexistent authentication mechanisms can allow unauthorized access to IoT devices and networks. Without proper authorization protocols in place, Malicious attackers can exploit vulnerabilities to compromise the integrity of the IoT ecosystem.

DOI: 10.1201/9781003477327-22

Firmware and Software Vulnerabilities: IoT devices rely on firmware and software to function, and these components can contain vulnerabilities that hackers can exploit. Manufacturers must frequently update and patch their devices to address security flaws, but many devices lack the capability for regular updates, leaving them perpetually vulnerable.

Supply Chain Risks: The global supply chain for IoT devices is complex, involving multiple vendors, components, and assembly points. This complexity increases the potential for supply chain attacks, where Malicious attackers compromise devices during manufacturing or distribution.

Lack of Security Standards: While efforts are underway to establish security standards for IoT devices, there is still a lack of universal standards and regulations. This fragmentation makes it difficult for consumers to make informed choices and for manufacturers to adhere to consistent security practices.

Botnets and DDoS Attacks [2]: Compromised IoT devices are often used to form botnets, which can be leveraged for large-scale Distributed Denial of Service (DDoS) attacks. These attacks can disrupt internet services, causing significant economic and operational damage.

Human Factor: Human negligence or lack of awareness can also pose a significant security challenge in IoT. Users often fail to change default passwords, update firmware, or follow best practices, inadvertently leaving devices vulnerable to attacks.

22.2   KEY CHARACTERISTICS OF IoT

- **Connectivity**: IoT devices are connected to the internet or other networks, allowing them to communicate and share data.
- **Sensors**: IoT devices are equipped with sensors that can collect various types of data, such as temperature, humidity, location, motion, and more.
- **Data Processing**: The data collected by IoT devices can be processed and analyzed either locally on the device or in the cloud [3] to derive meaningful insights.
- **Automation**: IoT devices often include the capability to automate actions based on the data they collect. For example, a thermostat can adjust the temperature based on occupancy and weather data.
- **Remote Control**: Many IoT devices can be remotely controlled or monitored via a smartphone app or a web interface.

22.3   IMPORTANCE OF SECURITY IN IoT

Data Protection: IoT devices collect and transmit vast amounts of data, including sensitive and personal information. Ensuring the security of this data is essential to protect individuals' privacy and prevent data breaches. Unauthorized access or data leaks can have severe consequences, including identity theft, financial losses, and reputational damage.

Safety Concerns: Many IoT applications are directly tied to safety-critical systems, such as autonomous vehicles, medical devices, and industrial control systems. Inadequate security measures in these applications can lead to accidents, injuries, or even loss of life.

Preventing Unauthorized Access: Unauthorized access to IoT devices can result in various malicious activities, including device manipulation, Data manipulation, and unauthorized control. Strong security measures are necessary to prevent unauthorized individuals or entities from gaining control over IoT devices.

Network Security: IoT devices are often part of larger networks, and compromising one device can potentially provide attackers with a foothold to launch broader attacks on networked systems. Robust security measures are crucial to safeguard the integrity of entire networks.

Malware and Botnets: Insecure IoT devices are susceptible to malware infections and can be harnessed into botnets for large-scale cyberattacks, such as DDoS attacks. Strengthening security can help mitigate the risk of devices becoming part of botnets.

Data Integrity: Ensuring the integrity of data collected and transmitted by IoT devices is vital. Manipulated or tampered data can lead to incorrect decisions and actions, particularly in applications like healthcare, manufacturing, and critical infrastructure.

Compliance and Legal Requirements: Many regions have introduced regulations, such as the General Data Protection Regulation (GDPR) in Europe, that impose legal requirements on data handling and security. Non-compliance can result in significant fines and legal consequences.

Business Continuity: Businesses and organizations that rely on IoT systems need to ensure the continuity of their operations. Security breaches or disruptions in IoT systems can lead to downtime, financial losses, and damage to an organization's reputation.

Reputation Management: Security incidents involving IoT devices can harm the reputation of manufacturers and service providers. Building and maintaining trust with customers and users requires a commitment to security.

Future-Proofing: As the IoT ecosystem continues to expand, it becomes a more attractive target for cybercriminals. Implementing robust security measures from the outset can help future-proof IoT solutions against evolving threats.

22.4   SECURITY VULNERABILITIES IN IoT

Weak Authentication and Passwords: Many IoT devices ship with default usernames and passwords that are rarely changed by users. Weak or easily guessable passwords make it simple for attackers to gain unauthorized access to devices.

Inadequate Firmware and Software Security: Manufacturers may not prioritize security in the development of firmware and software for IoT devices. Lack of updates and patches for known vulnerabilities can leave devices susceptible to attacks.

Unauthenticated Firmware Updates: IoT devices often receive firmware updates without proper authentication, allowing attackers to push malicious updates to compromise devices.

Insecure APIs and Interfaces: Weak or insecure application programming interfaces (APIs) and web interfaces can be exploited by attackers to gain control of devices or extract sensitive data.

Physical Access Vulnerabilities: Physical tampering with IoT devices, such as unsecured ports or lack of tamper-evident seals, can lead to unauthorized access or compromise.

Lack of Device Identity Management: IoT devices may not have robust identity management mechanisms, making it challenging to verify the authenticity of devices and their communications.

Default and Weak Network Configurations: Some IoT devices have insecure network configurations or use outdated and vulnerable network protocols.

Denial of Service (DoS) Attacks: IoT devices can be targeted in DoS attacks, causing them to become unresponsive or unavailable.

Privacy Concerns: Data collected by IoT devices may not be adequately protected, leading to privacy violations and potential misuse of sensitive information.

Third-Party Integrations and Supply Chain Risks: Integrations with third-party services and components can introduce security vulnerabilities. Supply chain risks, including compromised hardware or software during manufacturing, can impact device security.

Lack of Security Standards: The absence of universal security standards [4] and best practices can result in inconsistent security measures across IoT devices.

Limited Resources in Constrained Devices: Resource-constrained IoT devices may struggle to implement robust security measures due to limitations in processing power, memory, and energy resources.

Human Error and Social Engineering: Users and administrators may inadvertently expose IoT devices to security risks through misconfigurations or by falling victim to social engineering attacks.

Legacy and Outdated Devices: Older IoT devices may no longer receive updates or support from manufacturers, leaving them vulnerable to known vulnerabilities.

22.5 AUTHENTICATION AND ACCESS CONTROL

22.5.1 AUTHENTICATION METHODS USED IN IoT [5]

Authentication is the process of verifying the identity of a user or device trying to access a system or resource. In IoT, authentication is essential to establish trust and prevent unauthorized access.

- **Password-Based Authentication**: Users or devices provide a username and password to authenticate. However, strong passwords are crucial, and password policies should be enforced.

- **Token-Based Authentication**: Devices or users receive unique tokens (e.g., API keys, OAuth tokens) that they present for authentication. Tokens are more secure than passwords and can be easily revoked.
- **Biometric Authentication**: Some IoT devices, such as smartphones and fingerprint scanners, use biometrics like fingerprints, facial recognition, or retinal scans for authentication.
- **Certificate-Based Authentication**: Devices and users are issued digital certificates that can be verified by a certificate authority (CA) or a trusted third party.
- **Multi-factor Authentication (MFA)**: MFA combines multiple authentication methods, such as something you know (password), something you have (token), and something you are (biometric), to enhance security.
- **Device Identity Authentication**: Devices can authenticate each other using unique identifiers (e.g., MAC addresses and device certificates) to ensure that only trusted devices can communicate.

22.5.2 Access Control in IoT

Access control refers to the policies and mechanisms that determine what resources or actions a user or device is allowed to access. In IoT, access control is crucial for safeguarding sensitive data and ensuring the integrity of IoT networks.

Access Control Mechanisms

- **Role-Based Access Control (RBAC)**: Users or devices are assigned roles with specific permissions. Access is granted based on the user's role.
- **Attribute-Based Access Control (ABAC)**: Access decisions are based on attributes such as user identity, device type, location, and time of access.
- **Policy-Based Access Control**: Access policies are defined based on predefined rules, and access decisions are made according to these policies.
- **Access Tokens**: Access tokens are used to grant temporary access to specific resources. Tokens can have an expiration time, limiting access duration.
- **Dynamic Access Control**: Access control rules can adapt to changing conditions, such as the context of the user or device.
- **Fine-Grained Access Control**: Access control can be granular, allowing for precise control over individual resources and actions.

22.6 DATA PROTECTION AND ENCRYPTION

22.6.1 Data Protection in IoT [6, 7]

Data protection in IoT involves ensuring that data is handled, stored, and processed in a way that preserves its confidentiality, integrity, and availability. The following are some key considerations for data protection in IoT.

- **Data Classification**: Classify data based on its sensitivity and importance. Identify which data requires stronger protection measures.

- **Data Retention Policies**: Define data retention policies to specify how long data should be kept and when it should be securely deleted.
- **Data Minimization**: Collect only the data that is necessary for the intended purpose. Avoid unnecessary data collection to reduce the risk of exposure.
- **Data Lifecycle Management**: Implement processes for managing data throughout its lifecycle, including secure disposal of data when it is no longer needed.
- **Secure Data Storage**: Use encryption and access controls to protect data stored on IoT devices, gateways, and cloud servers.

22.6.2 ENCRYPTION IN IoT [6–8]

Encryption is a fundamental technique for protecting data in transit and at rest. It involves encoding data in such a way that only authorized parties can decipher it. In IoT, encryption plays a crucial role in securing communication and stored data. The following are some aspects of encryption in IoT.

- **Data in Transit Encryption**: Encrypt data as it is transmitted between IoT devices, gateways, and servers. Transport Layer Security (TLS) and Datagram Transport Layer Security (DTLS) are commonly used encryption protocols for securing communication.
- **End-to-End Encryption**: Implement end-to-end encryption to ensure that data is protected from the point of origin to the final destination. This prevents eavesdropping along the communication path.
- **Data at Rest Encryption**: Encrypt data when it is stored on IoT devices, local storage, or cloud servers. This protects data even if physical access to storage media is compromised.
- **Key Management**: Implement secure key management practices, including key generation, distribution, rotation, and storage. Protect encryption keys from unauthorized access.
- **Data Integrity Protection**: Use cryptographic techniques, such as digital signatures and message authentication codes (MACs), to verify the integrity of data and detect any tampering.
- **Secure Boot and Firmware Updates**: Ensure that firmware updates and device boot processes are secure by using secure boot mechanisms and code signing.

22.7 IoT DEVICE MANAGEMENT AND PATCHING

IoT device management and patching [9, 10] are critical aspects of maintaining the security, functionality, and reliability of IoT deployments.

22.7.1 IoT DEVICE MANAGEMENT

Definition: IoT device management refers to the process of provisioning, monitoring, configuring, updating, and maintaining IoT devices throughout their lifecycle.

Lifecycle Stages: Device management covers various stages, including onboarding, deployment, monitoring, maintenance, and end-of-life disposal.

Key Functions

- **Provisioning**: Enrolling devices into the network, assigning unique identifiers, and configuring initial settings.
- **Configuration**: Remote configuration of device settings, including network parameters and application-specific settings.
- **Software Updates**: Managing firmware and software updates to address vulnerabilities and bugs, and add new features.
- **Security Management**: Implementing security measures such as authentication, access control, and encryption.
- **Remote Control**: Enabling remote control and troubleshooting of devices when necessary.
- **Data Collection**: Gathering data from devices for analytics, diagnostics, and optimization.
- **Alerts and Notifications**: Setting up alerts and notifications for abnormal device behavior or security incidents.
- **IoT Device Management Platforms**: Many organizations use IoT device management platforms (IoT Device Management as a Service or IoT DMaaS) to streamline device management tasks. These platforms provide centralized control and automation.

22.7.2 IoT Device Patching

Definition: IoT device patching involves updating the device's software, firmware, or operating system to address security vulnerabilities and bugs, and improve device performance.

Importance of Patching

- **Security**: Patches often include fixes for known security vulnerabilities, reducing the risk of exploitation by attackers.
- **Reliability**: Patching can improve the stability and reliability of IoT devices by resolving software-related issues.
- **New Features**: Patching may introduce new features and capabilities to enhance device functionality.
- **Regulatory Compliance**: Many industries and regions require regular patching to comply with security and privacy regulations.

22.8 REGULATORY COMPLIANCE AND PRIVACY

Regulatory compliance and privacy [11, 12] in the IoT are critical aspects that require careful attention due to the potential for data breaches and privacy violations in IoT ecosystems. Various laws and regulations around the world aim to safeguard individuals' privacy and ensure that IoT devices and applications adhere to certain standards.

General Data Protection Regulation (GDPR): GDPR [13] is a comprehensive data protection regulation in the European Union (EU). It applies to IoT devices and services that process the personal data of EU residents. Organizations must obtain explicit consent to collect and process personal data, implement data protection by design and by default, and provide individuals with the right to access, rectify, and erase their data. Non-compliance can lead to significant fines.

California Consumer Privacy Act (CCPA): CCPA [14] is a privacy law in California, USA, that grants California residents certain rights over their personal information. IoT companies doing business in California need to comply with CCPA by providing transparency about data collection, allowing consumers to opt out of data sales, and ensuring data security.

HIPAA (Health Insurance Portability and Accountability Act): For IoT devices and applications in the healthcare sector, HIPAA [12] in the United States sets strict rules for the protection of patients' health information. IoT solutions must ensure the confidentiality, integrity, and availability of healthcare data and implement safeguards to prevent data breaches.

IoT Cybersecurity Improvement Act: In the United States, this legislation focuses on IoT security [15]. It requires federal agencies to establish minimum security standards for IoT devices purchased or used by the government. Compliance with these standards can influence IoT device manufacturers and suppliers.

Industry-Specific Regulations: Certain industries, such as automotive, energy, and industrial IoT, may have specific regulations or standards that apply to their IoT devices and systems [12]. For example, the automotive sector may have safety and cybersecurity standards for connected vehicles.

Data Minimization: IoT devices should practice data minimization, collecting only the data necessary for their intended purpose [16]. Storing excessive data poses privacy risks and may lead to regulatory non-compliance.

Security by Design: IoT developers should follow the principle of "security by design," considering security measures throughout the development lifecycle. This includes regular security assessments and updates to address vulnerabilities.

Consent and Transparency: IoT providers should obtain clear and informed consent from users before collecting and processing their data. Transparency about data practices, including what data is collected and how it's used, is essential.

Data Encryption: Encrypting data in transit and at rest helps protect it from unauthorized access. This is particularly important for IoT devices that transmit sensitive data over networks.

IoT Device Lifecycle Management: Proper lifecycle management, including timely security updates and patches, is crucial for maintaining the security and compliance of IoT devices. Manufacturers should consider end-of-life disposal and data erasure procedures.

Privacy Impact Assessments: Conduct privacy impact assessments to evaluate and mitigate privacy risks associated with IoT deployments. This process helps identify potential privacy concerns and take steps to address them.

22.9   THREATS AND ATTACK SCENARIOS

The IoT introduces a wide range of new threats and attack scenarios [2, 11, 17–21] due to its interconnected nature and the proliferation of devices. These threats can have serious consequences, including data breaches, privacy violations, and disruption of critical services.

Unauthorized Access

- **Scenario**: Unauthorized users gain access to IoT devices, networks, or data.
- **Impact**: Intruders can control or manipulate devices, steal sensitive data, or disrupt operations.
- **Mitigation**: Implement strong authentication mechanisms, use unique credentials for each device, and regularly update passwords.

Device Compromise

- **Scenario**: Attackers compromise IoT devices, either through vulnerabilities or weak security controls.
- **Impact**: Compromised devices can be used for malicious purposes, such as launching DDoS attacks or infiltrating networks.
- **Mitigation**: Regularly update device firmware, apply security patches, and use hardware-based security features when available.

Data Interception

- **Scenario**: Attackers intercept data transmitted between IoT devices and backend systems.
- **Impact**: Sensitive data, such as personal information or critical commands, can be stolen or manipulated.
- **Mitigation**: Encrypt data in transit using strong encryption protocols and implement secure communication channels.

Distributed Denial of Service Attacks

- **Scenario**: IoT devices are hijacked to launch DDoS attacks, overwhelming target systems or networks.
- **Impact**: Targeted services become unavailable, causing operational disruptions and financial losses.
- **Mitigation**: Implement network segmentation, rate limiting, and anomaly detection to mitigate DDoS threats.

Malware and Botnets

- **Scenario**: Malicious software infects IoT devices, turning them into a botnet for coordinated attacks.
- **Impact**: Botnets can execute various attacks, including spam, DDoS, and data theft.
- **Mitigation**: Use reputable anti-malware solutions, isolate infected devices, and maintain up-to-date security measures.

Physical Attacks

- **Scenario**: Attackers physically tamper with or damage IoT devices.
- **Impact**: Disruption of device functionality, data theft, or unauthorized access.
- **Mitigation**: Implement tamper-evident hardware designs and physical security controls.

Supply Chain Attacks

- **Scenario**: Attackers compromise IoT devices during manufacturing or distribution.
- **Impact**: Infected devices may enter the market, posing a risk to users and organizations.
- **Mitigation**: Strengthen supply chain security, verify device integrity, and conduct thorough security audits.

Insider Threats

- **Scenario**: Malicious insiders or employees with access to IoT systems misuse their privileges.
- **Impact**: Unauthorized access, data theft, or sabotage of IoT systems.
- **Mitigation**: Implement access controls, monitor user activities, and conduct background checks.

Inadequate Authentication

- **Scenario**: Weak or default authentication credentials are exploited by attackers.
- **Impact**: Unauthorized access to devices, networks, or data.
- **Mitigation**: Enforce strong authentication practices, and encourage users to change default passwords.

Lack of Security Updates

- **Scenario**: Manufacturers do not provide security updates or patches for IoT devices.
- **Impact**: Devices remain vulnerable to known exploits and vulnerabilities.
- **Mitigation**: Choose devices from reputable manufacturers with a history of providing updates and support.

Privacy Violations

- **Scenario**: IoT devices collect and transmit personal data without user consent or adequate protection.
- **Impact**: Breach of user privacy, potential legal consequences, and reputational damage.
- **Mitigation**: Implement data protection measures, obtain user consent, and follow privacy regulations.

22.10   SECURITY BEST PRACTICES FOR IoT

Security practices in IoT [22–26] are essential to protect the vast and diverse ecosystem of interconnected devices from potential threats and vulnerabilities.

- Device Authentication and Authorization
- Encryption
- Regular Updates and Patch Management
- Secure Boot and Firmware Integrity
- Network Segmentation
- Intrusion Detection and Prevention
- Physical Security
- Privacy by Design
- Access Control and Role-Based Permissions
- Security Audits and Testing
- Logging and Monitoring
- Vendor Security Assessment
- Incident Response Plan
- User Education and Awareness
- Regulatory Compliance
- Lifecycle Management
- Collaboration and Knowledge Sharing

22.11   SUMMARY

The IoT has ushered in transformative possibilities but also brings forth a host of issues and security challenges. These issues encompass a wide range of concerns that need to be addressed Proliferation of Devices, Diverse Ecosystem, Resource Limitations, Data Privacy, Weak Authentication and Authorization, Firmware and Software Vulnerabilities, Supply Chain Risks, Lack of Security Standards, Botnet and DDoS Attacks, and Human Negligence. To mitigate these challenges, a comprehensive approach to IoT security is essential. This approach includes robust authentication and encryption, regular software updates, secure boot processes, network segmentation, intrusion detection, physical security measures, privacy protection, access control, audits, compliance with regulations, and collaboration among industry stakeholders. By addressing these issues and challenges, the IoT ecosystem can be made more secure and reliable.

REFERENCES

1. S. G. H. Soumyalatha, "Study of IoT: understanding IoT architecture, applications, issues and challenges," *International Journal of Advanced Networking & Applications*, vol. 478, 2016.
2. N. Tariq, M. Asim, Z. Maamar, M. Z. Farooqi, N. Faci, and T. Baker, "A mobile code-driven trust mechanism for detecting internal attacks in sensor node-powered IoT," *Journal of Parallel and Distributed Computing*, vol. 134, pp. 198–206, 2019.

3. H. Yi and Z. Nie, "Side-channel security analysis of UOV signature for cloud-based Internet of Things," *Future Generation Computer Systems*, vol. 86, pp. 704–708, 2018.

4. K. Zhang, X. Liang, R. Lu, and X. Shen, "Sybil attacks and their defenses in the Internet of Things," *IEEE Internet of Things Journal*, vol. 1, no. 5, pp. 372–383, 2014.

5. N. C. Winget, A. R. Sadeghi, and Y. Jin, "Invited: can IoT be secured: emerging challenges in connecting the unconnected," in *Proceedings of the 53rd Annual Design Automation Conference*, pp. 1–6, New York, USA, 2016.

6. G. S. Hukkeri and R. H. Goudar, "IoT: issues, challenges, tools, security, solutions and best practices," *International Journal of Pure and Applied Mathematics*, vol. 120, no. 6, pp. 12099–12109, 2019.

7. Q. D. La, T. Q. S. Quek, J. Lee, S. Jin, and H. Zhu, "Deceptive attack and defense game in honeypot-enabled networks for the Internet of Things," *IEEE Internet of Things Journal*, vol. 3, no. 6, pp. 1025–1035, 2016.

8. N. Zhang, R. Wu, S. Yuan, C. Yuan, and D. Chen, "RAV: relay aided vectorized secure transmission in physical layer security for Internet of Things under active attacks," *IEEE Internet of Things Journal*, vol. 6, no. 5, pp. 8496–8506, 2019.

9. M. López, A. Peinado, and A. Ortiz, "An extensive validation of a SIR epidemic model to study the propagation of jamming attacks against IoT wireless networks," *Computer Networks*, vol. 165, article 106945, 2019.

10. N. Sharma, M. Shamkuwar, and I. Singh, "The history, present and future with IoT," in *Internet of Things and Big Data Analytics for Smart Generation*, pp. 27–51, Springer, 2019.

11. A. Gopi and M. K. Rao, "Survey of privacy and security issues in IoT," *International Journal of Engineering & Technology*, vol. 7, no. 2.7, p. 293, 2018.

12. R. H. Weber, "Internet of things: privacy issues revisited," *Computer Law and Security Review*, vol. 31, no. 5, pp. 618–627, 2015.

13. K. Hamid, M. W. Iqbal, A. U. R. Virk et al., "K-Banhatti Sombor invariants of certain computer networks," *Computers Materials & Continua*, vol. 73, no. 1, pp. 15–31, 2022.

14. M. B. M. Noor and W. H. Hassan, "Current research on Internet of Things (IoT) security: a survey," *Computer Networks*, vol. 148, pp. 283–294, 2019.

15. P. Zhang, S. G. Nagarajan, and I. Nevat, "Secure Location of Things (SLOT): mitigating localization spoofing attacks in the Internet of Things," *IEEE Internet of Things Journal*, vol. 4, no. 6, pp. 2199–2206, 2017.

16. N. Aleisa and K. Renaud, "Privacy of the Internet of Things: a systematic literature review," in *Proceedings of the 50th Hawaii International Conference on System Sciences*, pp. 1–10, Hilton Waikoloa Village, Hawaii, 2017.

17. A. K. Mishra, A. K. Tripathy, D. Puthal, and L. T. Yang, "Analytical model for sybil attack phases in Internet of Things," *IEEE Internet of Things Journal*, vol. 6, no. 1, pp. 379–387, 2019.

18. A. Raoof, A. Matrawy, and C.-H. Lung, "Routing attacks and mitigation methods for RPL-based Internet of Things," *IEEE Communications Surveys & Tutorials*, vol. 21, no. 2, pp. 1582–1606, 2019.

19. A. S. Genadiarto, A. Noertjahyana, and V. Kabzar, "Introduction of Internet of Thing technology based on prototype," *Jurnal Informatika*, vol. 14, no. 1, pp. 47–52, 2018.

20. B. Xu, W. Wang, Q. Hao et al., "A security design for the detecting of buffer overflow attacks in IoT device," *IEEE Access*, vol. 6, pp. 72862–72869, 2018.

21. C. Li, Z. Qin, E. Novak, and Q. Li, "Securing SDN infrastructure of IoT–fog networks from MitM attacks," *IEEE Internet of Things Journal*, vol. 4, no. 5, pp. 1156–1164, 2017.

22. A. Tewari and B. B. Gupta, "Security, privacy and trust of different layers in Internet-of-Things (IoTs) framework," *Future Generation Computer Systems*, vol. 108, pp. 909–920, 2020.

23. D. Yin, L. Zhang, and K. Yang, "A DDoS attack detection and mitigation with software-defined Internet of Things framework," *IEEE Access*, vol. 6, pp. 24694–24705, 2018.

24. H. Yan, Y. Wang, C. Jia, J. Li, Y. Xiang, and W. Pedrycz, "IoT-FBAC: function-based access control scheme using identity-based encryption in IoT," *Future Generation Computer Systems*, vol. 95, pp. 344–353, 2019.

25. J. Moon, I. Y. Jung, and J. H. Park, "IoT application protection against power analysis attack," *Computers and Electrical Engineering*, vol. 67, pp. 566–578, 2018.

26. O. O. Bamasag and K. Youcef-Toumi, "Towards continuous authentication in Internet of Things based on secret sharing scheme," in *Proceedings of the WESS'15: Workshop on Embedded Systems Security*, pp. 1–8, Amsterdam, Netherlands, 2015.

23 Vulnerabilities in Internet of Things and Their Mitigation with SDN and Other Techniques

S. Shalini

Dayananda Sagar Academy of Technology and Management, Visvesvaraya Technological University, Bengaluru, India

S. Sheela

Global Academy of Technology, Visvesvaraya Technological University, Bengaluru, India

Shabeen Taj

Government Engineering College, Visvesvaraya Technological University, Bengaluru, India

Madhumala R. Bagalatti

Dayananda Sagar Academy of Technology and Management, Visvesvaraya Technological University, Bengaluru, India

23.1 INTRODUCTION

A lot of modern technologies are being used recently. Modern technologies such as internet-enabled cameras, tracking/recording of the current location, microphones, or websites are leading to security breaches. Internet of Things (IoT) threats have made online banking very dangerous, and medical data is no longer secured well, files are vulnerable to attacks, chats won't remain confidential, etc. The threats have become so strong that they are very difficult to mitigate. The communication through the primary network device is no longer safe, and a back network is needed as a backup. Cheaper IoT devices available on the internet are not safe and high-end products with better security [15] techniques are therefore required.

Devices have to be updated frequently to ensure security. Strong authentication passwords have to be used. Firewalls and antivirus are not effective many times.

DOI: 10.1201/9781003477327-23

Some good practices such as IoT security analysis can be implemented to predict and detect threats. They also help in identifying anomalies. Artificial intelligence (AI) and machine learning (ML) concepts also help in avoiding threats. They predict threats and find a way to attack the threats.

23.2 SOFTWARE-DEFINED NETWORKS

A large number of problems can be resolved by rendering intelligence to the network. This intelligence can be done using Software-Defined Networks (SDNs). They enable the network in a way that data transfer can be done efficiently without packet loss, and protect data integrity [1, 16]. By separating the control plane and the data plane forwarding processes in individual networking devices, central administration of SDNs is achieved. SDNs streamline operations. They auto-discover and auto-configure networks and switches. SDNs help in faster troubleshooting. They help monitor the health of networks using network analytics and machine learning. SDNs also help in policy-based segmentation and help in automating WAN and cloud network domains.

The IoT faces a lot of problems in different modes of hybrid communication technologies. There are also a few problems and challenges [1, 12] in security. SDN is an approach to control communication effectively.

23.3 DIFFERENCES BETWEEN SOFTWARE-DEFINED NETWORKING AND TRADITIONAL NETWORKING

The differences between SDN and normal networks are a lot. Some of them are given in Table 23.1.

Nowadays, IoT security methods incorporating AI and ML are used. Since networks are expanding, IoT devices produce a lot of data. Massive sets of data can be analyzed with AI and ML, allowing the machines to teach themselves, retain what they learned, and hence improve the capabilities of IoT. These mitigation techniques using the above-mentioned technologies, such as AI and ML, can be easily implemented using SDNs in networks. The implementation of SDNs makes networks more adaptable, reliable, and flexible.

TABLE 23.1
Difference between the Two Types of Networks

Software-Defined Networking	Traditional Networking
A virtual networking strategy known as software-defined network (SDN) is used.	A conventional network uses the outdated traditional networking method.
SDN is the centralized control.	Distributed control describes an older network.
The network is programmable.	The network is non-programmable.
The open interface is the SDN.	A traditional network is a closed interface.
The control plane and data plane are detached in SDNs.	The control plane and data plane are positioned on the same plane in a conventional network.

23.4 SDN ARCHITECTURE

The SDN controller [7] provides efficient data forwarding and resource allocation. As shown in Figure 23.1, SDN has three planes: the control plane, the application plane, and the data plane [2]. Each switch has a control plane as well as a data plane when the traditional network is taken into account. Different switches' control planes exchange topological information among themselves. Therefore, a forwarding table may be created. The control plane is moved from the switch to the SDN controller, a centralized component, with the help of an SDN. Therefore, a network administrator may arrange traffic from a single platform without having to communicate to each switch individually. The switch's data plane determines a packet's forwarding action based on entries made in flow tables which the controller has previously assigned. Match fields (such as the header of the packet and input port number) and instructions make up a flow table [1]. The entries in the flow table's match fields are first compared with the packet. Then, the corresponding flow's orders are carried out. The instructions may include port-by-port packet forwarding. The switch either forwards or drops the packet based on the flow entry.

SDNs work using algorithms to make the configuration of the device and management. Without taking the size of the network into consideration, an SDN may scale operations to meet changing network needs. In order to configure the network based on the patterns of traffic flow or other factors, IT administrators may enhance the SDN's capabilities. Various businesses may improve their operations globally because of network programmability. When each component is set separately, it would be difficult to create a consistent network-wide state. However, this is possible using SDNs.

Network operations use SDNs that are logically centralized. Networks may now be operated using APIs to maintain a constant level of overall functionality, control, and efficiency since they are no longer dependent on restrictions. In the previous architectures, manual configuration was common. A hardware-based setup makes it

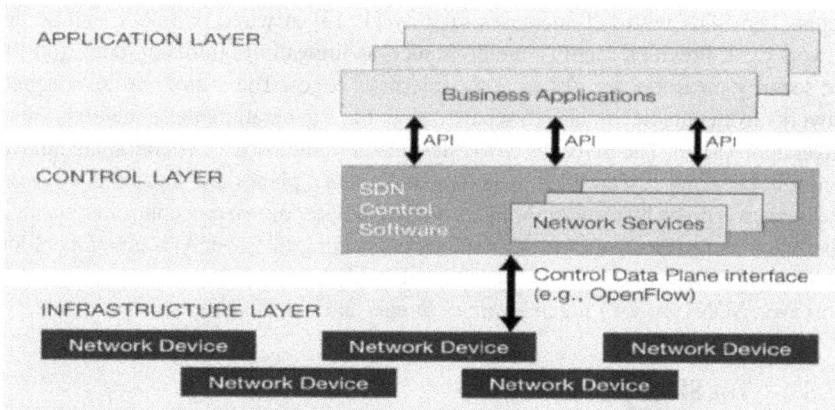

FIGURE 23.1 SDN and IoT. (*Source*: IIT Kharagpur (2020).)

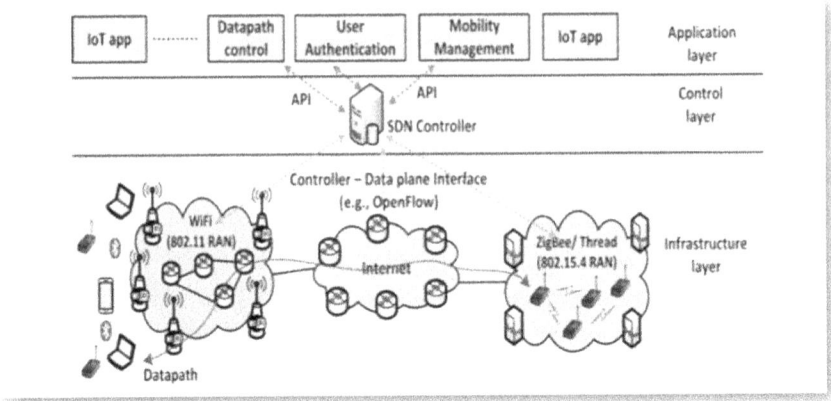

FIGURE 23.2   SDN architecture. (*Source*: https://www.opennetworking.org/images/stories/downloads/white-papers/wp-sdn-newnorm.pdf.)

challenging to find and fix problems. The functionality of three levels in an SDN architecture [2] is shown in Figure 23.2.

- **Application layer**: Numerous network applications, including load balancing, firewalls, and intrusion detection, are included.
- **Control layer**: The SDN controller lies in this layer. It enables hardware abstraction for the applications built on it.
- **Infrastructure layer**: The actual data flow is carried out by physical switches that contain the data plane.

23.5   IoT AND SDN

The SDN provides many advantages and gives the potential to overcome many constraints. An SDN with IoT increases security [1, 13] in wired, wireless, and ad-hoc networks [5]. Previous security methods such as firewalling, intrusion detection [9], and security measures are placed at the internet's edges. These are used to safeguard networks from outside threats. To secure the next IoT generation, these methods are no longer appropriate. The SDN is a newly developed framework for encouraging innovation in networking research [2]. The data and control planes are separated. Network intelligence is given by programming. A modern device known as a controller manages this switch using the OpenFlow protocol by connecting to it over a secured OpenFlow channel. The flow entries may be updated, added, or removed by the controller. Moreover, SDNs are very much aware of threats and react fast to security threats.

23.5.1   THE SDN FRAMEWORK

The SDN framework provides an intelligent framework. The SDN performs the following functions:

- Works very efficiently with different server platforms.
- Works with an open operating system along with various hardware.
- Utilizes standardized protocols to interact with different operating systems or control platforms.
- Network-reachable device security is provided by the SDN framework. After connecting to the OpenFlow switches, the SDN has a globally available network based on the information obtained through the OpenFlow protocol [10]. Additionally, the SDN can:
  - Perform network discovery, utilizing the Link Layer Discovery Protocol (LLDP).
  - Collect information in the connected network.

Prior network protocols are built for high degrees of scalability, massive traffic volumes, and mobility. Figure 23.3 depicts a heterogeneous network and different kinds of nodes in each area. Heterogeneous connections with a variety of SDN domains are included in an SDN architecture for the IoT. Connecting to and communicating with other SDN border controllers is the responsibility of these controllers. This architecture's expansion is predicated on the idea of equal interaction across controls while utilizing the current security measures. There are distinct maintenance processes and security guidelines for each SDN domain.

The network can be effectively protected against outside threats while still being managed by the controller. By introducing security policies into OpenFlow switches, security can be maintained. Regarding the framework of the next-generation internet, even greater security standards are required, such as the ability to identify objects, network devices, and users connected to users utilizing both types of communications [3]. Additionally, the actions of both individuals and objects establish trust boundaries that must be monitored. Software verification is combined with accounting procedures. To satisfy the requirements of next-generation internet architecture, present security mechanisms do not, however, provide these security protocols. However, the methods for network access control and security already in use served as inspiration for how the SDN-based architecture is designed [9].

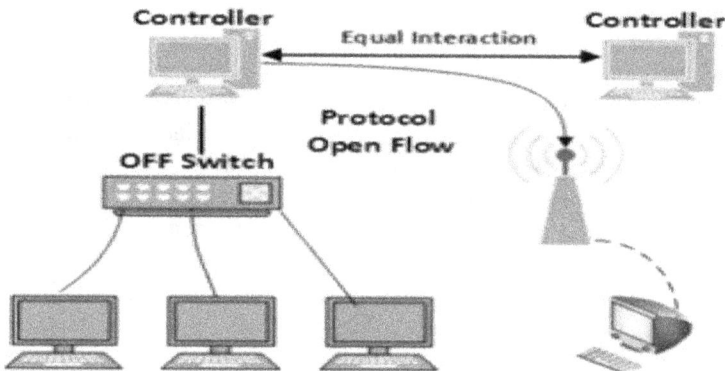

FIGURE 23.3 SDN architecture in IoT. (*Source*: March 2018, Intelligent Systems Design and Applications.)

To explain the framework, a solution in which a controller has security is presented. Second, the resources on each control platform may be used with this approach in many controllers. A secure IoT model may be built thanks to the distributed architecture, which links all SDN domains together through border controllers. Because of the lack of network infrastructure, network access control and worldwide traffic monitoring cannot be provided by the typical ad-hoc architecture. These security constraints are removed by the architecture suggested in this study, which also makes it possible to apply security policies and build networks dynamically. Our objective is to achieve the highest synchronization of SDN controllers inside a secure perimeter while providing granular network access control and network infrastructure point monitoring [2]. The SDN controllers initially validate the network devices before providing network access and resources. The controller restricts the switch ports that are connected to users when an OpenFlow [10] secure connection is made between the switch and the controller. The controller then only permits traffic for user authentication. Depending on the user's level of permission after verification, the needed flow entries will be pushed by the controller to the SDN table.

23.6 TYPES OF VULNERABILITIES

There are different types of vulnerabilities [4] faced by the IoT and its devices. Some of the vulnerabilities are mentioned below.

23.6.1 MAN-IN-THE-MIDDLE ATTACKS

Man-in-the-Middle (MitM) attacks happen when a hacker affects the communication between systems by inserting a rogue node in between two genuine ones or by attacking the IoT network's communication protocols. In order to compromise an IoT system, hackers may change the traffic flow, modify the network topology, establish false identities, and produce harmful and false information [11].

23.6.2 SOFTWARE VULNERABILITIES

- **Weak or hardcoded passwords**: Many passwords are very easy to tell and socially available [7].
- **Not updating the process or mechanism**: Since many IoT applications and devices are invisible on the network, admins exclude them from updates. Also, IoT devices may not even have an update mechanism incorporated into them due to age or purpose, meaning admins can't update the firmware regularly.
- **Not secured network services and ecosystem interfaces**: Each IoT [11] app connection has the potential to be compromised, either through an inherent vulnerability within themselves or because it is vulnerable to attack. That includes any gateway, router, modem, external web app, API, or cloud service connected to an IoT application.
- **Outdated or unsecured IoT app components**: Many IoT apps depend on third-party frameworks and libraries when they are being developed. If they are out-of-date, have known vulnerabilities, or aren't tested before being put in a network, they might pose security risks [11].

- **Unsecured data storage and data transfer**: IoT apps and other linked objects and systems may store and send a large variety of data kinds. They must be properly secured.

23.6.3 Data Leaks from IoT Systems

Data leakages occur because of inside problems. They usually do not occur because of cyber-attacks [4]. Let us go over a few of the most typical reasons for data leaks.

- **Worst infrastructure**: Bad infrastructure can expose information. Having outdated software can potentially expose data.
- **Social scams**: Data leaks are caused by cyber-attacks, although criminals often use a variety of methods to do it. The data breach will subsequently be used by the criminal to launch more cyber-attacks. To provide one example, spam attacks may be effective in gaining someone's login details, which might result in a bigger data breach.
- **Bad password policies**: People often use the monotonous password across many accounts because it's easier to remember. Even something as simple as jotting down login information in a notebook might lead to a data breach.
- **Missing devices**: A potential data breach occurs when the employee of a certain company misjudges a device containing sensitive data. A criminal might access the device's content, which may lead to identity fraud or data insecurity.
- **Very old data**: People may lose track of data as firms grow. Increase and infrastructure changes may accidentally expose earlier data.

23.6.4 Malware Attacks

Malware is malicious software intended to enter your computer or other device and cause harm [6]. IoT technology is vulnerable to malware attacks, according to experts [11]. Because they lack security, these devices are constantly online. Because of this, hackers may easily access them. These malware assaults are constantly being used by hackers.

23.6.5 Cyber-Attacks on IoT

Bad communication is the main risk associated with IoT. Device-to-device data communications are vulnerable to cyber parties capturing them. Threat actors could be able to obtain personal data in this way.

23.7 MITIGATION TECHNIQUES FOR IoT ATTACKS

There are several methods to avoid attacks such as the following:

- MitM attacks can be reduced using proper encryption mechanisms, and TLS/SSL setup needs to be done in a proper way.
- The hackers need to be hired.
- Employers need to be made aware of leakage in data.

- The resources on the cloud need to be used to update the device and focus on network quality so that cyber-attacks can be minimized.
- Server maintenance needs to be done well.
- The usage of default passwords can be avoided.
- Mobile device management concepts can be used to provide security remotely.
- Network scanning and identifying all the IoT components connected.
- The network needs to be designed in such a way that security is incorporated.
- Authentication based on identity needs to be used.

SDN and Raspberry Pi: All the above techniques can be easily implemented by installing SDN in Raspberry Pi [14] Since Raspberry Pi is a portable pocket-sized computer, it can be made a server as well. It can also act as a mobile server. Security management can be done remotely. All the above-mentioned types of hazards can be overcome. Since SDNs give intelligence to the networks, they can avoid 90% of the above-mentioned problems. SDNs allow keeping track of packets entering the network, leaving the network, and packet loss. Since Raspberry Pi is a portable processor, it can be carried anywhere and can be used as a switch as shown in Figure 23.4. The security threats can also be gauged [14].

SDN Controller: SDN is a reliable technology. There are a number of ways that an attacker utilizes to attack the SDN controller. A rule-based approach to mitigate these attacks on the SDN domain [2] is used. Open Source SDN controller is incorporated to develop the rule-based security application. This can be used for real-time attack scenarios.

As shown in Figure 23.5, the Policy Manager is installed in the SDN which in turn is installed in Raspberry Pi [14]. This helps mitigate the attacks [8]. Installing a controller in the SDN helps mitigate all types of vulnerabilities to the maximum extent. For every type of vulnerability, a policy can be installed in the SDN controller. Since the controller is programmable, the controller is set up with different network policies.

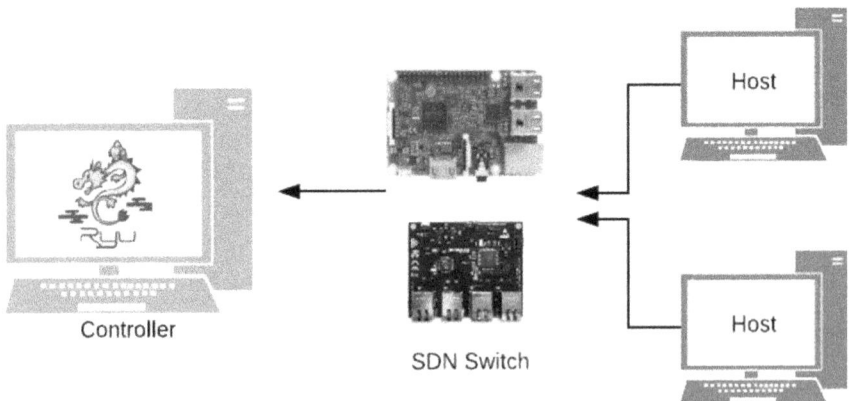

Controller

SDN Switch

Host

Host

FIGURE 23.4 SDN in Raspberry Pi connected to networks. (*Source*: Sinan Karakaya (2018).)

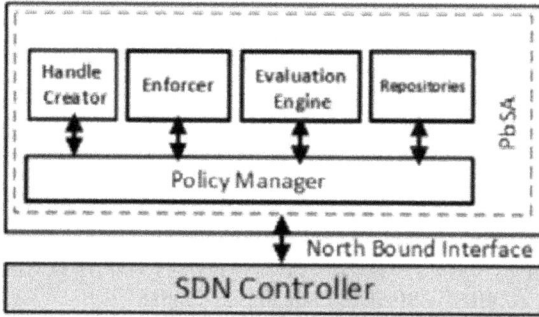

FIGURE 23.5   Policy manager in SDN controller to mitigate attacks in IoT devices. (*Source*: Xuekai Du (2016).)

23.8   SUMMARY AND FUTURE WORK

IoT advancements are being hindered by security issues for deployments. Making sure trusted parties can connect with devices and services inside safe ecosystems is a fundamental IoT security component.

In this chapter, the possible threats to SDN controllers and other communication devices in SDNs are presented. Then a rule-related security application is developed. This application is good enough to defend and control the SDN domain behavior. An SDN controller in Raspberry Pi works against threats. Apart from anomaly detection, it also enables proper traffic-forwarding techniques to avoid congestion.

The installed controller detects cyber-attacks, mitigates their bad impacts on the network performance, and ensures correct data delivery of traffic scenarios. The multiple layers of responsibilities, authorizations, and information required by these complex contexts cannot be handled by the protocols alone. Organizations may therefore offer their devices to meet these objectives more securely and cheaply by implementing a strong, managed SDN service. The above-mentioned concepts are achievable with the assistance of an SDN. Introducing AI and ML into SDN controllers will bring more productivity and the attacks can be mitigated efficiently.

REFERENCES

[1] Aglan, M.A., Sobh, M.A., Bahaa-Eldin, A.M. "Reliability and Scalability in SDN Networks," in *2018 13th International Conference on Computer Engineering and Systems (ICCES)*, Cairo, Egypt, 2018, pp. 549–554, doi: 10.1109/ICCES.2018.8639201

[2] Flauzac, O., González, C., Hachani, A., Nolot, F. "SDN Based Architecture for IoT and Improvement of the Security," in *IEEE 29th International Conference on Advanced Information Networking and Applications Workshops*, Gwangju, Korea (South), 2015, pp. 688–693, doi: 10.1109/WAINA.2015.110

[3] Bukar, U.A., Othman, M. "Architectural Design, Improvement, and Challenges of Distributed Software-Defined Wireless Sensor Networks," *Wireless Pers Commun* **122**, 2395–2439, 2022, doi: 10.1007/s11277-021-09000-2

[4] Singh, S., Jayakumar, S.K.V. "A Study on Various Attacks and Detection Methodologies in Software Defined Networks," *Wireless Pers Commun* **114**, 675–697, 2020, doi: 10.1007/s11277-020-07387-y

[5] Remondo, D. "Wireless Ad Hoc Networks: An Overview," in Kouvatsos, D.D. (ed) *Network Performance Engineering*. Lecture Notes in Computer Science, vol. 5233. Springer, Berlin, Heidelberg. doi: 10.1007/978-3-642-02742-0_31

[6] Hajizadeh, M., Phan, T.V., Bauschert, T. "Probability Analysis of Successful Cyber Attacks in SDN-Based Networks," in *IEEE Conference on Network Function Virtualization and Software Defined Networks (NFV-SDN)*, Verona, Italy, 2018, pp. 1–6, doi: 10.1109/NFV-SDN.2018.8725664

[7] Sinha, M., Bera, P., Satpathy, M. "DDoS Vulnerabilities Analysis in SDN Controllers: Understanding the Attacking Strategies," in *International Conference on Wireless Communications Signal Processing and Networking (WiSPNET)*, Chennai, India, 2023, pp. 1–5, doi: 10.1109/WiSPNET57748.2023.10134518

[8] Zolotukhin, M., Hämäläinen, T., Immonen, R., "Curious SDN for Network Attack Mitigation," in *IEEE 21st International Conference on Software Quality, Reliability and Security Companion (QRS-C)*, Hainan, China, 2021, pp. 630–635, doi: 10.1109/QRS-C55045.2021.00096

[9] Cheleng, P.J., Chetia, P.P., Das, R., Singha, B.C., Majumder, S. "A Survey on the Latest Intrusions and Their Detection Systems in IoT-Based Networks," in Mishra, M., Kesswani, N., Brigui, I. (eds) *Applications of Computational Intelligence in Management & Mathematics*. ICCM 2022. Springer Proceedings in Mathematics & Statistics, vol. 417. Springer, Cham, doi: 10.1007/978-3-031-25194-8_6

[10] Rotsos, C., Sarrar, N., Uhlig, S., Sherwood, R., Moore, A.W. "OFLOPS: An Open Framework for OpenFlow Switch Evaluation," *Lecture Notes in Computer Science*, vol. 7192, 2012, Springer, Berlin, Heidelberg, doi: 10.1007/978-3-642-28537-0_9

[11] Patel, A.B., Sharma, P.R., Randhawa, P. "Internet of Things (IoT) System Security Vulnerabilities and Its Mitigation," *Security and Privacy in Cyberspace*. BlockchainTechnologies. Springer, 2019, Singapore, doi: 10.1007/978-981-19-196028

[12] Dey, S., Bhale, P., Nandi, S. "ReFIT: Reliability Challenges and Failure Rate Mitigation Techniques for IoT Systems," in *Innovations for Community Services. I4CS 2020. Communications in Computer and Information Science*, vol. 1139. Springer, 2020, Cham, doi: 10.1007/978-3-030-37484-6_7

[13] Shu, X., Yao, D. "Data Leak Detection as a Service," in *Security and Privacy in Communication Networks. SecureComm*. Lecture Notes of the Institute for Computer Sciences, Social Informatics and Telecommunications Engineering, vol. 106. Springer, 2012 Berlin, Heidelberg, doi: 10.1007/978-3-642-36883-7_14

[14] Babayigit, B., Karakaya, S. and Ulu, B., "An Implementation of Software-Defined Network with Raspberry Pi," in *26th Signal Processing and Communications Applications Conference (SIU), 2018*, Izmir, Turkey, 2018, pp. 1–4, doi: 10.1109/SIU.2018.8404546

[15] Shu, X., Yao, D. "Data Leak Detection as a Service," in Keromytis, A.D., Di Pietro, R. (eds) *Security and Privacy in Communication Networks*. SecureComm 2012. Lecture Notes of the Institute for Computer Sciences, Social Informatics and Telecommunications Engineering, vol. 106. Springer, Berlin, Heidelberg, doi: 10.1007/978-3-642-36883-7_14

[16] Wani, A., Revathi, S. "Analyzing Threats of IoT Networks Using SDN Based Intrusion Detection System (SDIoT-IDS)," in *Smart and Innovative Trends in Next Generation Computing Technologies*. NGCT 2017. Communications in Computer and Information Science, vol. 828, Springer, Singapore, doi: 10.1007/978-981-10-8660-1_4

24 Intrusion and Malware Detection and Prevention Techniques in IoT

Ramanpreet Kaur and Parveen Singla
CEC-CGC, Landran, Mohali, India

Jaskirat Kaur
Punjab Engineering College (Deemed to be University), Chandigarh, India

Balram Kumar
CEC-CGC, Landran, Mohali, India

24.1 INTRODUCTION

The contribution of numerous researchers has conceptualized the advancements in Artificial Intelligence (AI) which has brought scientific revolution as a result of computing technologies. It has become essential to practically every sphere of human effort, including technology, engineering, business, and education.

The use of smart devices that are connected to one another has expanded significantly in recent years and is now referred to as the Internet of Things (IoT). To make daily tasks easier, IoT devices may be accessible from any place, including our home, business unit, and transportation. These intelligent gadgets are employed in a variety of fields, including healthcare, home automation, transportation networks, smart grids, and smart cities [1]. These smart devices offer distinctive qualities like smaller size, weight, lighter protocols, and lower consumption of power, that make them more adaptive. The Web of Things has become more adaptable as a result of the increased release of smart devices into the market and the erosion of consumer confidence in detection technology [2].

With their benefits and drawbacks, internet-connected devices pose a greater risk for cyber threats and attacks. It is more vulnerable to digital risks since there are so many of these devices, and they are diverse and heterogeneous. As of now, there are no established security measures to ensure these gadgets' digital safety.

DOI: 10.1201/9781003477327-24

The world has gone through an innovative phase in the last three decades, one that signaled a quick change in society's focus from an industrialized, machine-based economy to one centered on technology-enabled devices. The advancements in science and technology provide many barriers, especially when it comes to digital technology, which dates back to the 1980s, and where there are now more than 40,000 viruses in existence [3].

In the present era, antivirus software and other security measures play a significant role in dealing with malicious attacks. The AI-based techniques in IoT must aim at protecting data of different domains such as healthcare, finance, and aviation. Additionally, various techniques need to aim to protect both crucial data from malware attacks along with security liabilities and data leakage through unauthorized access [4].

From the viewpoint of artificial intelligence, the present chapter aims at a collection of malware detection and prevention approaches in IoT. An analysis of AI-based approaches for malware detection and prevention in IoT and their challenges in implementation will be studied.

24.2 MALWARE: ORIGIN AND FLOW

Malware is a term used to refer to harmful software or applications that are not just found on computers but also on the internet and in related sectors. A computer virus, worm, ransomware, or rootkit is an example of malicious software, which also includes scripts, active content, and intrusive software that aim to damage the targeted computer systems, programmers, mobile, and web applications. Researchers claim that malware attacks can target IoT systems easily as these devices are constantly online and lack security. Some of the major malware that have been studied so far are discussed here [5].

Malware is capable of infecting networks and devices and is made to harm users, networks, or targeted equipment. Depending on the nature of the target and type of malware, this damage expresses itself to the destination in many ways. Malware occasionally poses a threat; however, it can also have quite minor and non-harmful consequences.

IoT malware searches a variety of open ports for IoT services including Telnet, FTP, and SSH. IoT malware uses force attacks to take control of IoT equipment. Malicious apps can also be downloaded through drive-by downloads, which secretly and automatically install harmful software on a user's device.

Another popular method of disseminating malware is through phishing assaults, in which perilous links or attachments are concealed in emails that seem to be regular messages and distributed to unwary recipients

24.3 TYPES OF MALWARE

This section studies different forms of malware with their distinctive characteristics [6]. Some of them are presented in Figure 24.1. A virus is the most common type of malware that may be set to run automatically and spread by infecting other programs or files.

A worm may replicate itself without a host programmer and frequently spreads without assistance from the malware's developer. In order to gain access to a system,

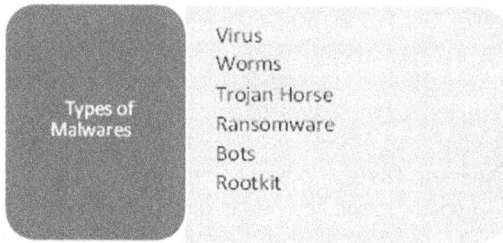

FIGURE 24.1   Types of malware.

a trojan horse is constructed to appear to be a reliable piece of software. Trojans have the ability to perform their malicious actions if activated after installation. Without the user's knowledge, spyware is able to collect information and personalized data about a person or device. A user's machine is infected with ransomware, which encodes their data. The person has to pay ransom to the cyber criminals for decoding the data.

A rootkit allows the victim to become the system administrator. After the software is installed, the programmer gives threat actors privileged access or root access to the machine. Nearly all computer operations are recorded by keyloggers, also known as system monitors. Emails, websites visited, programs run, and keystrokes all fall under this category. Practically, every computer activity is recorded by keyloggers, also known as system monitors. Emails, websites visited, programs run, and keystrokes all fall under this category [7].

24.4   ARTIFICIAL INTELLIGENCE

Artificial intelligence is the capacity of machines, especially computer systems, to imitate cognitive processes in humans. Expert systems, machine learning, speech recognition, natural language processing, and vision are a few examples of specific AI applications as mentioned in Figure 24.2. The term "weak AI" or "narrow AI"

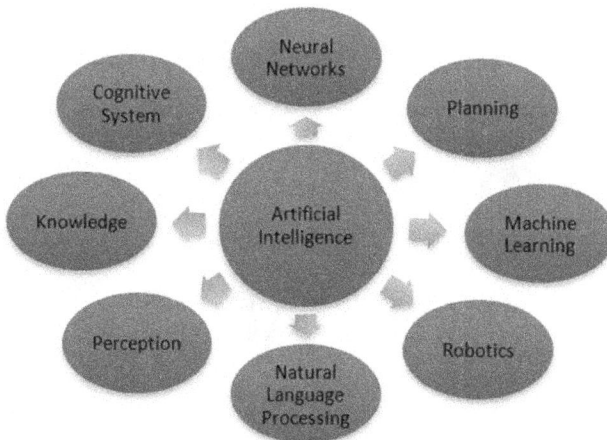

FIGURE 24.2   Applications of AI.

describes an AI system that has been created and trained to do a single task. Weak artificial intelligence is used by commercial robotics and virtual personal assistants like Siri from Apple. Artificial General Intelligence (AGI), sometimes known as strong artificial intelligence, is a term used to describe computer programs that can mimic the cognitive functions of the human brain (AGI). Strong AI systems can use Fuzzy logic to transfer knowledge from one domain to another and come up with a solution on their own when presented with an unexpected job [8, 9]. With increased public awareness of AI, vendors have rushed to highlight how their products and services use it.

Learning: The objective of this branch of artificial intelligence programming is to collect data and create the rules that will allow the data to be turned into knowledge.

Syllogism: The basic objective of this branch of AI programming is to select the optimum method to generate a given outcome.

Self-correction: A facet of AI programming that is constantly improving algorithms to ensure they deliver the most accurate results.

24.5 MALWARE DETECTION TECHNIQUES

To commit specific unlawful activities, threat actors use malicious software, or malware. There are many different types of malware as shown in Figure 24.3, and each one has a distinct function. For instance, trojans and spyware both seek to permanently take over computer systems, whereas ransomware encrypts data and demands a fee from its owners [9]. The process of identifying, excluding, alerting, and responding to malware threats involves the application of procedures and technologies. The use of simple malware detection methods like "check summing", "application allow listing", and "signature-based detection" can aid in locating and minimizing known

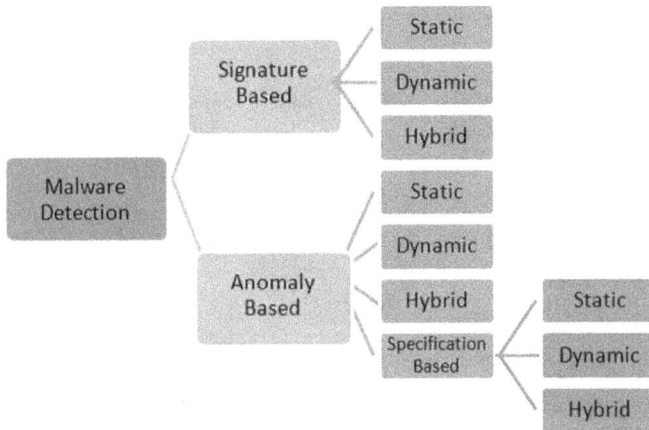

FIGURE 24.3 Malware detection types.

risks. Advanced malware detection programmers actively search for and find new and unknown malware threats using AI and machine learning [10, 11].

Signature-based detection: An identifiable digital trail, or signature, is left by the software creators of a secured system that is used in signature-based detection. When antivirus software scans a piece of software, it detects the signature and compares it to signatures of known malware [12].

An antivirus-supplier-led security research team frequently updates the extensive database of recognized malware signatures utilized by antivirus software. This database is regularly updated and synced with safe devices, ensuring the most recent version is available.

An antivirus program halts the activity and either deletes or places any software that matches a known signature. It is crucial to use this quick and efficient method to identify malware as the first line of defense. The calculation of CRC checksums is a step in this method's signature analysis process. Check summation is a method for ensuring the integrity of files. Check summing addresses the primary drawback of signature-based detection, which is the accumulation of a large database producing false positives.

In order to get around signature-based identification measures, hackers typically use polymorphic malicious advertising. Hackers commonly encrypt non-constant keys that act as substitutes for various command sets for decryption in the virus code.

When the security team discovers a malicious signature, the virus is no longer detectable due to the deletion of the code fragment. Other malicious code detection methods such as the variable code lacks a detectable signature.

Statistical analysis: To ascertain whether a file is infected, the frequency of processor commands is assessed.

Cryptanalysis: recognized plaintext—Cryptanalysis employs a mathematical methodology to decode viruses to decode some of the virus's overall body to reassemble the algorithm and keys used by the decryption application.

Reduced masks: Using components from the encrypted virus body, the malware detection team may be able to obtain static code without the need for an encryption key. The mask or signature of the infection can be seen in the static code that is produced.

Heuristics: To find unusual activity, a malware detection team searches and examines behavioral data. The team is tasked with looking for malicious code linked to dubious actions, such as serving a code to tens of thousands of users in a matter of minutes.

Application allow listing: Utilizing attack signatures is the opposite of using "application allow" lists, also referred to as whitelists. Instead of defining which applications it should block, everything outside of the permitted programs is forbidden by the antivirus program, which keeps a list of them.

Because it is based on an analysis of the behavior of possible processes, AI/ML malware detection is also referred to as "behavioral" detection. When a file or process displays abnormal behavior that exceeds the threshold set by these algorithms for hazardous activity, it is classified as malicious.

Though behavioral analysis is useful, it occasionally fails to detect malicious software or misclassified harmless programs as potentially harmful. Training procedures for AI and ML are equally vulnerable to assault. Attackers have occasionally used specially created artifacts to train behavioral analysis tools to mistakenly classify malicious software as safe.

Modern malware detection technologies: Several companies no longer use traditional antivirus as their main malware detection technique. On the other hand, sophisticated security organizations routinely implement endpoint protection platforms (EPPs) and endpoint detection and response tools to combat malware.

Endpoint protection platform: Employee workstations, servers, and cloud resources are just a few types of endpoints where EPPs are employed. In their capacity as the first line of defense, they recognize threats and neutralize them before they have a chance to damage crucial assets. EPPs use a range of techniques to identify and stop malware [12, 13].

Static analysis: In order to recognize well-known malware strains and decide whether to allow or reject apps that administrators have marked, EPPs use standard static analysis techniques.

Inspection in a sandbox: EPPs have the ability to run questionable applications separately from the main operating system. With the help of this, it is feasible to "detonate" a file, watch how it behaves, and determine whether it truly poses a risk. Endpoint reaction to detection (EDR).

Security teams are given the ability to notice and respond to assaults on endpoint devices via EDR solutions, which are a complement to EPP solutions. EDR enables when EPP fails to get rid of a threat. Security analysts can spot indicators of an attack and investigate them to verify a security problem in EDR. EDR enables proactive endpoint inspection and suitable data search for breach indicators [11].

Using firewalls, email security, network segmentation, and intrusion prevention systems (IPS), to mention a few, is one technique to organize a response to a malware threat. Many EDR implementations integrate EPP features.

24.6   HOW AI HELPS?

However, when combined with the modeling of both undesirable and desirable behaviors, even the most complex virus can be defeated using AI; this is a strong and effective weapon. While AI is not yet capable of automatically identifying and resolving every malware or cyber threat incident, the following are some ways AI can help your security team combat malware [12, 13].

Security solutions that rely on models of bad behavior, such as signatures, are easily overcome by advanced malware due to its intelligence. This shift from trapping to hunting, or from emulating negative behavior to modeling positive behavior, is required. Malware authors today are quite skilled at producing limited- or single-use malware that is never discovered by businesses that develop signatures. Without signatures, traditional bad-behavior model-based detection techniques are entirely useless.

Modern hunting systems that use good-behavior models rather than signatures can find evasive malware much more quickly. Products that use good behavior modeling can find many varieties of malware that a signature-based tool will miss. The following types of anomalies are discovered by malware-hunting technologies because they actively monitor behavior.

- Overuse of some resources (such as CPU or memory).
- Connections to hosts with which the infected target has never previously connected.
- Unusually high data transfer to a distant host and weird login times.
- The use of software that hasn't been used before, like compilers or network analysis tools.

AI will play a bigger and more significant role as malware detection techniques are developed. By adding the automatic application of bad-behavior models, these errors can be significantly reduced; nevertheless, false positives are regularly produced by the good-behavior models most often connected to AI.

Even though it's hard to envisage a world without people working to tackle evasive malware, recent advances in AI will significantly reduce our dependence on human labor.

24.7 MALWARE DETECTION USING AI

Modern systems are seriously threatened by the diversity and ongoing growth of malware, and new countermeasures have been created as a result of the failure of current security measures to keep up with hackers' creativity and competence [10, 13].

The field of AI is also developing swiftly, and new developments could lead to excellent results in a variety of application areas. These developments will be essential for creating potent anti-malware programs that can outperform the limitations of the current preventative technologies. This section discusses the outcomes of the AI-based malware detection method as well as any potential drawbacks.

Shanxi Li and colleagues created a malware classifier using a graph convolutional network that can adapt to the many characteristics of malware. From the malicious code, the approach initially generates an API call sequence.

The principal component analysis and Markov chain techniques are used to design a classifier based on a graph convolutional network. It is constructed by extracting the feature map of a directed cycle graph. The technique also evaluates and compares its effectiveness. The evaluation results indicate that the approach with the best accuracy has an FPR and accuracy that are higher than those of other methods currently in use, which is 98.32%.

Moreover, Long Wen and Haiyang Yu suggested a rapid machine-learning-based method to identify unknown malware on Android devices. The Android malware detection architecture makes use of machine learning. The increased detection rate and lower mistake detection in the demonstration led to better performance.

24.8 IoT MALWARE

The IoT is a system of interconnected, individually numbered things. Without human involvement, the gadgets may communicate with one another through a network. The components of the IoT system are things with distinctive identities that easily link to the information network through intelligent interfaces. Networked and portable IoT devices are frequently used in IoT systems that have several applications in business, industry, smart cities, healthcare, and the environment. IoMT (Internet of Medical Things) devices are IoT gadgets utilized in the medical field. These are necessary for health monitoring and intervention, hospital-connected medical equipment, and wearable monitoring medical devices. Because of this, IoMT device and system security is crucial and demands stringent virus detection procedures [14]. Sensors that measure temperature and humidity are examples of IoT devices used in environmental and agricultural applications. These gadgets are frequently placed far away and run on batteries. Hence, an energy-efficient and computationally efficient malware detection approach is required to maximize battery life.

Security devices and other equipment are supported by IoT systems in smart cities. Important information is captured by cameras, so strong security measures are required to prevent unwanted access. To maintain an efficient Industrial IoT operation and to ensure the safety of the workforce in these circumstances, the IoT system must be protected from malware. These IoT objects might be both real-world objects and digital ones, but they all work together to form an operational IoT system.

24.8.1 IoT CHARACTERISTICS AND CHALLENGES

A device's capacity to link to other devices and/or the cloud is referred to as "interconnectivity". Connectivity is required in order to allow remote device operation and to access the data that the IoT device's sensors have collected. The heart rate of the patient is monitored by an IoMT device, for instance, which is remotely controlled. collection and transmission of real-time health indicators to a cloud-based data center. Because of this, it is crucial to secure this connection in order to protect critical data. The major characteristics and challenges for IoT-enabled devices are shown in Figure 24.4.

In an IoT ecosystem, physical and virtual devices can trade services while continuing to abide by each type's restrictions. Since no person or central processor is in charge of managing the connectivity between different IoT devices, this might be extremely dangerous. A rogue device could start to disrupt other devices by installing harmful data if it poses as an approved IoT device. The volume of data being produced by IoT devices is unparalleled, and their number is growing quickly. Between 25 billion and 50 billion IoT devices are anticipated to exist by 2025. Because of the sheer size, massive-scale networks have significant issues with regard to data privacy and integrity. Large real-time data points are generated by IoMT-based applications like COVID-19, for example, and are saved in the cloud. However, when the volume of freshly created data increases, the network is put under more stress, which may lead to some misunderstandings [13, 14].

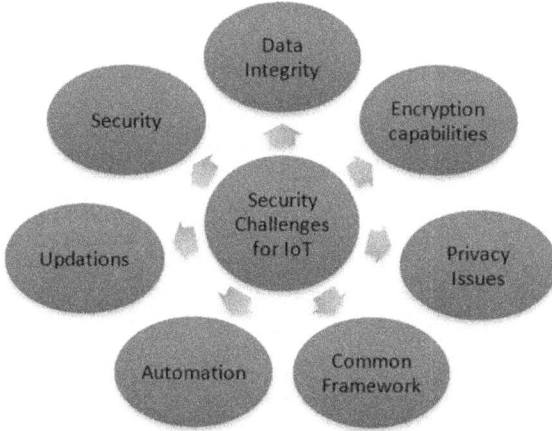

FIGURE 24.4 Security challenges for IoT.

It is exceedingly challenging to issue certificates to every IoT item because there are so many connected devices and there isn't a single global root certificate authority. IoT network-layer security is also challenged by DNS, which is used to describe things' properties and identify them. In this circumstance, there is a danger to data integrity due to attacks such as DNS cache poisoning or man-in-the-middle. With the use of misleading information, this exploit sends traffic to malicious websites [14].

The security problems in the IoT make malware attacks a problem. The major obstacle to real-time malware detection is antivirus software. Conventional security solutions, however, have not been successful in preventing assaults and do not offer the IoT with decentralized and trustworthy security safeguards. Because a device must be both secure and energy-efficient, the adoption of security measures is limited by concerns with battery size and expected durability. Table 24.1 presents a summary of various IoT detection techniques studied so far.

24.8.2 MALWARE IN IoT

The methods for malware detection for IoT have been modified from those covered in the preceding section. For instance, SVELTE, an intrusion detection tool based on signatures and anomalies, has been deployed in order to protect the IoT from attacks based on the IPv6 routing protocol. A time-consuming simulation method that is expensive computationally is required for a behavioral-based or specification-based approach to IoT security [12].

It is difficult for IoT systems to adopt the two main AI options for safeguarding the IoT, specification-based or behavior-based methodologies. The authors of [14] reviewed the newest developments in AI/ML techniques in IoT security. Despite developments in IoT-related AI techniques, the authors claimed that security is still fragile when used in actual IoT systems because 80% of the information was used to train the module. In addition, by highlighting the challenges and algorithmic limitations, the authors published a survey of AI technologies that enhance IoT security.

TABLE 24.1

Pros and Cons of the IoT Detection Techniques

Techniques	Pros	Cons
Signature-based detection	Effective against known malware.	Ineffective against new or unknown malware.
Heuristic-based detection	Can detect new or unknown malware.	May produce false positives, leading to unnecessary alerts or actions.
Behavior-based detection	Can detect malware based on unusual behavior patterns.	May require significant processing power and resources leading to higher costs.
Anomaly detection	Can detect unusual patterns or deviations from normal behavior.	May produce false positives or false negatives, leading to potential security risks.
Access control	Can restrict access to devices and data.	May be inconvenient for users and limit functionality.
Firmware updates	Can fix known vulnerabilities and security flaws.	May not be available for all devices or be timely in case of new threats.

IoT devices will be used by fraudsters to distribute malware payloads more frequently as long as companies and consumers keep connecting gadgets to the internet without taking the appropriate security precautions. In the first half of 2019, compared to the first two quarters of 2018, there were 55% more IoT threats, according to SonicWall. In the first half of 2019, around 100 million assaults on IoT devices were discovered by a security firm, underscoring the ongoing risk posed by unprotected IoT devices. The Russian antivirus business Kaspersky asserted that in the first six months of 2019, it had discovered 106 million threats emanating from 267,000 different IP addresses [14].

This number of attacks—from just 69,000 IP addresses—was more than nine times higher than the 12 million that were reported for the first quarter of 2018 [13]. According to the authors, the rising inclination of customers to purchase smart home solutions without conducting adequate research into security measures is a key factor fueling this trend. For all the aforementioned reasons, malware attacks pose major security risks to the IoT, necessitating the adoption of a dedicated IoT security solution.

Based on the characteristics and design of the IoT, security is best achieved by constructing a distributed, dynamic, adaptive, and self-monitoring system. This motivates us to research Artificial Immune System (AIS) products and how they may be applied to safeguard the IoT against malware attacks [15].

24.9 SUMMARY

This chapter provides an insight into the basics of malware and its types. An effective understanding of the various kinds of malware and AI-based techniques to detect and prevent these malware have been presented. Also, with the advancement in IoT-based communication, ensuring the security of IoT devices is also a crucial task. These IoT devices need to be protected from malware. The AI-based techniques to detect and protect malware in IoT has also been discussed. Out of the techniques being studied, it is clear that signature-based techniques are not able to identify newly generated malware files. It is also concluded that there would be a probability of malware attacks on IoT devices even after applying advanced preventive techniques of malware.

REFERENCES

1. Mendez, D.M., Papapanagiotou, I., and Yang, "Internet of Things: Survey on security and privacy", 2017. ArXiv:1707.01879.
2. Tama, B.A. and Rhee, K.H., "A random forest and PSO-based feature selection integration for IoT network anomaly detection", 2018 *MATEC Web Conf.*, 159, 10503.
3. Naeem, H., Ullah, F., Naeem, M.R., Khalid, S., et al., "Malware detection in the industrial Internet of Things based on a hybrid image visualization and deep learning model", *Ad Hoc Networks*, 2020, 105, 102154.
4. Moti, Z., Hashemi, S., Karimipour, H., Dehghantanha, A., Jahromi, A.N., Abdi, L., and Alavi, F., "Generative adversarial network to discover unknown Internet of Things - Malware", *Ad Hoc Networks*, 2021, 122, 102591.
5. Humayun, M., Jhanjhi, N., Alsayat, A., and Ponnusamy, V., "Internet of Things and ransomware: Evolution, mitigation and prevention", *Egypt. Inform. J.* 2021, 22, 105–117.
6. Yan, Q., Huang, W., Luo, X., Gong, Q., and Yu, F.R., "A multi-level DDoS mitigation framework for the industrial Internet of Things", *IEEE Commun. Mag.*, 2018, 56, 30–36.
7. Kumar, S. "MCFT-CNN: Malware classification with fine-tune convolution neural networks using traditional and transfer learning in Internet of Things", *Future Gener. Comput. Syst.*, 2021, 125, 334–351.
8. Chaganti, R., Ravi, V., and Pham, T.D., "Deep learning based cross architecture Internet of Things malware detection and classification", *Comput. Secur.*, 2022, 120, 102779.
9. Madan, S., Sofat, S., and Bansal, D., "Tools and techniques for collection and analysis of Internet-of-Things malware: A systematic state-of-the-art review", *Comput. Inf. Sci. J. King Saud Univ.* 2022, 34, 9867–9888.
10. De Albuquerque, V.H.C., Lisboa, C.O., Munoz, R., da Costa, K.A., and Papa J.P., "A survey of machine learning-based intrusion detection systems for the Internet of Things", *Computer Networks*, 2019, 151, 147–157.
11. Xiao, L., Wan, X., Lu, X., Zhang, Y., and Wu, D., "Machine learning-based IoT security measures: How does AI help secure IoT devices?", *IEEE Signal Process. Mag.*, 2020, 35, 41–49.
12. Diro, A.A. and Chilamkurti, N., "Distributed attack detection system for the Internet of Things using a deep learning approach", *Future Gener. Comput. Syst.*, 2018, 82, 761–768.
13. Yeo, M., Koo, Y., Yoon, Y., Hwang, T., Ryu, J., Song, J., and Park, C., "Detecting malware using a convolutional neural network and flow", in *The Proceedings of International Conference on Information Networking (ICOIN)*, Thailand, January 10–12, 2018.
14. Naveed, M., Arif, F., Usman, S.M., Hadjouni, M., Elmannai, H., Hussain, S., Ullah, S.S., Umar, F., and Anwar, A., "Using a Deep Learning-Based Framework, feature extraction and classification of intrusion detection in networks", *Wireless Commun. Mob. Computing*, 2022, 2215852.
15. Kamal, P., Sharma, R., Gupta, A., and Kumar, G., "Mitigating Gray Hole attack in Mobile Adhoc Network using Artificial Intelligence Mechanism", *Int. J. Innovative Technol. Exploring Eng.*, 2019, 8, 640–645.

25 5G and 6G Security Issues and Countermeasures

T. R. Reshmi

Society for Electronic Transactions and Security (SETS), Chennai, India

Kunal Abhishek

Centre for Development of Advanced Computing (C-DAC), Patna, India

25.1 INTRODUCTION: BACKGROUND AND DRIVING FORCES

The evolution and technological advancements of mobile communication were initiated as a contemporary effort for strategic applications. Later on, new wireless communication network adoption started in various applications and improvised versions came out every decade since 1980. To date, five generations of wireless communication networks have evolved and the sixth-generation technology is in testing phases for deployment across the world [1]. The attractive spectacle of 2G which followed the 1G was successful in delivering voice services. The data communication which started along with voice communications was not successful in 3G and 4G networks. These cellular networks in the last five years have enabled data-driven applications such as multimedia, multiplayer services, and high-definition streaming media. The number of smartphone subscribers, their data traffic exchanges, and subscriptions for the data have increased exponentially after 3G. Similarly, the evolution of the Internet of Things (IoT) also introduced massive increases in network traffic. Hence, the surge of the network traffic caused a significant expansion of wireless network capability and the development and expansion of networks. The evolution of the features of 1G to 6G mobile wireless communication is very drastic in each phase and is shown in Figure 25.1. The following sections also detail the features and surge of evolution in each generation.

First-Generation Mobile Communication (1G): The first-generation technology is a simple analog system with data speeds of up to 2.4 kbps and focused on its usage for voice communications. The technology uses frequency modulation (FM) and frequency division multiple access (FDMA) technologies and has a 30-kHz bandwidth. Considering North America's advanced mobile phone system (AMPS), Scandinavia's Nordic mobile

DOI: 10.1201/9781003477327-25

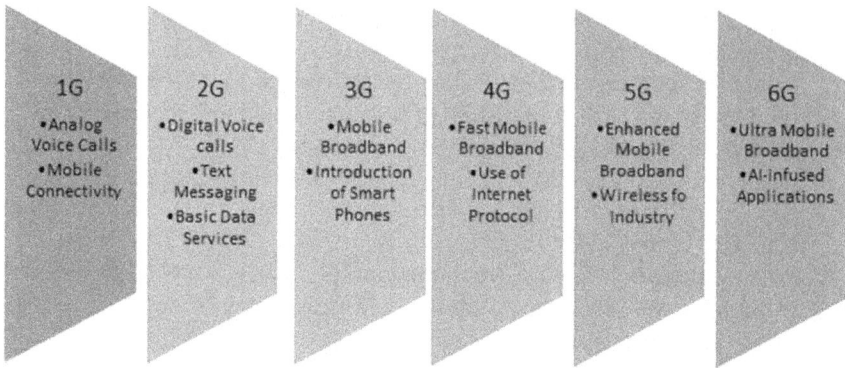

FIGURE 25.1 Evolution of generations of mobile communication networks.

telephone (NMT), the United Kingdom's total access communications system (TACS), and Japan's total access communications system (JTACS) as the principal users, 1G was declared in 1970. The released generation has many shortcomings such as (i) security concerns as it couldn't support encryption, (ii) scalability issues because of the use of FDMA technology, (iii) lack of transfer procedures and insecure base station power radiation, (iv) doesn't support services other than voice, and (v) variants in systems came up to formulate a consistent international standard. With the evolution of cellular communication, the 2G standardization tried to overcome the issues in 1G.

Second-Generation Mobile Communication (2G): The global mobile communication (GSM) was the first second-generation system introduced in the 1990s. GSM uses a bandwidth of 200 kHz for voice communications and uses Gaussian minimum frequency shift keying (GMSK) modulation and time division multiple access (TDMA) transmission technology. The feature sets of this generation include (i) a unified international mobile communication standard that promoted the development of the technology, (ii) provision for improved services, (iii) support network security through encrypted numbers, (iv) increased system capacity, and (v) the longer mobile phone battery life. The general packet radio service (GPRS) technology is classified as 2.5G, which raised the data rate by up to 50 kbps and used GSM's packet switching and circuit switching technologies. It also used the transmission and modulation similar to GSM. GPRS is the first phase that supports enhanced data (EDGE) in a GSM environment. EDGE is a radio technology that allows data at a rate of up to 200 kbps. The EDGE technology also used a transmission mechanism similar to GSM but it employed eight phase-shift keying (8PSK) that provides a higher data rate in narrow coverage areas and GMSK modulation for reliable wide coverage.

Third-Generation Mobile Communication (3G): The 3G technology focused on delivering fast internet access and employed wideband code division multiple access (WCDMA) and high-speed packet access (HSPA) technology to improve video and audio transmission capabilities. The two protocols high-speed downlink packet access (HSDPA) and high-speed uplink packet access (HSUPA) improve network performance. The developed HSPA (also known as HSPA+) is an upgraded third-generation partnership project (3GPP) standard used internationally in 2010 and 3.9G long-term evolution (LTE) was developed with more features. LTE was later called the 4G technology by the International Telecommunication Union (ITU) and 3GPP.

Fourth-Generation Mobile Communication (4G): LTE is a wireless access technology with expandable transmission bandwidths of up to 20 MHz using orthogonal frequency division multiplexing (OFDM) which allows multi-antenna transmission. LTE uses multiple input multiple output (MIMO) technology that allows higher data rates and multi-stream transmission to achieve improved spectral efficiency, connection quality, and radiation pattern changes for signal gain. Interface arrays are created using the antenna's adaptive beamforming techniques and the technology boosts data speeds through mobile to 100 megabits per second (Mbps). The 4G systems have improved the communication networks with a comprehensive and consistent solution based on the Internet protocol (IP). The three primary research topics were investigated in-depth to improve services of 4G mobile networks such as densification of the network, improved spectral efficiency, and carrier aggregation. Many outcomes of this research gave satisfactory improvements. But to satisfy the growing needs of the service requirements, the network technology evolved.

Fifth-Generation Mobile Communication (5G): The 5G is the opted standard defined by the ITU and ought to meet three common scenarios: enhanced mobile broadband (eMBB) Gb/s data rate, ultra-reliable low-latency communication (URLLC), and massive machine challenge technology communication (mMTC). There are eight key performance indicators (KPI) which ought to show the performance of 5G [2]. The highlighted features are MIMO, sophisticated coding and modulation, millimeter-wave communications, ultra-dense networks (UDN), non-orthogonal multiple access, flexible frame structure, dual connectivity architecture, and adaptability to other wireless technologies. There is exponential growth of data traffic due to the substantial growth in the number of connected devices. The adoption of 5G has improved and extended to accomplish high throughput, higher system capacity, greater spectrum efficiency, lower latency, and coverage wider and deeper to support high speed of movement [3].

The growth of innovative applications like virtual reality/augmented reality (VR/AR), self-driving cars, integrated 3D communications requires high data rates and lower latency than 5G networks. These challenges are thought to be the key driving factors for the realization of 6G communication systems [4]. With these foreseen growth and requirements for the evolving technologies, 6G communication technologies are designed and getting standardized.

The ITU and 3GPP have updated and enlarged 6G based on the visualization and expansion of 5G that requires 100 times the data throughput of 5G, higher system capacity, low latency, more efficient and higher spectrum, and more coverage. To promote the evolution of intelligent life and industrial automation, 6G is expected to enable faster movement to serve the Internet of Everything (IoE). The 6G with IoT is expected to generate vast volumes of data with its use in edge computing, cloud computing, artificial intelligence, and blockchain. The 6G is expected to have the ability to actualize intelligence gathered and to support ubiquitous computing in a better way.

25.2 THE IMPACT OF 5G AND BEYOND ON IoT NETWORKS

The wide range of radio frequencies creates more connectivity as assured in the 5G and beyond networks. The range of support includes lower and higher ranges. The services like television signals, Wi-Fi signals, and phone signals work in the lower band whereas the other higher range bands are used for services that require higher bandwidth for faster transmission such as high-definition video streaming and web browsing. The virtual networks created by the concept of network slicing in 5G cater to the specific needs of the user. The set of unique resources is allocated based on the needs of each network. The Service Level Agreement (SLA)-specific factors like speed, and capacity act as the decision factors for the resource allocation. Hence, 5G systems are deployed quickly in case of need for only a few functions allowing faster performances.

The application where IoT devices are used is increasing day by day. These IoT devices use sensors, applications for communication, and communicating interfaces to exchange packets. These devices are highly hooked on the features required such as accuracy, consistency, and security to assure reliability and performance [5–7]. Therefore, the 5G technology is promising for IoT to ensure the reliability and performance in its applications and is represented in Figure 25.2. The upgrades that 5G and beyond can offer in IoT applications are listed below.

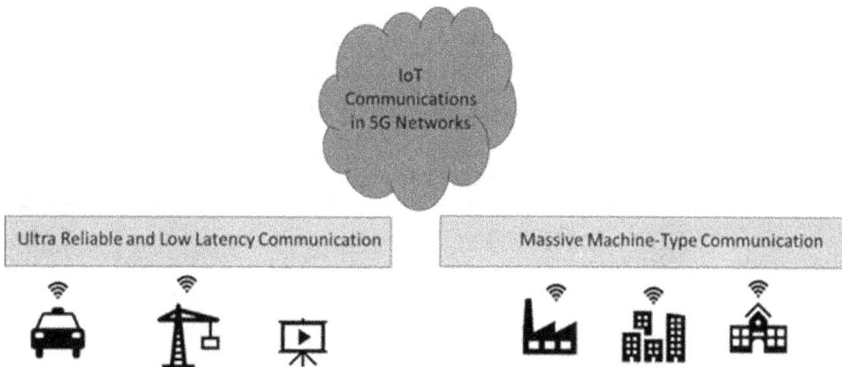

FIGURE 25.2 IoT communications in 5G networks.

- **Faster Transmission**
  The 5G technology has evolved to ensure very high speeds for transmissions. The devices can quickly communicate and share data effectively. Therefore, the applications will run better and do seamless data exchanges from sensors and devices. Such reliability assurance in 5G makes connections between IoT devices more reliable.
- **Increase in Device Capacity**
  There are billion devices predicted in 2030, the exponential growth of these networks is supported by the evolving 5G and 6G technologies. The higher bandwidth in these technologies will consequently increase the device capacity for device-to-device and low-power wide area networks.
- **Low Latency**
  The data transfer happens in real-time with greater accuracy in low-latency applications. The lagging is caused due to issues or problems in guaranteed precision in the delivery of packets in networks. The advancement of 5G ensures a convenient and safer way of exchanging packets with considerably low latency.

  The improvements in communication technologies and support for massive IoT in 5G technologies have given a better and new phase for IoT [8–11]. The scaling up of large amounts of IoT devices is enhancing all sorts of industries. Some industry-specific IoT applications are given below.

  1. **Manufacturing and Factories**: The reliable use of robotic technology and advancements in network technologies can bring down labor and improve the efficiency of manufacturing and factories. The downtime is hence managed well with more accurate sensors.
  2. **Transportation**: The sensors placed in various roadside units and automated vehicles rely on real-time traffic information and driver assistance for safer and smarter drives on roads. The 5G relies on this data accurately to make the driving experience safer and smarter.
  3. **Surveillance**: The intervention of 5G has improved video surveillance for asset protection. Extended wireless connectivity, more accurate videos, alerts on time, etc., are the features that improve the surveillance that can be assured by 5G.
  4. **Asset Tracking**: The IoT sensors deployed and used to monitor the delivery and management of goods in asset tracking systems give promising performance in logistics. Better communication with the cloud, sensing the location and climate change, and moving parts of logistics gave accurate and smart asset tracking.
  5. **Live Streaming and Broadcasting**: The 5G-supported devices allow users to seamlessly view live content and broadcasts on their devices. The broadcasters can send the feeds directly to production hubs and thereby do efficient streaming at a low cost. The faster data transmission and low latency in live video give a better user experience to the users.
  6. **Home Automation**: The fast local or remote access to the home automation system through intelligent resources over 5G is expected to boom very rapidly in the coming years. The greater security offered in 5G can overcome the technical glitches in the home automation industry.

25.3 SECURITY AND PRIVACY RISKS IN 5G NETWORKS

The vulnerabilities in Network Function Virtualization (NFV), Software-Defined Networks (SDN), cloud techniques, and interoperability of networks are the major concerns in 5G networks. The security and privacy concerns in 5G are addressed by many mechanisms [12] and special attention is given to the vulnerabilities identified in mechanisms such as authentication, access control, malicious behavior monitoring, encryption, and data transmission. The traditional IoT used cloud storage which faced many security and privacy issues such as identity thefts, location privacy, attacks due to node compromise, attacks with layer removal/addition, forward and backward security issues, and problems due to semi-trusted and malicious cloud. In 5G edge-based IoT, traditional security and privacy issues still exist. There are also many other specific security and privacy problems emerging for the characteristics of edge paradigms. In various use cases of IoT, the IoT devices don't have conventional hardware protocols, software updates, or security configurations required for the secured life cycle of the device. Edge computing allows physical objects to autonomously communicate over 5G networks for many service applications. For example, IoT applications in devices such as microwaves, air conditioners, and cameras can detect widespread sensitive personal information collected by these devices [13]. The personal sensitive information collected and transmitted by endpoints has privacy risks. The attacker intercepting exchanged information can collect, integrate, and do analysis to illicitly use the sensitive personal information archived. There is much research focusing on privacy-preserving algorithms in IoT. A few popular strategies of privacy metrics such as k-anonymity and differential privacy are employed for addressing the privacy issues. Still, the studies confirm that integrating and analyzing the encrypted data in the database collected from multiple sources can be used to determine and identify information about a particular individual.

25.4 SECURITY CONCERNS AND RESEARCH DIRECTIONS IN 5G-IoT AND BEYOND

IoT security includes security of all elements such as devices, cloud, software, mobile applications, network interfaces, encryption algorithms, authentication, and physical security techniques [14]. The weaknesses in the data security of IoT applications, as stated in the Open Web Application Security Project (OWASP), are due to insecure web interfaces, insufficient authentication/authorization, insecure software/firmware, privacy concerns, insecure network services, lack of transport encryption, insecure cloud interface, insecure mobile interfaces, insufficient security configurability, and poor physical security, IoT application security, and endpoint security. The cellular IoT faces many signaling attacks such as identification threats, location privacy attacks, data interception and tapping, and denial of service.

The 5G-IoT, with the sudden exponential growth in the attack surface on mesh networks connected by IoT devices, faces serious security concerns. The traditional hub-and-spoke configuration has totally changed in the 5G-IoT with the edge computing techniques. Being in a meshed environment, any weak link can

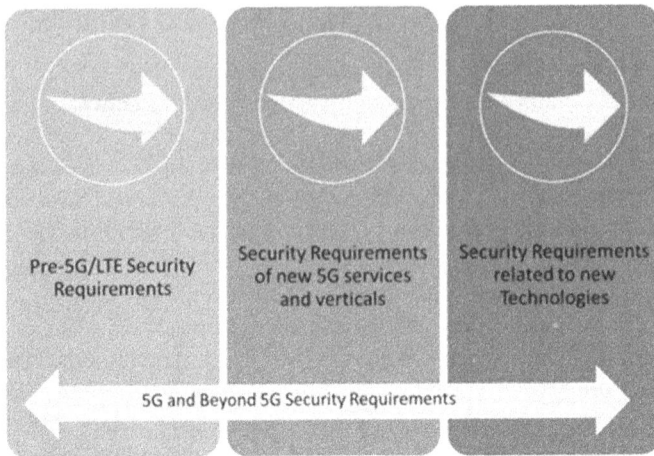

FIGURE 25.3   Security requirements of 5G and beyond 5G networks.

become a vulnerable point in 5G-IoT. The issues listed in Figure 25.3 show the broad categorization of security concerns. There are many different concerns, some of which are detailed below.

Access Control Mechanisms: The security of all individual links connecting from edge to edge, across IoT edge, across core enterprise networks, and on local offices or multiple public clouds is required to have some access control mechanisms. Therefore, any entity connected to the enterprise ecosystem must be identified, verified, and confirmed in their states; thereby, all requests for access to resources will be verified, validated, and authenticated.

Network Segmentation: Elastic and adaptive security mechanisms are required for combining edge-to-edge hybrid systems with traditional security strategies. The network segmentation is proven to have vulnerabilities for a 5G world. New segmentation strategies are required to navigate local, remote resources that mix segments that are in its control or not. Hence, there is a requirement to manage the complexity of multiple co-managed systems to implement 5G networks and public cloud services for IoT Services.

New Security Architecture: The new security architecture for 5G-IoT requires threat intelligence sharing, event data correlation, and automated incident response. The development and adoption of a comprehensive security architecture with machine learning, artificial intelligence, and automation can bridge the gap between the detection and the mitigation of security issues in 5G-IoT.

Interoperability across Security Tools: A new open 5G security standard with interoperable security tools, API adoption across vendors, and agnostic management tools to see security events and orchestrate security policies is another requirement to overcome the security challenges in 5G-IoT.

25.5 SECURITY AND PRIVACY RISKS IN 6G NETWORKS

The 6G technologies inherit many of the security features of 5G technologies with or without improvements. There are many defense techniques to protect from jamming and eavesdropping but when multiple adversaries collect scattered signals, it is challenging to prevent them as there comes the complexity of attack re-verification during hand-off. There are also privacy concerns in 6G. The Access Points (AP) working in THz follow the user's motion in precision of centimeters that improves the link connectivity and exposes the locations by creating privacy concerns. These compromised APs can also become potential surveillance devices in the hands of the attacker.

The remarkable upgrade on 6G networks is in the physical layer. The THz communications and ultra-massive MIMO antenna technologies are the major changes. But these enhancements have proved to be vulnerable to physical layer attacks specifically, jamming and pilot contamination attacks. Therefore, enhancing the security of THz communications and relevant technologies is an open research issue focused by many researchers. There are solutions built with AI/ML models to minimize the secrecy outage probabilities and improve secrecy rates. The quantum-safe encryption and key distribution are expected to be first applied to 6G. As a complete transition to THz technologies requires many years, the vulnerabilities of mm-wave and massive MIMO technologies are still in researchers' focus for ensuring 6G security. The security concern of vRAN and cloud paradigms (C-RAN) is seen as an area to be addressed. The vulnerability testing of vRAN API and the complexity of vRAN is very much required. The flaw or bug in this shared infrastructure of vRAN/C-RAN can cause big chaos in the network.

The concept of predictability and timely processing has a major influence on the availability-integrity principles of 6G security. The security architecture and relevant technologies of 6G are designed in such a way that they are simplified and transparent. AI/ML plays a major role in the 6E IoE. As the network coverage will be expanded substantially with 6G, the diversity of applications and services with different protection requirement goals will also be evolving, which poses challenges to the underlined architecture features of 6G. The initiatives to bring in concepts such as slicing, open encryption standards, open security orchestration, and open authentication protocols are researching these areas in 6G. 6G requires the capability of backward compatibility maintenance in network access authentication and mutual authentication between the serving network and the home network, which opens the door to exposing many vulnerabilities. Applications from untrusted sources can threaten the trust of networks and be a real danger to business assets. The major applications focused and feature sets of 5G and 6G technologies are given in Table 25.1.

TABLE 25.1

Comparison of Security Requirements of 5G and 6G Technologies

Technology	Features	Applications	Security Requirement
5G Technology	Software-Defined Networks	Smart Grid	High Confidentiality and Integrity
	Connected Things	Telehealth	Zero-Trust Security
		Smart Factory	Subscriber Privacy
		Smart Agriculture	Lightweight Security
		Unmanned Aerial Vehicle	Real-time Security
		Augmented Reality	Energy Efficiency
			Security Cloudization
6G Technology	Network Edge Intelligence	Extended Reality Digital Twin	Ultra-High Confidentiality and Integrity
	Connected Intelligence	Smart Medical Nano-Robot	Zero-Touch Automation Security
		Autonomous Driving	Subscriber Privacy
		Tactile Internet	Ultra-Lightweight Security
		Holographic Telepresence	Real-Time Security
		Space–Air–Sea Communication	Energy Efficiency
			Security Automation

25.6   SECURITY CONCERNS AND RESEARCH DIRECTIONS OF 6G IN IoT

The new authentication model and cryptographic schemes for communication security are a requirement of 6G technology. The evolved 6G-AKA, quantum-safe cryptography schemes are the top candidates for this specific security requirement in 6G. The security models used in 6G inherit some security models from 5G, mainly the unified authentication platform which is used for both open and access network agnostic. The 6G-AKA protocol is used to certify the clear roles of the Authentication Server Function (AUSF) or Security Anchor Function (SEAF) and selects the authentication in cross-slice communications. The identity of an endpoint hence needs to be verified by 6G-AKA. Other technologies such as quantum communication, quantum-safe cryptography schemes, and open authentication protocols for non-3GPP Access Networks such as Satellite, Sea, and Enhanced EAP-TLS for mutual authentication for core network components such as Blockchain are some other areas of research focus in 6G.

25.7   THE FUTURE OF IoT

The researchers foresee Social IoT (SIoT) consisting of connected intelligent, inter-related objects that work together to share information and services with the user as the next evolution of IoT. The proof of concept of SIoT is in development without a clear definition. SIoT is seen as a human network configured at various levels such as personal, group, or public to connect to objects through internet technology. The personal SIoT consists of objects that can be connected by a single individual, for example, smart devices for health and fitness, sleeping habits, and workout monitors. The intelligent objects that users can access based on controllers providing permission comprise the group SIoT. The devices are allowed to access and communicate with other user objects based on roles and permission sets in group SIoT. For example, the individuals in a fitness club can exchange information about their health monitoring devices with each other and interact with each other through smart devices. The public SIoT might include users and their intelligent objects connected to others to share information through publically available networks. The IoT-enabled car or devices that can connect to objects in public networks to get real-time updates on traffic conditions is an example of a public IoT.

The evolution of IoT-enabled objects poses various privacy and security challenges. A device that connects to the IoT network would be identified with a massive source of data which raises privacy- and security-related issues [15, 16]. There is an increasing importantance and need for qualitative research to evaluate these unstructured nature of the data getting generated for IoT. There are limitations in the quantitative methods in evaluating unstructured data coming out of these devices. There are various researches going on for the analyses of these mixed data for better performance and security practices.

The 5G and beyond-IoT ensures the ubiquitous computing of devices which makes it intelligent and paradigm for new attacks. The layered approach of IoT security must cover all the physical security, secured firmware, and gateway or network perimeter security. There is no single or fits-all solution to ensure the performance and security needs of IoT networks.

REFERENCES

1. Solyman, Ahmed Amin Ahmed, and Khalid Yahya. "Evolution of wireless communication networks: From 1G to 6G and future perspective." *International Journal of Electrical and Computer Engineering* 12, no. 4 (2022): 3943.
2. Liberg, Olof, Marten Sundberg, Eric Wang, Johan Bergman, Joachim Sachs, and Gustav Wikström. *Cellular internet of things: from massive deployments to critical 5G applications*. Academic Press, 2019.
3. Ramezanpour, Keyvan, Jithin Jagannath, and Anu Jagannath. "Security and privacy vulnerabilities of 5G/6G and WiFi 6: Survey and research directions from a coexistence perspective." *Computer Networks* 221 (2023): 109515.
4. Senbagavalli, G., T. Kavitha, Aruna Ramalingam, and V. A. Velvizhi "6G with TeraHertz Communications." In *Handbook of research on design, deployment, automation, and testing strategies for 6G mobile core network*. edited by Kumar, D. Satish and G. Prabhakar, and R. Anand, 218–247. Hershey, PA: IGI Global, 2022.
5. Von Butovitsch, Peter, David Astely, Christer Friberg, Anders Furuskär, Bo Göransson, Billy Hogan, Jonas Karlsson, and Erik Larsson. "Advanced antenna systems for 5G networks." Ericsson white paper (2018).
6. New, S. I. D. on Support of reduced capability NR devices, document RP-193238, RAN# 86, Ericsson, 3GPP, Dec. 2019.

7. Service requirements for next generation new services and markets, 3GPP TS22.261, 2019.

8. 5G for automation in industry, 5G ACIA White Paper, 2019.

9. 3GPP, D. "System architecture for the 5G system (5GS)." 3rd Generation Partnership Project (3GPP), Technical Specification (TS) 23.501 (2020).

10. Sachs, J., K. Wallstedt, F. Alriksson, and G. Eneroth. "Boosting smart manufacturing with 5G wireless connectivity." *Ericsson Technology Review* 2 (2019): 1–12.

11. Farkas, Janos, B. Varga, G. Miklós, and J. Sachs. "5G-TSN integration meets networking requirements for industrial automation." *Ericsson Technology Review* 96, no. 7 (2019): 45–51.

12. Reshmi, T. R., and M. Azath. "Improved self-healing technique for 5G networks using predictive analysis." *Peer-to-Peer Networking and Applications* 14, no. 1 (2021): 375–391.

13. Simplifying the 5G ecosystem by reducing architecture options, Ericsson Technology Review, 2018.

14. Liu, Liyuan, and Meng Han. "Privacy and security issues in the 5g-enabled Internet of Things." In *5G-Enabled Internet of Things*, pp. 241–268. CRC Press, 2019.

15. Wang, Minghao, Tianqing Zhu, Tao Zhang, Jun Zhang, Shui Yu, and Wanlei Zhou. "Security and privacy in 6G networks: New areas and new challenges." *Digital Communications and Networks* 6, no. 3 (2020): 281–291.

16. Kim, Jin Ho. "6G and Internet of Things: A survey." *Journal of Management Analytics* 8, no. 2 (2021): 316–332.

Index

Pages in *italics* refer to figures and pages in **bold** refer to tables.

For Product Safety Concerns and Information please contact our EU
representative GPSR@taylorandfrancis.com
Taylor & Francis Verlag GmbH, Kaufingerstraße 24, 80331 München, Germany

www.ingramcontent.com/pod-product-compliance
Lightning Source LLC
Chambersburg PA
CBHW060812220326
41598CB00022B/2597